A MANUAL

OF

AQUATIC PLANTS

A MANUAL

OF

AQUATIC PLANTS

By

NORMAN C. FASSETT

With Revision Appendix by

EUGENE C. OGDEN

THE UNIVERSITY OF WISCONSIN PRESS

Madison · Milwaukee · London

Published 1940, 1957

The University of Wisconsin Press
Box 1379, Madison, Wisconsin 53701
The University of Wisconsin Press, Ltd.
70 Great Russell Street, London

Printings 1957, 1960, 1966, 1969, 1972

Printed in the United States of America
ISBN 0-299-01450-9; LC 57-6593

PREFACE

The aim of this book is to make possible the identification of aquatic plants in sterile as well as in flowering or fruiting condition. An effort has been made to render such identification as simple as possible, but it must be remembered that the aquatics include many of the plants in groups that are most difficult for the taxonomist.

For present purposes, an aquatic is defined as a plant that may, under normal conditions, germinate and grow with at least its base in the water and is large enough to be seen with the naked eye. Under some conditions almost any plant may be found in the water. If a lake or river has been at a low level for some time and then rises to its normal level, many plants not ordinarily considered as aquatics will be found in the water. Bogs, which are often saturated, are excluded from this work, as are small woodland brooks, waterfalls, tidal, salt and brackish waters. With the field thus circumscribed, it is still very indefinite. The author is certain that no two individuals would make the same list of species; it is probable that the same individual would not make identical lists at different times. So it is frankly admitted that the present list of species is highly subjective; the goal has been to treat such plants as the aquatic biologist will be likely to find, but it is unavoidable that in many places plants will be found in the water which have been omitted.

The region covered is from Minnesota to Missouri and eastward to the Gulf of St. Lawrence and Virginia. In the last-named state many plants not heretofore known from this area are being discussed in a series of articles in *Rhodora* by Prof. M. L. Fernald. Most of these plants are omitted. The list of discoveries in this significant region is not yet complete, and anyone working in southeastern Virginia will in any event find it necessary to consult Professor Fernald's papers. Inclusion here of his findings would add to the bulk and difficulty of this book without materially aiding workers in that area.

The preparation of this book in its present form would not have been possible but for the assistance of two graduate students, Mr. John L. Blum and Miss Elizabeth A. Chavannes, who drew most of the plates, in large part without recompense. Some of the plates were drawn by Mr. Russell Stevens, Dr. Stephen Kliman, and, through cooperation with the Milwaukee Public Museum and the WPA, by Miss Emmeline O. Krause. Assistance on special problems has been given by Mr. C. C. Deam, Dr. E. C. Ogden, Prof. M. L. Fernald, Mr. Neil Hotchkiss, Dr.

H. S. Conard, Dr. F. J. Hermann, Dr. R. I. Evans, Dr. J. W. Thomson, Jr., Dr. Pauline Snure and Dr. Carl Epling.

To the Wisconsin Alumni Research Foundation the writer is indebted for the aid of research assistants and the opportunity for months of uninterrupted study, both of which were made possible by grants from research funds.

NORMAN C. FASSETT.

UNIVERSITY OF WISCONSIN,
   April, 1940.

PREFACE TO THE REVISED EDITION

Shortly before his death, Dr. Fassett arranged with the University of Wisconsin Press for a revised edition of his *Manual of Aquatic Plants*, the revision to appear as an appendix bringing the nomenclature into agreement with present-day usage. Dr. Fassett then invited me to undertake its preparation in the likely event that he would not live to complete it. Unfortunately, our plans to work together on this could not materialize and suggestions by letter were of a very general nature. So, while I have attempted to follow his wishes and have been guided by his numerous publications, the final decisions are my own and I must accept all responsibility.

The Revision Appendix is designed to serve as a revision of pages 3–341. Its primary purpose is to bring the nomenclature into agreement with present-day usage as indicated by Fernald's revision of *Gray's Manual*, Gleason's revision of the Britton and Brown *Illustrated Flora*, Muenscher's *Aquatic Plants of the United States*, and by recent local floras and monographic treatments, including several by Fassett.

Dr. Fassett's custom of using the specific name to include all typical (from the standpoint of nomenclature) subspecific categories is followed. Thus *Rhynchospora macrostachya* means variety *macrostachya* and forma *macrostachya* but does not include var. *colpophila* or any other segregated varieties and forms considered to be worthy of recognition here. Some minor forms, which I find scarcely recognizable, have been retained but many newly described ones have not been added; this makes a slight, but unimportant, disuniformity. In general, treatments before 1940 and available to Dr. Fassett were not considered. No attempt has been made to change his determinations as to what is or is not an aquatic. Important range extensions within this manual area have been added but those outside the area have not been mentioned. Fassett's custom, not my own, has been followed where both are permissible. Thus *Saggittaria Eatoni* is not changed to *Saggittaria eatonii*. A determined, though often difficult, attempt has been made to keep technical terms to those in the Glossary.

It will be readily understood that a complete revision cannot be handled as an appendix; in numerous cases the rewriting of portions would result in a somewhat different text from that as modified here.

EUGENE C. OGDEN.

June, 1956.

CONTENTS

MANUAL OF AQUATIC PLANTS

INTRODUCTION

The simplest way to identify a plant is to examine a set of pictures until one is found that resembles it. This becomes not practical when a large number of species is involved. Moreover, many kinds of plants are superficially alike, and even if the pictures show the differences the user must be directed as to what points to observe. It is on this principle that these keys are based. The text is essentially a set of directions for looking at the pictures. Part I is based as far as possible on superficial characters for the identification of sterile specimens, and by it a plant may be run to species or to genus or to family; the lack of uniformity results from the fact that some groups can be identified more closely in the sterile condition than can others.

For the identification of aquatic plants a lens and a millimeter scale are necessary. A lens magnifying six or eight times is sufficient for most work; in a few cases a microscope is called for. If much identification is to be done, a binocular microscope is very helpful. If possible it should be mounted so that light may be thrown upon the object from above or from below, for venation is often best seen by transmitted light.

Five categories of names are used here. The family (*Alismaceae, Onagraceae, Cruciferae*) includes a number of genera whose floral or fruiting structure has certain factors in common. These genera may be very diverse in appearance. A genus (*Potamogeton, Nuphar, Proserpinaca*) includes a number of species, usually similar in flower and fruit, and often of somewhat uniform appearance. A species (*Potamogeton pectinatus, P. natans, P. gemmiparus*) is a group of individuals so similar that they are naturally classed together. It may be subdivided into variety (abbreviated var.) and form (forma, or f.). Varieties are phases of a species differing usually on minor points and often occupying different parts of its range. They frequently intergrade. A form is a phase of a species that may occur anywhere throughout the range and is often found with the typical plant. A form often represents a response to habitat: the Pickerelweed commonly has stout leaves with more or less heart-shaped blades standing straight in the air, but when submersed it may have instead narrow ribbon-like leaves resembling those of Wild Celery. Again, the emersed leaves may be as broad as long and heart-

1

shaped on some plants and long and narrow on others. These sporadic variations are treated as forms.

The personal names (Pursh, Britton, Fernald) following plant names refer to the botanists who named them. These are often abbreviated (L., Raf., Willd.); the abbreviations are the same as those in the seventh edition of Gray's Manual, and are explained on pages 28–30 of that book. As in Gray's Manual and in Britton and Brown, two accents are used, the grave (`) to indicate a long vowel, and the acute (´) for a short vowel.

Names of plants follow the International Code of Botanical Nomenclature. When a name differs from the one used in Gray's Manual, seventh edition,[1] or in Britton and Brown, second edition,[2] the name used in these works appears as a synonym, with the symbol "G" for the former and "B" for the latter. If the name is in neither book, a reference to the appropriate description or discussion is given following the treatment of the genus or of the family.

[1] ROBINSON, B. L., and M. L. FERNALD: A Handbook of the Flowering Plants and Ferns of the Central and Northeastern United States and Adjacent Canada. American Book Company, (New York 1908).

[2] BRITTON, N. L., and ADDISON BROWN: An Illustrated Flora of the Northern United States, Canada and the British Possessions, 3 vol., Charles Scribner's Sons, (New York 1913).

PART I

GENERAL KEY

This key is designed for the identification of plants, whether flowering or sterile. In some cases an unknown can be placed in the proper species, whereas in others it can be taken to genus or family only; in any event a reference is given for the detailed treatment in Part II.

Procedure is as follows. On this page are described 17 categories, into one (or more) of which the unknown should fit. Under each category reference is given to another page, on which will be found categories for further subdivision. In most cases there is a series of pairs of choices, the two choices in each case bearing the same number: if the plant fits under the first, advance to the paragraph of which the number follows in parentheses; if under the second, advance as there directed. Proceed in this fashion until the name of the plant is reached, then refer to Part II.

1. Stems and leaves, or their divisions, limp, thread-like or ribbon-like, more than 10 times as long as broad (figures, p. 5)....................................**1**, p. 4
2. Stems erect or nearly so, naked, or with leaves more than 10 times as long as broad (figures, p. 13)..**1**, p. 12
3. Leaves flat, less than 10 times as long as broad, not lobed (figures, pp. 21, 23).. ....**1**, p. 20
4. Trees or shrubs...**1**, p. 30
5. Leaves deeply lobed (Figs. 2, 3, 5, 6, 7, p. 31)......................**11**, p. 30
6. Leaves compound, cut into several flat leaflets (figures, p. 33).........**1**, p. 32
7. Blades with lobes extending below the junction with the petiole, or with the petiole attached near the middle (figures, p. 35).......................**1**, p. 34
8. Plants floating, roundish, not over 6 mm. broad (figures, p. 167). DUCKWEED FAMILY (*Lemnaceae*), p. 164.
9. Long-stalked branching fronds (Fig. 14, p. 167). STAR DUCKWEED (*Lemna trisulca*), p. 168.
10. Plants 2-lobed or repeatedly 2-forked. See THALLOSE LIVERWORTS (*Ricciaceae*), p. 42, and RIVER WEED (*Podostemum*), p. 239.
11. Leaves 7 mm. or less long, scale-like or overlapping. See MOSSES and LEAFY LIVERWORTS (*Musci* and *Hepaticae*), p. 40, and WATER-VELVET (*Azolla*), p. 42.
12. Leafless stems supporting a yellow or purple spurred flower (Fig. 57, p. 312). BLADDERWORT (*Utricularia*), p. 309.
13. Hollow cylindrical jointed stems, each joint with a sheath (figures, p. 45). HORSETAIL (*Equisetum*), p. 42.
14. Cotton-like masses of very fine green threads. GREEN ALGAE (*Chlorophyceae*), p. 36.
15. Gelatinous irregular spheres or lumps. BLUE-GREEN ALGAE (*Cyanophyceae*), p. 40.
16. A leaf like a 4-leaved clover (Fig. 1, p. 43). PEPPERWORT (*Marsilea*), p. 42.
17. Clumps of ferns from large hard bases. ROYAL FERN (*Osmunda*), p. 42.

3

1. (From **1,** p. 3.) Leaves scattered on the stem (Figs. 1, 3)...................**2**

1. Leaves all from one point at the base of the plant (Fig. 5) or in bunches from a buried horizontal stem (Fig. 7)...**3**

    2. Leaves flat, ribbon-iike...**1,** p. 10

    2. Leaves, or their divisions, thread-like................................**8**

3. Leaves thread-like...**4**

3. Leaves flat, ribbon-like...**1,** p. 8

    4. Plants with a bulb-like base (Fig. 5) and no horizontal stem. Quillwort (*Isoetes*), p. 44.

    4. Plants with a horizontal stem under the soil (Fig. 7). The so-called "leaves" may actually be stems in some of the following..........................**5**

5. Horizontal stem scarcely thicker than the green thread-like leaves or stems. Needle Rush (*Eleocharis acicularis*), **22,** p. 135. A rare rush (*Juncus subtilis*), p. 180, may also key to this point.

5. Horizontal stem thicker than the thread-like green leaves or stems. Stouter erect stems, often bearing fruiting heads, may also be present; these can best be run down under **2,** p. 18...**6**

    6. Horizontal stem 2 mm. or more thick. Bayonet Rush (*Juncus militaris*). p. 180.

    6. Horizontal stem about 1 mm. thick....................................**7**

7. Green thread-like structures (leaves) somewhat flattened at base and cupped around each other. Water Bulrush (*Scirpus subterminalis*), p. 143.

7. Thread-like structures (stems) borne in close tufts, not flattened or cupped at base. Triangle Spike Rush (*Eleocharis Robbinsii*), p. 129.

    8. Stems with whorls of brittle branches (figures, p. 39). Stonewort (*Characeae*), p. 36.

    8. Stems without regular whorls of branches...............................**9**

9. Leaves with divisions having one side minutely spiny-toothed (Fig. 2). Coontail (*Ceratophyllum*), p. 210.

9. Leaves without spiny-toothed margins...................................**10**

    10. Plants in swiftly flowing water, making a moss-like growth clinging to rocks by suckers. River-Weed (*Podostemum*), p. 239.

    10. Plants with true roots in the soil.....................................**11**

11. Leaves simple (Figs. 1–3, p. 7)...**1,** p. 6

11. Leaves compound (Figs. 4–7, p. 7).......................................**5,** p. 6

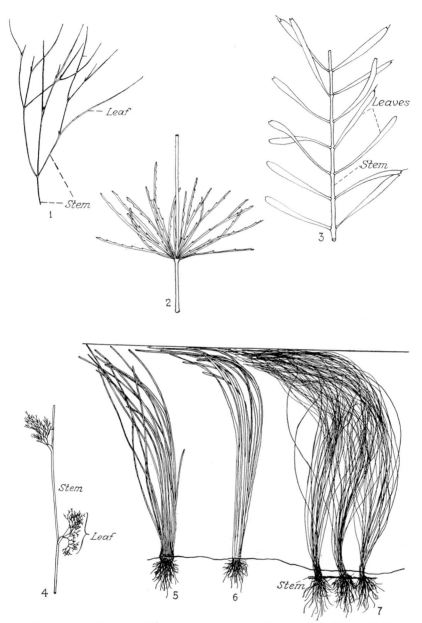

1. Scattered simple thread-like leaves on a branched stem, generalized.   2. **Cerato-phyllum**, leaf, × 4.   3. Flat, opposite, ribbon-like leaves on a stem, generalized.   4. **Ranunculus longirostris**, × ½.   5. **Isoetes**, generalized.   6. Ribbon-like leaves borne from one point, generalized.   7. Thread-like leaves or stems in clusters along a buried stem.

1. (From **11**, p. 4) Leaves opposite or in whorls (Fig. 3)........................**2**
1. Leaves borne singly on the stem (Fig. 1)....................................**3**
　　2. Base of leaf as in Fig. 2. BUSHY PONDWEED (*Najas*), p. 77.
　　2. Base of leaf as in Fig. 3. HORNED PONDWEED (*Zannichellia*), p. 75.
3. Stipule partly fused with the base of the leaf and partly free (Fig. 8). SAGO PONDWEED (*Potamogeton pectinatus* and related species), **5**, p. 55
3. Stipules nearly or quite free (Fig. 1), or absent or apparently so..............**4**
　　4. Base of leaf not inflated. PONDWEED (*Potamogeton*), p. 55.
　　4. Base of leaf apparently inflated due to the closely fused stipules (Fig. 10). WIGEON GRASS (*Ruppia*), p. 77.
5. (From **11**, p. 4) Leaf with one central axis (Figs. 9, 11)......................**6**
5. Leaf irregularly forking (Figs. 4–7)..**8**
　　6. Each leaflet consisting of a single thread (Fig. 9).........................**7**
　　6. Each leaflet cut into numerous thread-like divisions (Fig. 11). LAKE CRESS (*Neobeckia*), p. 235.
7. Stem bearing leaves for nearly its entire length. WATER MILFOIL FAMILY (*Haloragidaceae*), p. 261.
7. Leaves crowded along a few centimeters of the stem, which usually has a whorl of inflated branches. FEATHERFOIL (*Hottonia*), p. 279.
　　8. Leaves borne singly (Fig. 5)..**9**
　　8. Leaves opposite (Fig. 6)...**10**
9. Stems or leaves with scattered little bladders (Fig. 7). BLADDERWORT (*Utricularia*), p. 309.
9. Stems and leaves without bladders (Fig. 5). WATER CROWFOOT (*Ranunculus*), **2**, p. 218.
　　10. Leaves stalked, extending from opposite sides of the stem (Fig. 6). FANWORT (*Cabomba*), p. 217.
　　10. Leaves much divided, extending in all directions around the stem (Fig. 4). WATER MARIGOLD (*Megalodonta*), p. 320.

Rootstocks of Cattail (**Typha latifolia**).

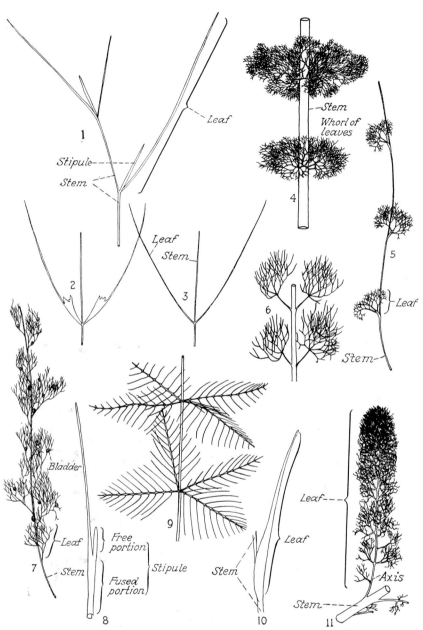

1. **Potamogeton**, diagrammatic.   2. **Najas,** showing dilated leaf bases, generalized.
3. **Zannichellia,** showing slender leaf bases, generalized.   4. **Megalodonta,** portion of stem and 2 nodes, × 1.   5. **Ranunculus,** generalized, × 1.   6. **Cabomba,** × 1.   7. **Utricularia,** generalized, × 1.   8. Leaf of **Potamogeton pectinatus,** diagrammatic.   9. **Myriophyllum,** generalized, × 1.   10. Leaf of **Ruppia,** generalized.   11. **Neobeckia,** × 1.

1. (From **3**, p. 4) Leaf with a ligule at junction of sheath and blade (Fig. 1).   GRASS
   FAMILY (*Gramineae*), p. 100, or sometimes SEDGE FAMILY (*Cyperaceae*), p. 122; see
   descriptions of these two families, p. 122.
1. Leaf without a ligule...2
   2. Leaves 2 mm. or less wide...3
   2. Leaves mostly 3 mm. or more wide..4
3. Horizontal stem, buried in the mud, about 1 mm. thick.   WATER BULRUSH
   (*Scirpus subterminalis*), p. 143.
3. Horizontal stem 5 mm. or more thick.   FLOWERING RUSH (*Butomus umbellatus*
   forma *vallisneriifolius*), p. 93.
   4. Leaf with a strong midrib (Figs. 2–4)...5
   4. Leaf with a weak midrib or none (Figs. 5, 6)..6
5. Roots hair-like, without cross markings; leaves with a broad central region showing
   many longitudinal and a few cross veins, and with 2 marginal zones the veins of
   which are few and obscure (Fig. 2).   WILD CELERY (*Vallisneria*), p. 99.   The
   venation is best seen with a lens, but after a little practice the 3 regions can be
   distinguished by the naked eye.
5. Roots thick as a broomstraw, with cross markings; leaves without 3 definite regions
   (Figs. 3, 4).   WATER PLANTAIN FAMILY (*Alismaceae*), p. 79, especially those
   following **19**, p. 91.
   6. Leaves veined as in Fig. 5.   PICKERELWEED, a submersed form (*Pontederia
      cordata* forma *taenia*), p. 173.
   6. Leaves veined as in Fig. 6.   BUR REED (*Sparganium*), p. 51.

Rosettes of **Sagittaria cristata**, uprooted from deep water by muskrats, and floating on the
surface.

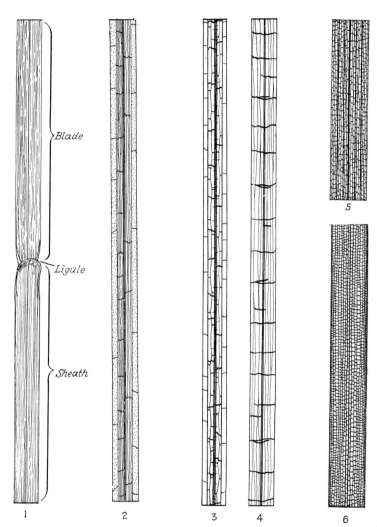

Portions of leaves, about natural size. 2–6 shown as by transmitted light. 1. Grass, generalized. 2. **Vallisneria.** 3, 4. **Sagittaria.** 5. **Pontederia cordata** forma **taenia.** 6. **Sparganium,** generalized.

1. (From **2,** p. 4) Leaves with little bunches of smaller leaves in their axils (Fig. 6). Bushy Pondweed (*Najas*), p. 77.

1. Leaves without bunches of leaves in their axils............................2

 2. Leaves opposite (Fig. 8), or whorled (Fig. 3, p. 25)......................3

 2. Leaves borne singly (Fig. 7)...5

3. Leaves whorled. Mare's-tail (*Hippuris*), p. 263.

3. Leaves opposite..4

 4. Stem with a narrow wing running down from each leaf base (Fig. 8). Water Purslane (*Didiplis diandra* forma *aquatica*), p. 255.

 4. Stem without a wing from each leaf base. Water Starwort (*Callitriche*), p. 241.

5. Each leaf with a ligule at the junction of the blade and sheath (Fig. 9); leaves usually opaque. Grass Family (*Gramineae*), p. 100; see also **24,** p. 63.

5. Leaves without a ligule, usually translucent so that a network of veins can be seen when it is held to the light and examined with a lens.........................6

 6. Leaf, when held to the light, showing a midrib (Figs. 2, 3, 4). Pondweed (*Potamogeton*), p. 55.

 6. Leaf without a midrib...7

7. Leaves 1 mm. or more wide above the base, rather abruptly pointed (Figs. 1, 5)...8

7. Leaves 0.5 mm. or less wide above the inflated base, tapered to a long point (Fig. 10, p. 7). Wigeon Grass (*Ruppia*), p. 77.

 8. Leaves without stipules, veined as in Fig. 1. Bur Reed (*Sparganium*), p. 51.

 8. Leaves with stipules, veined as in Fig. 5. Water Star Grass (*Heteranthera dubia*), p. 173.

Glyceria borealis with floating leaves (left foreground); Eleocharis palustris (right foreground); Typha latifolia (background).

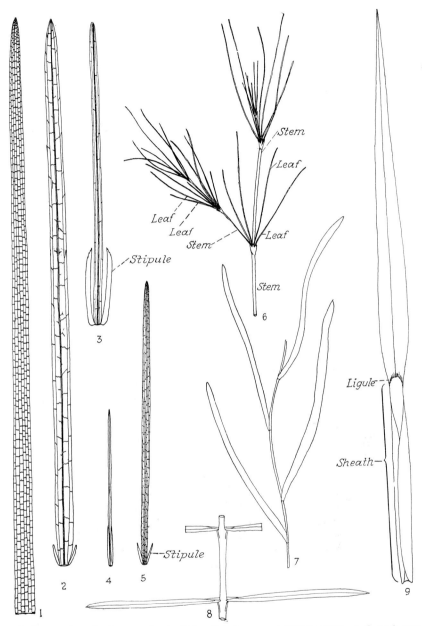

1. **Sparganium,** leaf, generalized, × 1. 2–4. Leaves of **Potamogeton,** each with two stipules at base. 5. **Heteranthera dubia,** leaf with two stipules at base. 6. **Najas,** part of plant, × about 1. 7. Plant with scattered ribbon-like leaves, diagrammatic. 8. **Didiplis diandra** forma **aquatica,** portion of stem, and leaves, × 2. 9. Grass, leaf, generalized; the ligule may vary in size and texture.

1. (From **2,** p. 3) Leaves represented by little bumps on the stem (Fig. 6). *Myrio-phyllum tenellum,* p. 263.

1. Plants not like Fig. 6..2

   2. Leaves, or naked green stems, all from one point (Figs. 4, 5, 8), or scattered along a horizontal buried stem (Figs. 1, 3)......................................3

   2. Leaves scattered on an erect stem (Figs. 2, 7)......................**9,** p. 18

3. Leaves or naked stems 1.5 dm. or more long...............................4

3. Leaves or naked stems 1 dm. or less long.........................**1,** p. 16

   4. Erect structures (usually leaves) flat...........................**1,** p. 14

   4. Erect structures (naked stems) round, square, or triangular in cross section ....**1,** p. 18

Juncus brevicaudatus.

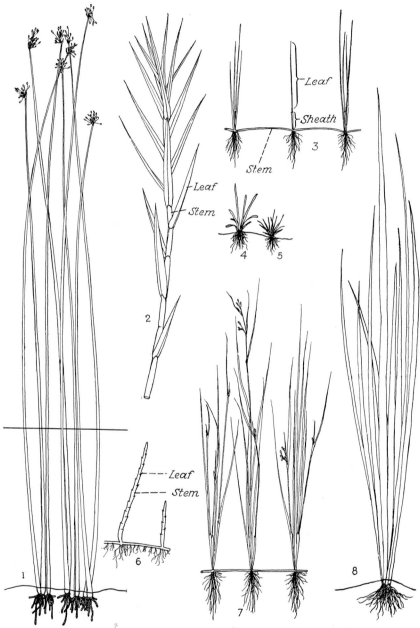

1. Bulrush (**Scirpus validus** or related species), × about $\frac{1}{10}$.   2. A sedge (**Dulichium**) with scattered leaves, × about $\frac{1}{4}$.   3. A sedge (**Eleocharis acicularis** or related species), × about $\frac{1}{2}$.   4, 5. Arrowhead (**Sagittaria**), Pipewort (**Eriocaulon**), or other rosette plants, generalized.   6. **Myriophyllum tenellum,** × $\frac{1}{2}$.   7. Sedges, generalized.   8. Sterile grass or sedge plant, generalized.

1. (From **4**, p. 12) Base of plant, where leaves enwrap one another, triangular in cross section (Fig. 7). Sedge Family (*Cyperaceae*), p. 122.
1. Base of plant not triangular..**2**
    2. Leaf with a ligule at the junction of blade and sheath (Fig. 2). Grass Family (*Gramineae*), p. 100.
    2. Leaf without a ligule; sheath present or absent.........................**3**
3. Leaf with a keel running down the center (Figs. 6, 8, 9).....................**4**
3. Leaf without a keel..**5**
    4. Crushed leaves with an aromatic odor. Sweet Flag (*Acorus*), p. 164.
    4. Crushed leaves without an odor. Bur Reed (*Sparganium*), p. 51.
5. Sheath abruptly narrowed to the blade (Fig. 3). Cattail (*Typha*), p. 49.
5. Sheath tapered to the blade (Fig. 4)....................................**6**
    6. Leaves 5 mm. or more wide. *Iris*, p. 180.
    6. Leaves less than 5 mm. wide...**7**
7. Leaves from a horizontal stem (Fig. 1). Flowering Rush (*Butomus*), p. 93.
7. Leaves in a close tuft, without a horizontal stem (Fig. 5). Yellow-eyed Grass (*Xyris*), p. 169.

Scirpus cyperinus var. **pelius.**

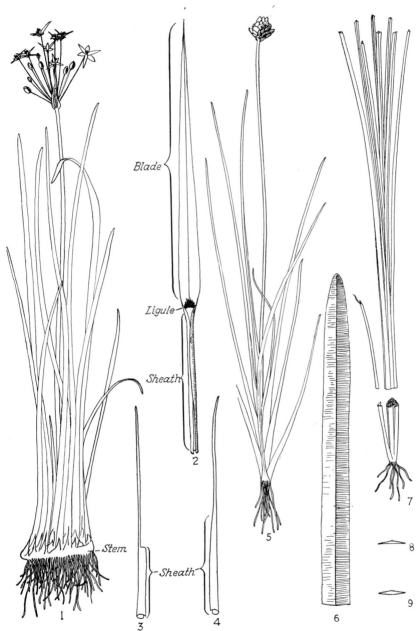

1. **Butomus,** × about ⅕. 2. Grass, leaf, generalized. 3. Leaf of **Typha,** diagrammatic, × ¹⁄₂₀. 4. Leaf with tapered sheath, diagrammatic. 5. **Xyris,** generalized. 6. **Acorus** or **Sparganium,** portion of leaf, generalized. 7. Sedge plant, diagrammatic. 8, 9. **Acorus** or **Sparganium,** leaves cut crosswise.

1. (From **3,** p. 12) Leaves consisting of 2 tubes joined side by side (Figs. 4, 7)· WATER LOBELIA (*Lobelia Dortmanna*), p. 319.
1. Leaves not consisting of 2 tubes...**2**
    2. Plant with an enlarged hard base (Fig. 5). QUILLWORT (*Isoetes*), p. 44.
    2. Plant without an enlarged base.....................................**3**
3. Leaves 1 cm. or more wide at base. WATER PLANTAIN FAMILY (*Alismaceae*), especially those following **20,** p. 91.
3. Leaves 3 mm. or less wide at base.................................**4**
    4. Tufts of leaves connected by fine horizontal stems (Figs. 1, 2, 6)...........**9**
    4. Tufts of leaves not connected by stems.............................**5**
5. Leaves with essentially parallel sides (Fig. 9). WATER STAR GRASS (*Heteranthera dubia* forma *terrestris*), p. 173.
5. Leaves tapering upward or downward **6**
    6. Leaves broadest at base and tapering upward (Figs. 8, 10)...............**7**
    6. Leaves broadest near or above the middle (Figs. 3, 11)..................**8**
7. Roots with cross lines or constrictions (Fig. 10). PIPEWORT (*Eriocaulon*), p. 169.
7. Roots without cross lines or constrictions (Fig. 8). AWLWORT (*Subularia*), p. 228.
    8. Leaves broadly rounded at tip (Fig. 11). MOUSETAIL (*Myosurus*), p. 227.
    8. Leaves pointed, or the tip slightly turned to one side (Fig. 3). YELLOW-EYED GRASS (*Xyris*), p. 169.
9. Erect green leaves or stems of essentially the same thickness throughout (Fig. 2), or somewhat enlarged and flattened upward (Fig. 1)......................**10**
9. Leaves thickest at base, tapering upward (Fig. 6).........................**12**
    10. Tufts of leaves connected by green runners about as thick as the leaves or their bases (Fig. 1). CREEPING SPEARWORT (*Ranunculus reptans*), **14,** p. 221. The leaves may be flat, as in Fig. 1, or thread-like.
    10. Tufts of leaves connected by thread-like white underground runners......**11**
11. Very common plants; erect green structures (stems) pointed at tip, or bearing a spikelet. SPIKE RUSH (*Eleocharis*, especially *E. acicularis*), **22,** p. 135.
11. Rare plants; leaves rounded at tip. MUDWORT (*Limosella*), p. 300.
    12. Leaves showing minute cross lines when held to the light and viewed with a lens..**13**
    12. Leaves opaque, not showing cross lines when held to the light. *Littorella*, p. 313.
13. Erect green structures (leaves) with the outer ones cupped around the inner ones at base. RUSH (*Juncus*, especially *J. pelocarpus* forma *submersus*), **19,** p. 179.
13. Erect green structures (stems) with the outer ones not cupped around the inner at base. SPIKE RUSH (*Eleocharis*, especially *E. acicularis*), **22,** p. 135.

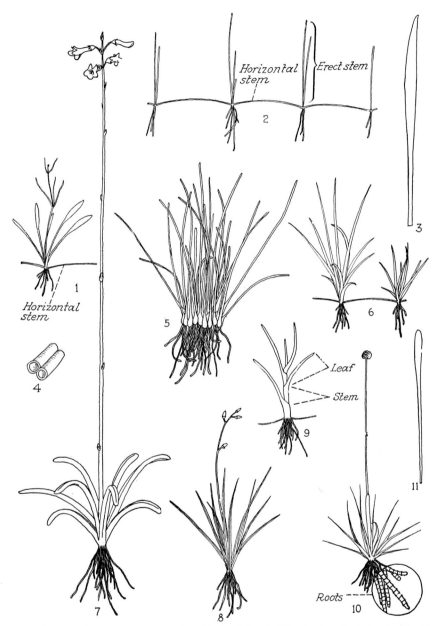

1. **Ranunculus reptans** var. **ovalis,** × about 1.   2. **Eleocharis acicularis,** × ½.   3. **Xyris,** leaf, generalized.   4. **Lobelia Dortmanna,** section of leaf, × 2.   5. **Isoetes,** generalized.   6. **Juncus pelocarpus** forma **submersus,** × 1.   7. **Lobelia Dortmanna,** × 1. 8. **Subularia,** × 1.   9. **Heteranthera dubia** forma **terrestris,** × 1.   10. **Eriocaulon,** × 1. with some roots enlarged.   11. **Myosurus,** leaf, generalized.

1. (From **4,** p. 12) Plants sterile...**4**
1. Plants with flowers at or near the tips of the stems........................**2**
    2. Flowers hidden behind overlapping scales (Figs. 1, 2, 5, 7, 12)............**3**
    2. Flowers visible, each with 3 scale-like sepals and 3 scale-like petals (Fig. 6).
        RUSH (*Juncus*), p. 174.
3. Spikelet of overlapping scales at the very tip of the stem, with no leaf continuing
    beyond it (Figs. 1, 2, 12). SPIKE RUSH (*Eleocharis*), p. 129.
3. Spikelets with a leaf extending beyond them (Figs. 5, 7). BULRUSH (*Scirpus*),
    p. 143.
    4. Stems 1 m. or more tall. BULRUSH (*Scirpus validus* or related species), **14,**
        p. 143.
    4. Stems less than 1 m. tall...**5**
5. Stems sharply angled...**6**
5. Stems nearly round in cross section....................................**7**
    6. Stems triangular in cross section (Figs. 5, 7). THREE-SQUARE (*Scirpus
        americanus* or related species), **6,** p. 143.
    6. Stems square in cross section (Fig. 1). SQUARE-STEM SPIKE RUSH (*Eleocharis
        quadrangulata*), **3,** p. 129.
7. Sheaths with edges often overlapping but not united to form a tube. RUSH
    (*Juncus*), p. 174.
7. Sheaths with edges united to form a tube about the stem (Figs. 2, 4)..........**8**
    8. Stems solitary, at intervals from a horizontal underground stem. SPIKE RUSH
        (*Eleocharis*, particularly *E. palustris* and its relatives), **5,** p. 131.
    8. Stems in bunches, from a tuft of roots. Some species of *Scirpus*, p. 143, and
        some of *Eleocharis*, p. 129; these are often difficult to distinguish in the sterile
        condition.
9. (From **2,** p. 12) Stem triangular in cross section (Figs. 4, 5, 7); the 3 angles may
    in some cases be so rounded as to make the stem appear nearly round in cross
    section. SEDGE FAMILY (*Cyperaceae*), p. 122.
9. Stem round in cross section...**10**
    10. Leaf with a ligule at the junction of sheath and blade (Fig. 3). GRASS
        FAMILY (*Gramineae*), p. 100.
    10. Leaf without a ligule...**11**
11. Leaves with sheaths at base (Figs. 2, 8, 9), or sometimes reduced to sheaths....**15**
11. Leaves without sheaths..**12**
    12. Leaves with file-like margins (Figs. 10, 11)...........................**13**
    12. Leaves with smooth margins..**17**
13. Leaves with 3 parallel veins (Fig. 10). GOLDENROD (*Solidago*), p. 339.
13. Leaves with faint veins or none branching from the midrib.................**14**
    14. Leaves 3 mm. or less wide (Fig. 11). BELLFLOWER (*Campanula*), p. 316.
    14. Leaves mostly 1 cm. or more wide. *Aster*, p. 341.
15. Leaves 3 mm. or less wide, or reduced to sheaths at the base of the stem. RUSH
    FAMILY (*Juncaceae*), p. 174.
15. Leaves 5 mm. or more wide...**16**
    16. Sheath abruptly contracted to the blade (Fig. 8). CATTAIL (*Typha*), p. 49.
    16. Sheath narrowed to the blade (Fig. 9). BUR REED (*Sparganium*), p. 51.
17. Leaves opposite. *Veronica*, p. 300.
17. Leaves scattered. *Lobelia*, p. 319.

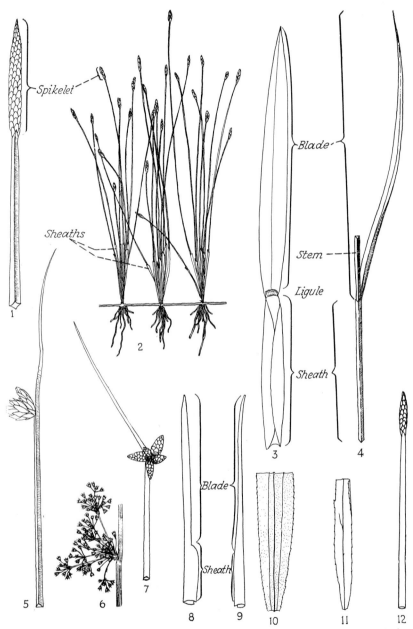

1. **Eleocharis quadrangulata**, summit of stem, $\times$ 1.   2. **Eleocharis**, generalized, showing horizontal underground stem, and sheaths.   3. Grass, leaf, generalized.   4. **Scirpus**, leaf and stem, generalized.   5. **S. americanus** or related species, summit of stem, $\times$ $\frac{1}{2}$.   7. **S. debilis** or related species, summit of stem, $\times$ 1.   6. **Juncus effusus**, inflorescence, $\times$ $\frac{1}{2}$.   8. **Typha**, leaf, $\times$ $\frac{1}{20}$.   9. **Sparganium**, leaf, generalized, $\times$ $\frac{1}{10}$.   10. **Solidago graminifolia**, basal part of leaf, $\times$ 2.   11. **Campanula aparinoides**, basal part of leaf, $\times$ 2.   12. **Eleocharis palustris** or related species, summit of stem, $\times$ 1.

1. (From **3,** p. 3) Leaves all from one point (Fig. 2) **2**

1. Leaves on an elongate stem (figures, p. 23) . . . .**1,** p. 22

    2. Leaves with 3 or more veins running from the base to the tip (Figs. 3, 5, 6, 8) . . .**3**

    2. Leaves with but one vein which runs from the base to the tip, and often with branch veins (Figs. 1, 4, 7) **8**

3. Leaf with stipules at the base of the petiole (Fig. 3). PONDWEED (*Potamogeton*), p. 55. These are short-stemmed forms of normally long-stemmed plants, which have been stranded on the mud. The most common is *P. gramineus* **16,** p. 57, and Fig. 14, p. 59.

3. Leaves without stipules **4**

    4. Central vein more conspicuous than the others (Figs. 6, 8) **5**

    4. Central vein not more conspicuous than the others (Fig. 5). GOLDEN CLUB (*Orontium*), p. 164.

5. Plants normally 1 dm. or more tall **6**

5. Plants 5 cm. or less tall **7**

    6. Leaves green beneath; long veins more conspicuous than the cross veins (Fig. 8). WATER PLANTAIN FAMILY (*Alismaceae*), p. 79.

    6. Leaves purple beneath; cross veins more conspicuous than the long veins (Fig. 6). FROGBIT (*Limnobium*), p. 99.

7. Common plants; rosettes of leaves connected by runners which are as thick as the petioles. CREEPING SPEARWORT (*Ranunculus reptans* var. *ovalis*), **14,** p. 221.

7. Very rare plants of northern and western regions; rosettes connected by white thread-like runners. MUDWORT (*Limosella*), p. 300.

    8. Blade about as broad as long (Fig. 4). WATER CHESTNUT (*Trapa*), p. 261.

    8. Blade several times as long as broad **9**

9. Blade 20 cm. or more long (Fig. 1). DOCK (*Rumex*), p. 198.

9. Blade 12 cm. or less long (Fig. 7). LANCE-LEAVED VIOLET (*Viola lanceolata*). p. 251.

Decodon verticillatus.

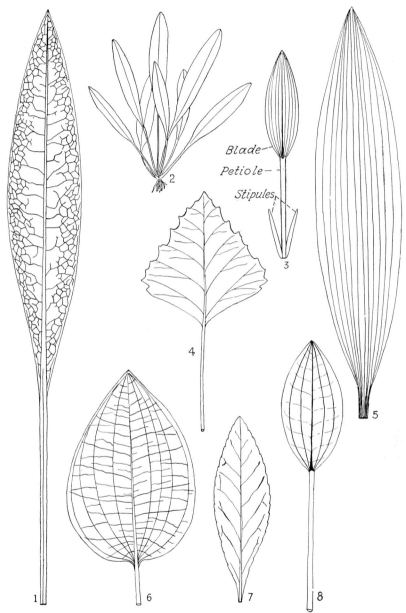

1. **Rumex,** leaf, $\times$ about $\frac{1}{4}$. 2. Plant with leaves from one point, generalized. 3. **Potamogeton,** emersed leaf. 4. **Trapa,** leaf, $\times$ about $\frac{1}{2}$. 5. **Orontium,** leaf, $\times \frac{1}{2}$. 6. **Limnobium,** leaf, $\times \frac{1}{2}$. 7. **Viola lanceolata,** leaf, $\times \frac{1}{2}$. 8. **Alismaceae,** leaf, generalized.

1. (From **1**, p. 20) Leaves opposite (Fig. 1) or whorled (Fig. 2)..................2
1. Leaves alternate (Fig. 3), or the uppermost pair sometimes opposite in the Pond-weeds...**1**, p. 26
    2. Leaf margins toothed or wavy (Figs. 4, 5, 10)...........................3
    2. Leaf margins not toothed (Figs. 6, 7, 8)................................7
3. Stems square in cross section. Mint Family (*Labiatae*), p. 286; some species of *Mimulus*, p. 305, also have square stems.
3. Stems round in cross section...4
    4. Pairs of leaves joined at base to encircle the stem (Fig. 10); leaves rarely separate in occasional individuals. Boneset (*Eupatorium perfoliatum*), p. 320.
    4. Leaves not joined at base...5
5. Leaves in whorls of 3–6. Joe-Pye Weed (*Eupatorium*), p. 320.
5. Leaves opposite...6
    6. Branch veins essentially straight, from the midrib nearly to the margin (Fig. 4). Beggar-ticks (*Bidens*), p. 325.
    6. Branch veins turning to run nearly parallel with the margin, often becoming lost in a network of veins (Fig. 5). Figwort Family (*Scrophulariaceae*), p. 300.
7. Leaves with numerous semitransparent dots visible with a lens when the leaf is held to the light. St. John's-wort (*Hypericum*), p. 245.
7. Leaves without semitransparent dots....................................8
    8. Petiole or base of leaf fringed (Fig. 9). Loosestrife (*Lysimachia*), p. 279.
    8. Petiole and leaf base not fringed...................................9
9. Leaves spattered with pin-point black dots, visible with a lens..............14
9. Leaves not spattered with black dots...................................10
    10. Leaf divided into blade and petiole (Figs. 7, 8)......................11
    10. Leaves without petioles...**1**, p. 24
11. Stem weak and thread-like, supported only by the water (Fig. 6); leaves round-tipped. Water Starwort (*Callitriche*), p. 241.
11. Stem usually 2 mm. or more thick, rigid enough to support itself; leaves pointed...12
    12. Juice milky. Milkweed (*Asclepias*), p. 285.
    12. Juice not milky...13
13. Blade of leaf about one-half as broad as long (Fig. 7). False Loosestrife (*Ludwigia*), p. 259.
13. Blade several times as long as broad (Fig. 8). Loosestrife Family (*Lythraceae*), p. 252.
    14. Dots irregular in shape, not raised above the surface of the leaf. Loosestrife (*Lysimachia*), p. 279.
    14. Dots nearly spherical, composed of waxy droplets. Hedge Hyssop (*Gratiola*), p. 307.

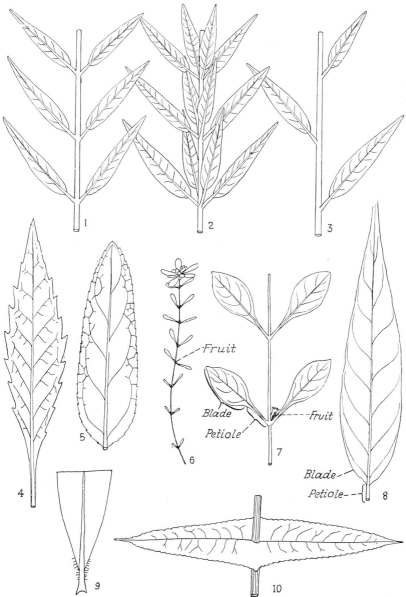

1. Opposite leaves, diagrammatic. 2. Whorled leaves, diagrammatic. 3. Alternate leaves, diagrammatic. 4. **Bidens,** leaf $\times$ ½. 5. **Scrophulariaceae,** leaf, diagrammatic. 6. **Callitriche,** $\times$ ½, showing floating rosette of uppermost leaves, and fruits in the axils of other leaves. 7. **Ludwigia palustris,** $\times$ ½, showing a fruit in one leaf axil. 8. **Decodon,** leaf, $\times$ ½. 9. **Lysimachia,** lower part of leaf, generalized. 10. **Eupatorium perfoliatum,** a pair of leaves united about the stem, $\times$ ¼.

1. (From **10**, p. 22) Leaves more than 2 at a level.............................**2**
1. Leaves 2 at each level...**4**
  2. Distance between whorls of leaves less than the length of the leaves (Figs. 1, 3)..**3**
  2. Distances between whorls of leaves greater than the length of the leaves (Figs. 60, 61, 62, p. 315). Bedstraw (*Galium*), p. 316.
3. Leaves 6–12 in a whorl (Fig. 3). Mare's-tail (*Hippuris*), p. 263.
3. Leaves 2–4 in a whorl (Fig. 1). Waterweed (*Anacharis*), p. 99.
  4. Leaves 7–15 cm. long. Water Willow (*Dianthera*), p. 313.
  4. Leaves 7 cm. or less long..**5**
5. Stems with sharp ridges or wings running down from the base of each leaf (Figs. 9, 10)..**6**
5. Stems without ridges from the leaf bases...............................**10**
  6. Leaves cut square across the base (Fig. 10)............................**7**
  6. Leaves tapered at base (Figs. 7, 9)...................................**9**
7. Leaves nearly as broad as long. Water Hyssop (*Bacopa*), p. 307.
7. Leaves several times as long as broad.................................**8**
  8. Leaves with branch veins in addition to the midrib, and often with file-like edges. Spiked Loosestrife (*Lythrum*), p. 252.
  8. Leaves with no veins in addition to the midrib (Fig. 10), or rarely with a few short faint veins, and with smooth edges. *Ammannia*, p. 255.
9. Leaves several times as long as broad, the side veins lacking or very obscure (Fig. 9)...**14**
9. Leaves not more than twice as long as broad, with conspicuous branch veins in addition to the midrib (Fig. 7). St. John's-wort (*Hypericum*), p. 245.
  10. Leaf bases joined in a little cup (Fig. 4). *Tillaea*, p. 239.
  10. Leaf bases not joined, although often touching.......................**11**
11. Bases of leaves embracing the stem (Fig. 8); some leaves usually in 3's or 4's. Waterweed (*Anacharis*), p. 99.
11. Leaves not embracing the stem, always in pairs.......................**12**
  12. Leaves pointed. Figwort Family (*Scrophulariaceae*), p. 300.
  12. Leaves blunt, rounded, or indented at tip............................**13**
13. Very common plants, most species with ribbon-like submerged leaves and a terminal rosette of broader leaves (Fig. 2); fruit in 2 one-seeded sections. Water Starwort (*Callitriche*), p. 241.
13. Very rare plants known from a few ponds in Maine, Wis., Ohio & Ill.; leaves narrowly oval (Fig. 5); fruit a pod with such papery walls that the seeds within are visible (Fig. 6). Waterwort (*Elatine*), p. 250.
  14. Plants 4 dm. or more tall, with creeping rootstocks; usually with a raceme of yellow flowers (Fig. 5, p. 280) or axillary bulblets (Fig. 10, p. 281). Swamp-candle (*Lysimachia terrestris*), **2**, p. 279.
  14. Plants mostly 3 dm. or less tall, without rootstocks; flowers borne singly in the leaf-axils (figures, p. 256). Water Purslane (*Didiplis diandra* forma *terrestris*, p. 255, and *Rotala*, p. 255.

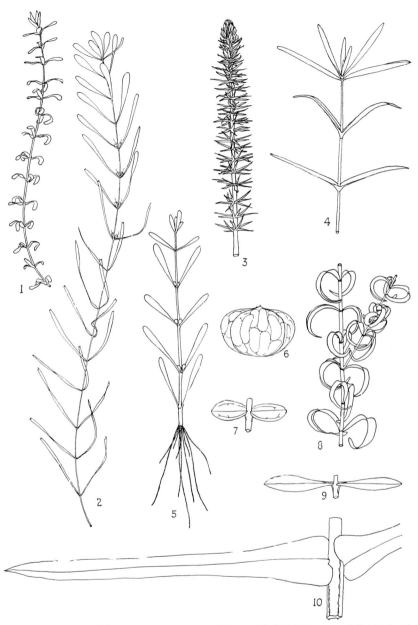

1. **Anacharis**, $\times \frac{1}{2}$. 2. **Callitriche**, generalized, with fruits in some of the leaf axils, $\times 1$. 3. **Hippuris**, upper part of plant, $\times \frac{1}{2}$. 4. **Tillaea**, upper part of plant, $\times 5$. 5. **Elatine triandra** forma **submersa**, $\times 1$. 6. **Elatine**, pod, generalized, $\times 25$. 7. **Hypericum boreale** forma **callitrichoides**, a node and 2 leaves, $\times 1$. 8. **Anacharis**, $\times 1$. 9. **Didiplis diandra** forma **terrestris**, a node and 2 leaves, $\times 2$. 10. **Ammannia coccinia**, node, leaf and base of another, $\times 2$, showing the somewhat roughened wings on the stem.

1. (From **1**, p. 22) Leaf margins toothed (nearly all figures on p. 27)............**2**

1. Leaf margins not toothed (all figures on p. 29).......................**1**, p. 28

    2. Leaves not petioled (Fig. 8)..**9**

    2. Leaves petioled (Figs. 1, 5, 9)......................................**3**

3. Leaves velvety beneath. Rose Mallow (*Hibiscus*), p. 245.

3. Leaves not velvety beneath..**4**

    4. Blade nearly as broad as long..**5**

    4. Blade several times as long as broad.................................**6**

5. Leaf coarsely toothed, the petiole with a swelling (Fig. 3). Water Chestnut (*Trapa*), p. 261.

5. Leaf with a wavy margin, the petiole not swollen (Fig. 4). Cress Family (*Cruciferae*), p. 228.

    6. Base of petiole embracing the stem (Fig. 2)..........................**7**

    6. Base of petiole not embracing the stem...............................**8**

7. Base of petiole with a collar-like sheath embracing the stem (Fig. 5). Dock (*Rumex*), p. 198.

7. Base of petiole without a sheath. Spearwort (*Ranunculus*), **13**, p. 221.

    8. Stems erect, or the plant standing mostly above the surface of the water; leaves all essentially alike...**11**

    8. Stems mostly prostrate in the water, the lower leaves tending to be more deeply lobed than the upper (Fig. 9) or even cut into thread-like leaflets. Mermaid Weed (*Proserpinaca*), p. 261.

9. Leaves with minutely saw-toothed edges, which feel like a file. Composite Family (*Compositae*), p. 320.

9. Leaves with edges smooth to the touch.................................**10**

    10. Each vein ending in a white gland at the leaf margin (Fig. 7). *Lobelia*, p. 319.

    10. Veins not ending in glands. Cress Family (*Cruciferae*), p. 228.

11. Leaf margins evenly and uniformly toothed............................**12**

11. Leaves with teeth larger or more coarse, sometimes grading into lobes, toward the base (Fig. 6). Cress Family (*Cruciferae*), p. 228.

    12. Base of petiole fringed with white hairs about as long as the width of the petiole. Bellflower (*Campanula*), p. 316.

    12. Base of petiole without long white hairs...........................**13**

13. Leaves with fine white hairs beneath (visible with a lens) which make them rough to the touch. Frog-fruit (*Lippia*), p. 286.

13. Leaves smooth and without hairs beneath (Fig. 1). Ditch Stonecrop (*Penthorum*), p. 239.

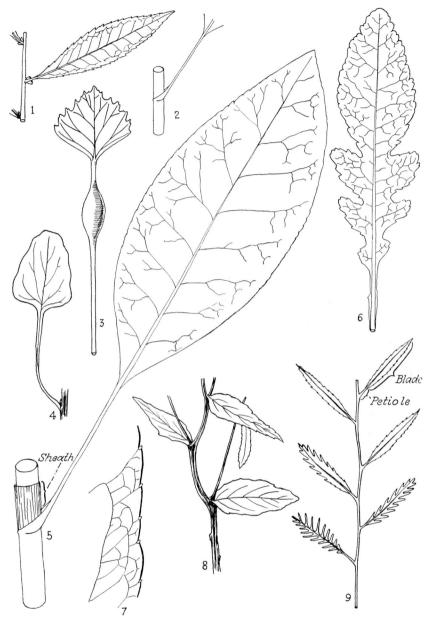

1. **Penthorum,** portion of stem, leaf and bases of two others, × ½. 2. Petiole with clasping base, diagrammatic. 3. **Trapa,** leaf, × ½. 4. **Neobeckia,** leaf, × ½. 5. **Rumex,** leaf, × ½. 6. **Rorippa palustris** var. **glabrata,** leaf, × 1. 7. **Lobelia cardinalis,** portion of leaf margin, × 5. 8. **Neobeckia,** part of stem and 4 leaves, × ½. 9. **Proserpinaca palustris,** part of plant, × ½.

1. (From **1,** p. 26) Leaves longer than the thickness of the stem **2**
1. Leaves represented by little bumps shorter than the thickness of the stem (Fig. 1). *Myriophyllum tenellum,* p. 263.
    2. Veins several, running from the base of the leaf to the apex without branching (Fig. 3) **13**
    2. Veins branching, or leaf with only the midrib showing **3**
3. Stem with a sheath from the base of the petiole (Fig. 10). Smartweed (*Polygonum*), p. 198.
3. Stem without sheaths **4**
    4. Stem with ridges or wings running down from the base of each leaf (Fig. 2). Loosestrife (*Lythrum*), p. 252.
    4. Stem without ridges or wings from the leaf base **5**
5. Leaves with 3 or more veins coming from near the base (Fig. 9). Spearwort (*Ranunculus*), **13**, p. 221.
5. Leaves feather-veined (Figs. 6, 7, 10) or with only the midrib **6**
    6. Plants covered with little bristly hairs. Forget-me-not (*Myosotis*), p. 285.
    6. Plants not hairy **7**
7. Stems slender and angled, the angles as well as the leaf margins with minute teeth which can be seen only with a lens. Bellflower (*Campanula*), p. 316.
7. Stems not angled and toothed **8**
    8. Leaf blades only a few millimeters long, and nearly as broad (Fig. 4). Water Cress (*Nasturtium officinale* forma *nanum*), p. 237.
    8. Leaves mostly 1 cm. or more long, much longer than broad **9**
9. Stems creeping, with clusters of narrow leaves (Fig. 5). Creeping Spearwort (*Ranunculus reptans* var. *ovalis*), p. 221.
9. Stems usually erect, the leaves not clustered **10**
    10. Leaves several times as long as broad **11**
    10. Leaves not over twice as long as broad. Water Pimpernel (*Samolus*), p. 279.
11. Leaves mostly 5 mm. or less wide, with whitish gland-like teeth (Fig. 7, p. 27). *Lobelia*, p. 319.
11. Leaves more than 5 mm. wide, without whitish glandular teeth **12**
    12. Leaves veined as in Fig. 7. Composite Family (*Compositae*), p. 320.
    12. Leaves veined as in Fig. 6. False Loosestrife (*Ludwigia*), p. 259.
13. Stems usually long and flexible, supported by the water (Fig. 8); stipules free from the petiole (Fig. 3). Pondweed (*Potamogeton*), p. 55.
13. Stems mostly stout; stipules fused with the petiole to form a sheath **14**
    14. Creeping or floating plants, the leaf blades not over 5 cm. long. Mud Plantain (*Heteranthera*), p. 173.
    14. Stems erect, usually with most of the leaves at the base, with blades 5–25 cm. long and commonly with basal lobes. Pickerelweed (*Pontederia*), p. 173.

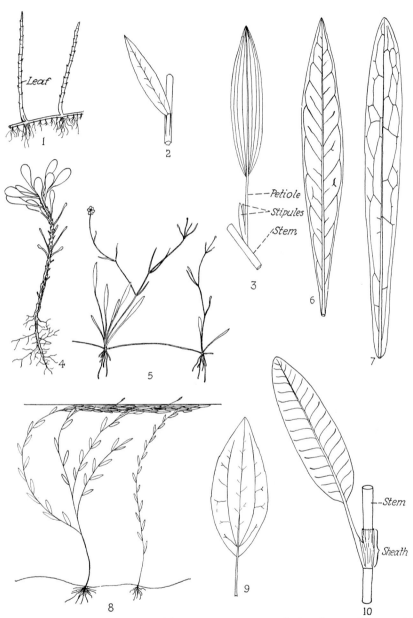

1. **Myriophyllum tenellum,** $\times \frac{1}{2}$. 2. **Lythrum alatum,** leaf and part of stem, $\times 1$. 3. **Potamogeton,** leaf, generalized, with stipules at base. 4. **Nasturtium officinale** forma nanum, $\times \frac{1}{2}$. 5. **Ranunculus reptans** var. **ovalis,** $\times \frac{1}{2}$. 6. **Ludwigia polycarpa,** leaf, $\times 1$. 7. **Boltonia asteroides,** leaf, $\times 1$. 8. **Potamogeton,** generalized, the stems reaching and partly floating on the surface. 9. **Ranunculus laxicaulis,** leaf, $\times 1$. 10. **Polygonum,** leaf, generalized.

1. (From **4**, p. 3) Leaves needle-like (Fig. 8). CYPRESS (*Taxodium*), p. 49.
1. Leaves flat and broad...**2**
   2. Leaves or buds fragrant when crushed. SWEET GALE (*Myrica*), p. 193.
   2. Leaves and buds not fragrant..**3**
3. Leaves pinnately compound. ASH (*Fraxinus*), p. 282.
3. Leaves simple...**4**
   4. Leaves toothed on the margin (Figs. 9, 10)...........................**8**
   4. Leaves not toothed (Figs. 1, 4)....................................**5**
5. Leaves 2.5 cm. or less long, covered beneath with minute brownish scales. LEATH-
   ERLEAF (*Chamaedaphne*), p. 276.
5. Leaves mostly 5 cm. or more long, not scaly beneath......................**6**
   6. Leaves opposite (Fig. 4); shrubs...................................**7**
   6. Leaves alternate (Fig. 1) or in clusters; trees. TUPELO (*Nyssa*), p. 276.
7. Leaves with minute close white hairs beneath. DOGWOOD (*Cornus*), p. 276.
7. Leaves without hairs beneath. BUTTONBUSH (*Cephalanthus*), p. 316.
   8. Leaves opposite (Fig. 10). SWAMP PRIVET (*Forestiera*), p. 282.
   8. Leaves alternate...**9**
9. Leaves at least 3 times as long as broad. WILLOW (*Salix*), p. 182.
9. Leaves less than twice as long as broad................................**10**
   10. Leaves broadly wedge-shaped at base, with many teeth (Fig. 9). ALDER
     (*Alnus*), p. 197.
   10. Leaves rounded or heart-shaped at base, with seldom more than 25 teeth on
     each side. WATER ELM (*Planera*), p. 197.
11. (From **5**, p. 3) Veins radiating (Figs. 5, 7)............................**12**
11. Veins branching from the midrib (Figs. 2, 6)...........................**13**
   12. Petiole attached near the center of the blade (Fig. 5). WATER PENNYWORT
     (*Hydrocotyle*), p. 270.
   12. Petiole attached at lower edge of the blade (Fig. 7). CROWFOOT (*Ranun-
     culus*), **18**, p. 221.
13. Vines. NIGHTSHADE (*Solanum*), p. 299.
13. Herbs...**14**
   14. Leaves scattered, or clustered at the base of the plant, the lobes usually with
     blunt or rounded teeth (Figs. 2, 3). CRESS FAMILY (*Cruciferae*), p. 228.
   14. Leaves opposite...**15**
15. Margins of lobes with sharp teeth (Fig. 6). BEGGAR-TICKS (*Bidens*), p. 325.
15. Margins of lobes not toothed, or lobes cut into a few coarse blunt teeth. *Leucospora*, p. 300.

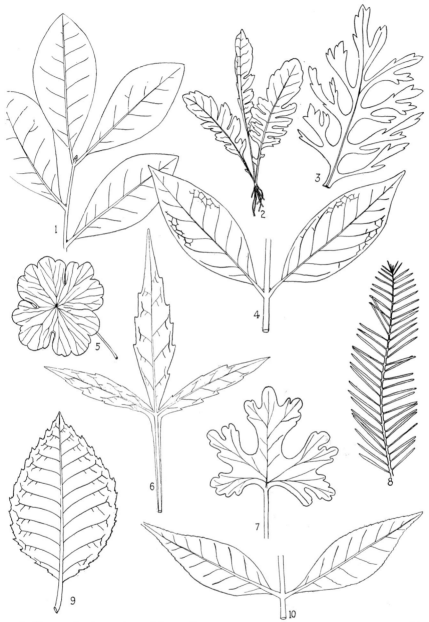

1. **Nyssa sylvatica,** twig, $\times \frac{1}{2}$. 2. **Rorippa palustris** var. **hispida,** basal leaves, $\times \frac{1}{2}$.
3. **Neobeckia,** submersed leaf, $\times 1$. 4. **Cephalanthus,** pair of leaves, $\times \frac{1}{4}$. 5. **Hydro-cotyle,** leaf, generalized. 6. **Bidens connata,** leaf, $\times \frac{1}{5}$. 7. **Ranunculus sceleratus,** leaf, $\times \frac{1}{2}$. 8. **Taxodium,** branchlet, $\times \frac{1}{2}$. **9. Alnus incana,** leaf, $\times \frac{1}{2}$. 10. **Forestiera,** pair of leaves, $\times \frac{1}{2}$.

1. (From **6,** p. 3) Trees. Ash (*Fraxinus*), p. 282.

1. Non-woody plants..**2**

   2. Divisions of leaves narrow, not toothed (Figs. 1, 3, 4)......................**3**

   2. Divisions of leaves toothed (Figs. 6, 10), or cut into smaller divisions (Figs. 2, 7)..**4**

3. Leaves crowded along a short portion of the stem (Fig. 1). Featherfoil (*Hottonia*), p. 279.

3. Leaves occupying most of the length of the stem (Figs. 3, 4). Water Milfoil Family (*Haloragidaceae*), p. 261.

   4. Leaves opposite (Fig. 8). Beggar-ticks (*Bidens*), p. 325.

   4. Leaves borne singly..**5**

5. Base of petiole expanded into a sheath (Figs. 5, 9); leaflets usually all about the same size..**6**

5. Base of petiole not expanded into a sheath; terminal leaflet usually (Figs. 10, 11), but not invariably (Fig. 2), the largest. Cress Family (*Cruciferae*), p. 228.

   6. Stem completely encircled by the base of the petiole (Fig. 9)...............**7**

   6. Stem not encircled by the base of the petiole (Fig. 5). Crowfoot (*Ranunculus*), **18,** p. 221.

7. Leaflets rounded at tip (Fig. 9). Cinquefoil (*Potentilla*), p. 241.

7. Leaflets pointed (Fig. 6) or cut into smaller divisions (Fig. 7). Parsley Family (*Umbelliferae*), p. 268.

Dulichium arundinaceum.

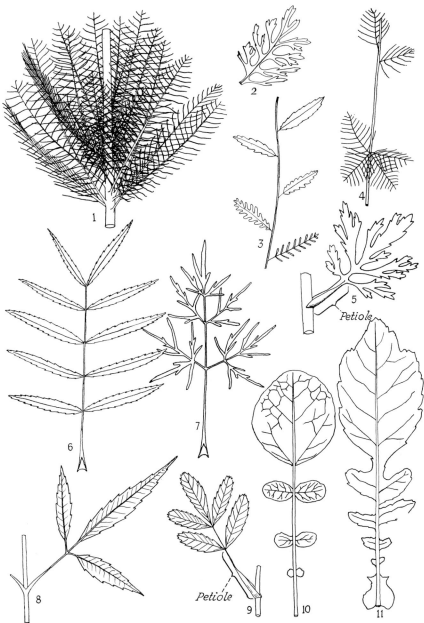

1. **Hottonia,** leaf-bearing portion of plant, $\times \frac{1}{2}$. 2. **Neobeckia,** submersed leaf, $\times \frac{1}{2}$. 3. **Proserpinaca,** showing divided leaves below and entire leaves above, $\times \frac{1}{2}$. 4. **Myriophyllum,** generalized, with 2 leaves scattered and 4 leaves whorled. 5. **Ranunculus,** leaf, generalized. 6. **Sium,** leaf, $\times \frac{1}{5}$. 7. **Sium,** submersed leaf, $\times \frac{1}{5}$. 8. **Bidens frondosa,** leaf, $\times \frac{1}{2}$. 9. **Potentilla palustris,** leaf, $\times \frac{1}{4}$. 10. **Nasturtium,** leaf, $\times \frac{1}{2}$. 11. **Rorippa palustris** var. **glabrata,** leaf, $\times \frac{1}{2}$.

  1. (From **7,** p. 3) leaf margin toothed......................................**2**
  1. Leaf margin not toothed..**3**
    2. Leaf almost round (Fig. 8).   Marsh Marigold (*Caltha*), p. 227.
    2. Leaf long-triangular, the lobes spreading (Fig. 7).   Rose Mallow (*Hibiscus militaris*), p. 245.
  3. Petiole attached in the middle of the blade (Fig. 1)........................**4**
  3. Petiole attached at the lower margin of the blade (Figs. 4, 6, 10).............**5**
    4. Leaf not more than 8 cm. wide.   Water Shield (*Brasenia*), p. 217.
    4. Leaf 15 cm. or more wide.   Lotus (*Nelumbo*), p. 217.
  5. Blade with a little finger-like tip (Fig. 5).   Arum Family (*Araceae*), p. 164.
  5. Blade without a finger-like tip...**6**
    6. Basal lobes pointed (Fig. 10).   Water Plantain Family (*Alismaceae*), p. 79.
    6. Basal lobes rounded...**7**
  7. Blade rounded at tip (Fig. 9)...**8**
  7. Blades mostly pointed at tip (Figs. 3, 11)................................**9**
    8. All radiating veins equally conspicuous, branching and recurving to unite with each other (Fig. 6).   Floating Heart (*Nymphoides*), p. 282.
    8. The central vein the heaviest, the others forking but not recurving (Fig. 9).   Water Lily Family (*Nymphaeaceae*), p. 210.
  9. Leaves with netted veins (Fig. 11).   Lizard's Tail (*Saururus*), p. 181.
  9. Leaves with parallel veins (Figs. 2, 3, 4)................................**10**
    10. Blades with additional cross veins (Fig. 4).   Frogbit (*Limnobium*), p. 99.
    10. Blades with only parallel veins.......................................**11**
  11. Stem erect, with most of the leaves at its base (Fig. 3); blades 5–20 cm. long. Pickerelweed (*Pontederia*), p. 173.
  11. Creeping or floating plants (Fig. 2), the blades rarely more than 5 cm. long. Mud Plantain (*Heteranthera*), p. 173.

Artificial planting of Arrowhead, Waterlily, Iris, etc.   Photograph by W. W. Chase, Soil Conservation Service.

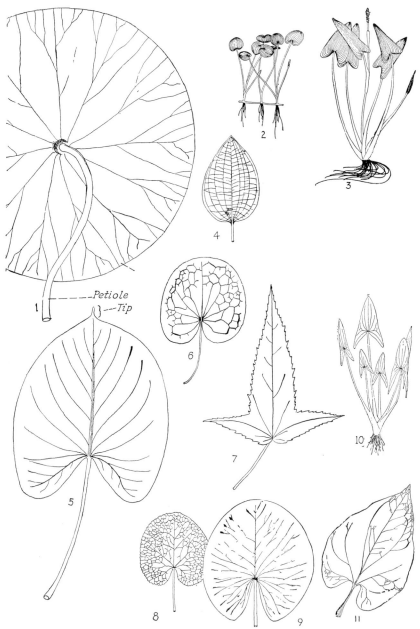

Petiole
Tip

1. **Nelumbo**, part of leaf, $\times \frac{1}{4}$.   2. Heteranthera reniformis, $\times \frac{1}{4}$.   3. Pontederia cordata, $\times \frac{1}{4}$.   4. **Limnobium**, leaf, $\times \frac{1}{4}$.   5. **Calla**, leaf. $\times \frac{1}{2}$.   6. **Nymphoides**, leaf. $\times \frac{1}{2}$.   7. **Hibiscus militaris**, leaf, $\times \frac{1}{2}$.   8. **Caltha**, leaf, $\times \frac{1}{4}$.   9. **Nuphar**, leaf, $\times \frac{1}{4}$.   10. **Sagittaria latifolia**, $\times \frac{1}{8}$.   11. **Saururus cernuus**, leaf, $\times \frac{1}{4}$.

PART II

DESCRIPTIVE TREATMENT

GREEN ALGAE   CHLOROPHÝCEAE
RED ALGAE   RHODOPHÝCEAE

These are shapeless masses of microscopic green threads. Sometimes they are attached to stones or to larger plants, and sometimes they float as bubble-filled scum. The Green Algae include a vast number of genera and species the identification of which is possible only with the aid of a microscope; a few of the common forms, however, can be recognized in the field with the use of a hand lens, but in all cases determinations should be verified by a microscopic examination.

1. Threads forming a net, visible to the unaided eye (Figs. 1, 2). **Hydrodíctyon,** the Water Net.

1. Threads not forming a net...2

    2. Threads with a silky, slippery feel; a mass, when picked up, tapering at its lower end to a corkscrew tip (Fig. 3)...................................3

    2. Threads of a cottony texture, not slippery...............................4

3. Chloroplast spiraled (Figs. 4, 5, 6). **Spirogýra.**

3. Chloroplast star-like (Fig. 7). **Zygnèma.**

    4. Plants floating, not attached...5

    4. Plants attached to stones or sticks....................................7

5. Threads freely branching (Figs. 9, 10, 12). **Cladóphora** and **Rhizoclònium.**

5. Threads branching slightly or not at all..................................6

    6. Threads about as thick as a hair (Fig. 12). **Rhizoclònium.**

    6. Threads finer than cotton fibers (Figs. 14, 15). **Oedogònium** and some of its relatives.

7. Plants forming a close felt-like mass. **Vauchèria.** Fig. 8.

7. Plants not close and thread-like...8

    8. Threads with a head-like cluster of branches (Figs. 11, 13, 16). **Batrachospérmum** and **Draparnàldia.**

    8. Threads without head-like clusters.....................................9

9. Fine unbranched threads in cold, constantly flowing water. **Ûlothrix.** Figs. 17, 18.

9. Fine or coarse threads, somewhat branched (Figs. 9, 10, 12). **Cladóphora** and **Rhizoclònium.**

Reference: The most useful comprehensive work on Algae is The Fresh-water Algae of the United States, by G. M. Smith, published by McGraw-Hill Book Company, Inc.

STONEWORTS, MUSKGRASS   CHARÀCEAE

This family of Algae is unique in its possession of cylindrical whorled branches. Each joint of the stem consists of a single cell. The plants

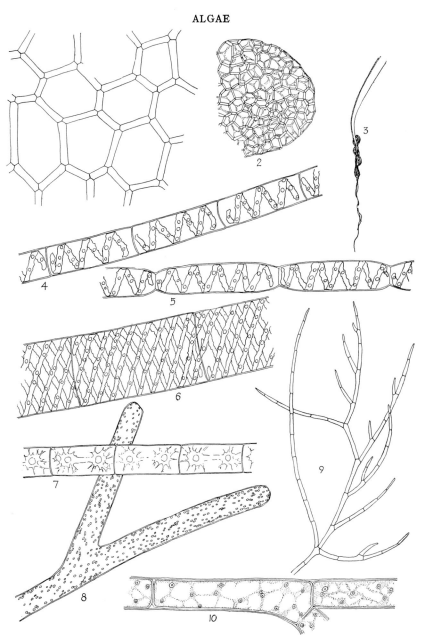

Hydrodictyon: 1. Portion of plant, × 5. 2. Portion of plant, × 1. **Spirogyra:** 3. Bit of plant held by forceps, × 1. 4, 5, 6. Different species, × 300. **Zygnema:** 7. Part of plant, × 300. **Vaucheria:** 8. Part of plant, × 100. **Cladophora:** 9. Part of plant, × 50. 10. Part of plant, × 300.

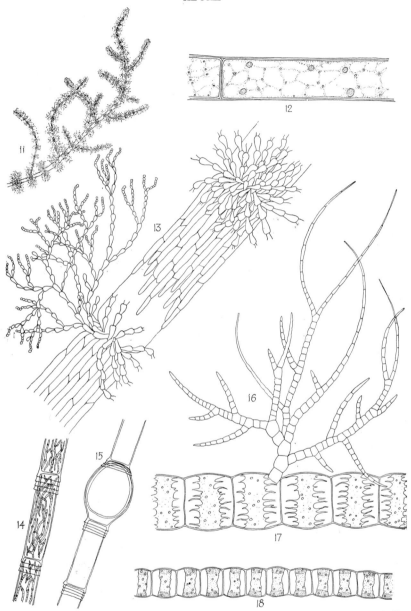

Batrachospermum: 11. Plant, × 5.   13. Part of plant, × 25.   **Rhizoclonium**: 12. Part of plant, × 300.   **Oedogonium**: 14, 15. Parts of plant, × 150.   **Draparnaldia**: 16. Part of plant, × 300.   **Ulothrix**: 17, 18. Parts of plants, × 300.

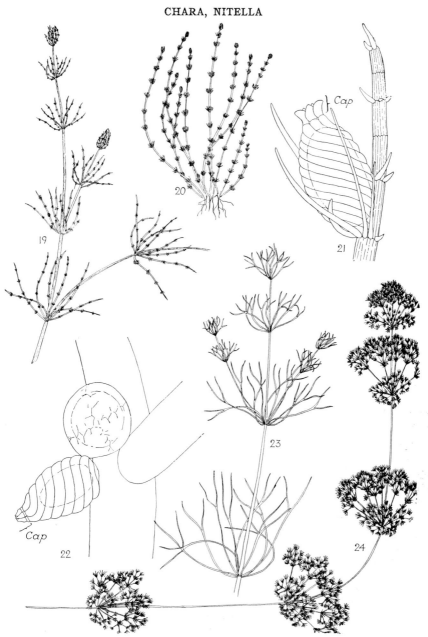

Chara: 21. Oögonium, × 45. **C. vulgaris:** 19. Plant, × 1. **C. aspera:** 20. Plant, × 1. **Nitella:** 22. Oögonium, × 45. **N. flexilis:** 23. Plant, × 1. **N. hyalina:** 24. Plant, × 1.

almost always occur in hard water and often have incrustations of lime. The two genera here described can sometimes be differentiated by use of a hand lens, but for certainty the tip of the oögonium (egg-bearing case) must be examined with a microscope. The illustrations (Figs. 19–24) show but a few of the many species.

1. Plants rough and harsh, appearing lined (Fig. 21) when seen with a lens; oögonium (Fig. 21) capped by 5 cells. **Chàra.** Figs. 19–21.
1. Plants smooth (Fig. 22); oögonium capped by 10 cells (Fig. 22). **Nitélla.** Figs. 22–24.

BLUE-GREEN ALGAE   CYANOPHÝCEAE

Gelatinous balls, of bluish-green color, from the size of a pinhead to that of a walnut or even larger, attached to rocks or to plants, sometimes free-floating and rolled by the waves in shallow water. Several genera occur, the commonest being *Nóstoc*, but their identification is entirely dependent on the use of a microscope.

Reference: Josephine Tilden, Minnesota Algae, University of Minnesota, 1910. The title is misleading; this treats the Blue-green Algae only, and includes all known in North America.

MOSSES   MÙSCI
LEAFY LIVERWORTS   HEPÁTICAE

Mosses and Liverworts (together comprising the Brýophytes) reproduce by microscopic spores borne in capsules (Figs. 33, 35). They are separated, technically, on the manner of opening of the capsule. For convenience, the Leafy Liverworts are here treated with the Mosses, which they superficially resemble, and the Thallose Liverworts are treated by themselves.

1. Plants with long stems floating out from the point of attachment.............**2**
1. Plants with rather short stems (about 5 cm.) forming a mat on soil or rocks.......**3**
  2. Leaves in 3 rows. **Fontinàlis.** Fig. 25. Water Moss. Common in flowing water; some species are among the largest mosses.
  2. Leaves in 2 rows vertically placed (Fig. 26), or loosely spiraled (Fig. 31)....**5**
3. Most stems creeping on the soil or rocks...................................**4**
3. Most stems erect. **Philonòtis.** Fig. 33.
  4. Leaves placed on the sides of the stem, giving a flattened appearance (Figs. 27, 28, 32). The Leafy Liverworts: **Plagiochìla, Chiloscỳphus,** and **Scapánia.**
  4. Leaves placed all around the stem so that the plant does not appear flattened. **Hypnàceae,** the Hypnum Family. Figs. 31, 37.
5. Leaves spreading from 2 sides of the stem (Fig. 26). **Físsidens.**
5. Leaves spreading on all sides of the stem (Fig. 31). **Drepanoclàdus.** Fig. 37.

References: A. J. Grout, Mosses with a Hand-lens, 3d ed. (1924). The simplest and most popular, yet accurate, book on mosses and some liverworts. Grout, Mosses with Hand-lens and Microscope (1903). A more complete and technical work. Grout, Moss Flora of North America. The most recent and complete work on mosses.

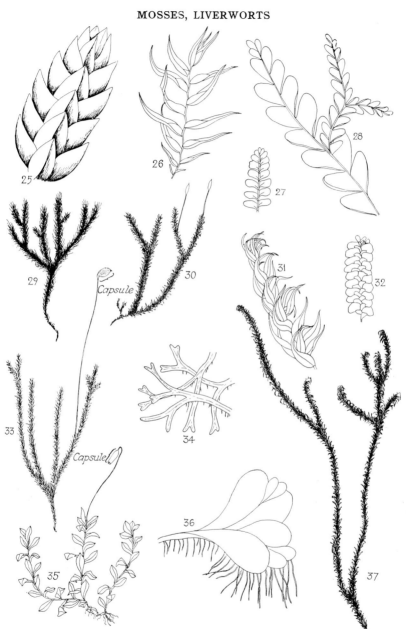

Fontinalis: 25. Branch, × 6. **Fissidens:** 26. Branch, × 6. **Plagiochila:** 27. Branch, × 2. **Chiloscyphus:** 28. Branches, × 2. **Grimmia:** 29. Plant, × 1. **Rhacomitrium:** 30. Plant, × 1. **Drepanocladus:** 31. Branch, × 6. 37. Plant, × 1. **Riccia:** 34. Plant, × 2. **Mnium:** 35. Plant, × 1. **Ricciocarpus:** 36. Plant, × 1. **Scapania:** 32. Branch, × 2. **Philonotis:** 33. Plant, × 1.

THALLOSE LIVERWORTS   RICCIÀCEAE

These may be recognized by their flat, equally forking stem-like bodies (thallus). Two species are commonly found in the water: **Rìccia flùitans** L. (Fig. 34), the Slender Riccia, which occurs in tangled masses just below the surface, and **Ricciocárpus nàtans** (L.) Corda (Fig. 36), the Purple-fringed Riccia, which floats on the water and looks much like a Duckweed but is 2-lobed with a furrow above and slender purplish scales beneath.

The text on Algae, Mosses and Liverworts has been contributed by Dr. Pauline Snure.

FLOWERING FERN FAMILY   OSMUNDÀCEAE
ROYAL FERN   Osmúnda

O. regàlis L., var. **spectábilis** (Willd.) Gray. Leaves 0.3–1.6 m. long, erect from a stout rootstock, divided into many oblong green leaflets; uppermost leaflets narrower, brown, and bearing spores. Moist woods, swamps and streams, rarely standing in the water; Nfd. to Sask., s. to W.I. & S.A.

MARSILEA FAMILY   MARSILEÀCEAE
PEPPERWORT   Marsílea

M. quadrifòlia L. Fig. 1. Rootstock slender, buried in the mud; petioles, when on land, 1–1.5 dm. long, longer when in the water; leaflets 4, fan-shaped; fruiting bodies roundish, black, on short branching peduncles. In quiet water, not common, probably all introduced.

SALVINIA FAMILY   SALVINIÀCEAE
WATER-VELVET   Azólla

A. caroliniàna Willd. Fig. 2. Leaves minute, about 0.5 mm. long, on branching, often radiating, stems; the entire plant 1–2 cm. in diameter, green or reddish, with a velvety appearance. Occurs floating on the water, often giving the whole surface of a pool a red coloring when seen from a distance, or stranded on the mud; Mass. to Minn., westward and southward.

FLOATING MOSS   Salvínia

S. rotundifòlia Willd. Fig. 3. Stem very slender, hidden by the leaves; leaves round, often somewhat folded along the midrib, 1 cm. or less long, thickly covered above with white erect forked hairs. Floating on the surface of the water; grown in aquariums, and rarely escaping to ponds. (*S. natans*, G, B.)

HORSETAIL FAMILY   EQUISETÀCEAE
HORSETAIL, SCOURING RUSH   Equisètum

Stems erect or reclining, hollow, jointed, without true leaves, but with a toothed sheath at each joint (Fig. 9), unbranched or with a whorl

MARSILEA, AZOLLA, SALVINIA, EQUISETUM

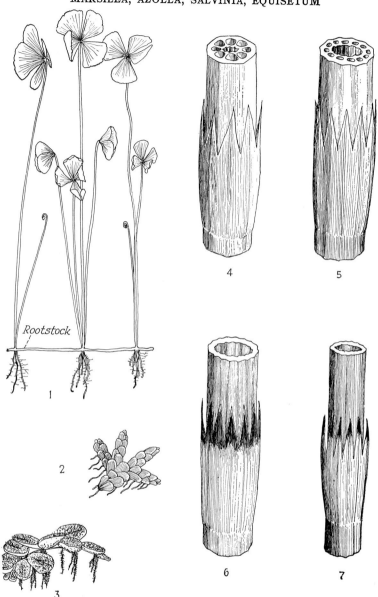

M. quadrifolia: 1. Habit, × ½. **A. caroliniana:** 2. Plant, × 3. **S. rotundifolia:** 3. Plant, × 1. **E. palustre** var. **americanum:** 4. Sheath, × 3. **E. litorale:** 5. Sheath, × 3. **E. fluviatile:** 6. Sheath, × 3. **E. fluviatile f. minus:** 7. Sheath, × 3.

of slender branches from each joint, sometimes with a terminal cone (Fig. 9) which bears spores.

1. Central cavity of the stem small, nearly equaled by the several side cavities (Fig. 4)..2
1. Central cavity one-half or more the diameter of the stem, the side cavities much smaller (Fig. 5) or absent (Fig. 6)...4
    2. Stem upright...3
    2. Stem trailing through the water, with a few upright branches (Fig. 10). **E. palústre** var. **americànum** forma **flùitans** Vict. A form in shallow running water.
    3. Branches few or none. **E. palustre** L., var. **americanum** Vict. Nfd. to Alaska, s. to Conn., N.Y., Ill. & Wash.
    3. Branches numerous, regularly whorled. **E. palustre** var. **americanum** forma **luxùrians** Vict. Occasionally found in the same range.
        4. Side cavities absent or so small as to be invisible to the naked eye (Fig. 6)...**5**
        4. Side cavities present (Fig. 5). **E. litoràle** Kuehl. Very variable in habit, and thought by some to be a series of hybrids cf **E. arvénse** with the preceding and the following.
5. Stem with branches (Fig. 8)...6
5. Stem without branches (Fig. 9)..7
    6. Branches 4–16 at each node, spreading in all directions (Fig. 8). **E. fluviátile** L. The commonest species in quiet water; Nfd. to Alaska, s. to Va., Neb. & Ore.
    6. Branches 1–2 at each node, all turned the same way. **E. fluviatile** forma **nàtans** (Vict.) Broun. A form in running water.
7. Stem stout, 3.5–7.5 mm. thick; teeth black throughout (Fig. 6). **E. fluviatile** forma **Linnaeànum** (Döll) Broun. With the typical form or by itself.
7. Stem slender, 1.5–3 mm. thick; teeth black only at the tip (Fig. 7). **E. fluviatile** forma **mìnus** (A. Br.) Broun. A form of drier places.

References: Victorin, Les Équisétinées du Québec, Contrib. Lab. Bot. Univ. Montréal, No. 9 (1927), and Broun, Index to North American Ferns (1938)—forms of *Equisetum;* Weatherby, Amer. Fern. Journ., **27**, 98–102 (1937)—nomenclature of *Salvinia.*

QUILLWORT FAMILY   ISOËTÀCEAE
Quillwort   Isòëtes

All species of Quillwort look very much alike, the plants always consisting of a number of awl-shaped leaves from a somewhat swollen base (Fig. 11). When the spores are mature, they may be found by removing a leaf base, on the inner surface of which will be found a spore-containing sac (Fig. 12). The leaves vary somewhat in length, ranging from 3–24, or sometimes to 70, cm., and are usually rigid, but sometimes flexible and floating toward the surface (Fig. 5, p. 2). There are two kinds of spores; the larger (megaspores) are just large enough to be seen with the naked eye, perhaps the size of salt crystals, and the smaller (microspores) appear as a fine powder. The species are separated almost entirely on the surface marking of the megaspores, for which a microscope is necessary.

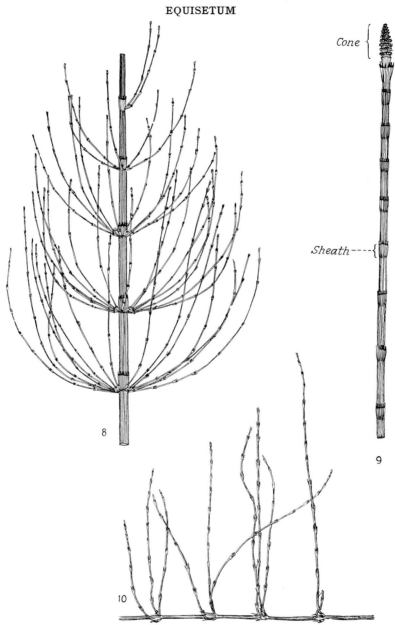

Cone {

Sheath ---- {

8

9

10

8. E. fluviatile, $\times \frac{1}{2}$.   9. E. fluviatile f. Linnaeanum, $\times \frac{1}{2}$.   10. E. palustre var. americanum f. fluitans, $\times \frac{1}{2}$.

I. **Braunii**: 11. Plant, × 1. 12. Inner side of leaf, × 2, showing spores. 13. Spore, × 60. I. **Eatoni**: 14. Spore, × 60. I. **riparia** var. **canadensis**: 15. Spore, × 60. I. **Tuckermani**: 16. Spore, × 60. I. **macrospora**: 17. Spore, × 60. I. **macrospora** f. **hieroglyphica**: 18. Spore, × 60. I. **foveolata**: 19. Spore, × 60. I. **Engelmanni**: 20. Spore, × 60.

1. Megaspores white..2
1. Megaspores black. **I. melanópoda** J. Gay. Low prairies, wet meadows, ditches, etc., submersed in spring; Ill. & Iowa to Tex.
  2. Megaspores with projections (Figs. 13–15)...............................3
  2. Megaspores with a network of ridges (Figs. 16–20)......................7
3. Projections spine-like (Fig. 13)...4
3. Projections flattened, crest-like (Figs. 14, 15)...............................5
  4. Leaves 10–35, rarely to 55, in number and 8–25 cm. long. **I. Braúnii** Dur. Fig. 11. Usually submersed, Nfd. to N.J., Ohio, Minn. & westward. (*I. echinospora* var. *Braunii*, G.)
  4. Leaves often as many as 75, and 12–15 cm. long. **I. Braunii** forma **robústa** (Engelm.) Pfeiffer. Vt. & Conn. (*I. Gravesii*, G, B.)
5. Crests heavy, crowded (Fig. 14); leaves 25–200 in number, and 10–60 cm. long, in spring forms, and 10–15 cm. long in summer forms, from a base sometimes 10 cm. thick. **I. Eatòni** Dodge. N.H., e. Mass. & N.J.
5. Crests less crowded, more delicate (Fig. 15); leaves 10–75 in number, from a smaller base..6
  6. Leaves 10–30 in number, and 9–30 cm. long. **I. ripària** Engelm. Maine to Del., w. to s. Ont.
  6. Leaves 15–75 in number, 10–45 cm. long. **I. riparia** var. **canadénsis** Engelm. Ottawa & vicinity; N.H. to N.J. (*I. Dodgei*, G, B.)
7. Plants always submersed..8
7. Plants of shallow water, exposed as the level drops........................10
  8. Megaspores 600–800 microns in diameter..............................9
  8. Megaspores 460–600 microns in diameter. **I. Tuckermàni** A. Br. Fig. 16. Labrador to Conn.
9. Spores marked as in Fig. 17. **I. macróspora** Dur. Nfd. to Mass., w. to Minn.
9. Spores marked as in Fig. 18. **I. macrospora** forma **hierogl**ý**phica** (Eaton) Pfeiffer. Lakes of n. Maine. (*I. hieroglyphica*, G, B.)
  10. Leaves less than 18 cm. long. **I. foveolàta** Eaton. Fig. 19. S.w. N.H.
  10. Leaves chiefly more than 18 cm. long. **I. Engelmánni** A. Br. Fig. 20. N.H. to Ga., w. to Ill. & Mo.

Reference: Pfeiffer, Ann. Mo. Bot. Gard. **9.** 79–232 (1922)—Monograph of the *Isoetaceae*.

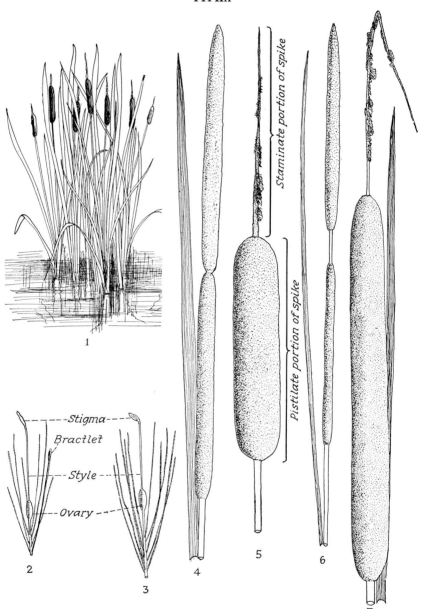

T. latifolia: 1. Colony, $\times \frac{1}{25}$. 3. Pistillate flower, $\times 3$. 4. Young flowering spike, $\times \frac{1}{2}$. 5. Fruiting spike with stamens withering, $\times \frac{1}{2}$. **T. angustifolia**: 2. Pistillate flower, $\times 3$. 6. Young flowering spike, $\times \frac{1}{2}$. **T. angustifolia** var. **elongata**: 7. Fruiting spike, $\times \frac{1}{2}$.

PINE FAMILY   **PINÀCEAE**
CYPRESS   **Taxòdium**

T. dístichum (L.) Richard. Bald Cypress. Tall straight trees, up to 18 dm. in diameter and 45 m. in height, the base usually much enlarged, often with tall slender "knees" from the roots, rising above the water; leaves needle-like, flat, dropping in the fall; cone nearly spherical. Damp soil and in the water; Del. to the Gulf of Mexico, north in the Mississippi basin to Mo. & s. Ind.

CATTAIL FAMILY   **TYPHÀCEAE**
CATTAIL   **Tỳpha**

Flowers borne in close cylindrical spikes, which consist of two portions; the pistillate, or female, portion is below, and the staminate, or male, portion is above (Figs. 4–6).

Cattails may appear in almost any wet place and are often the first invaders in a newly excavated pool. The underground stems spread extensively, so that a stand of Cattail an acre in extent may actually consist of but a few plants.

1. Stem stout; leaves flat; staminate and pistillate portions usually contiguous (Figs. 4, 5); pistillate portion dark-brown with black markings; pistillate flowers without bractlets (Fig. 3) and with flattened stigma; pollen grains (when seen under the microscope) in 4's. **T. latifòlia** L. Common Cattail. Forming extensive swales (Fig. 1), usually in shallow water, and very common. When it grows with the next, hybrids occur.
1. Stem slender; leaves somewhat rounded on the back; staminate and pistillate portions of the spike usually distant (Fig. 6), the interval being at least 5 mm.; pistillate portion light or cinnamon brown without black markings; pistillate flowers with bractlets (Fig. 2) and with thread-like stigma; pollen grains single............**2**
   2. Plant 1–1.5 m. tall; lower leaves 3–7 mm. wide; pistillate portion of spike 8–13 cm. long and 10–17 mm. thick in fruit (Fig. 6). **T. angustifòlia** L. Narrow-leaved Cattail. Atlantic Coast (where usually in brackish water) from s. Maine to Fla.; inland locally, usually in marl lakes.
   2. Plant 2–3.5 m. tall; lower leaves 9–15 mm. wide; pistillate portion of spike 15–30 cm. long and 20–23 mm. thick in fruit (Fig. 7). **T. angustifolia var. elongàta** (Dudley) Wiegand. Occasional in marl lakes and marshes.

Reference: Wiegand, Rhodora **26**, 1 (1924)—*T. angustifolia* var. *elongata.*

S. eurycarpum: 9. Upper part of plant, × ½. 10. Nutlet, × 4; the 2 stigmas have fallen as the fruit matured. **S. fluctuans:** 11. Nutlet, × 4. 12. Plant, × ½. **S. androcladum:** 13. Nutlet, × 4.

BUR REED FAMILY   **SPARGANIÀCEAE**
BUR REED   Spargànium

Leaves long and ribbon-like, erect (Figs. 9, 14, 18) or limp and floating partly on the surface (Fig. 12); flowers in spherical heads, the lower pistillate and the upper staminate (Fig. 14); fruit a nutlet, partly surrounded by the scale-like sepals.

The commonest and largest species is *S. eurycarpum*, and the floating species most apt to be found is *S. angustifolium*. Probably all the species produce sterile rosettes of limp ribbon-like leaves (see **Fig. 6**, p. 5, and **Fig. 6**, p. 9).

Mature fruit is necessary for the identification of most species, with the exception of *S. eurycarpum*, which is our only species with 2 stigmas (Fig. 9).

1. Stigmas 2 (Fig. 9); mature fruit cut nearly square across the top (Fig. 10). **S. eurycárpum** Engelm. Mucky shores; Que., N.S. & Maine to B.C., s. to Fla., Mo., Utah & Calif.
1. Stigma 1 (Fig. 19); mature fruit tapered to both ends......................2
   2. Staminate heads 2–20 on each plant.....................................3
   2. Staminate head only 1 on each plant....................................9
3. Leaves either erect or floating; sepals attached at the summit of the stipe (Fig. 13); beak slender, straight or slightly curved..................................4
3. Leaves always floating (Fig. 12); sepals attached near the base of the stipe (Fig. 11); fruit with a stout strongly curved beak (Fig. 11). **S. flúctuans** (Morong) Robinson. Floating-leaf Bur Reed; Broad Ribbon-leaf. Cold, usually soft, water; Nfd. to Minn., s. to Conn., Pa. & Wis.
   4. Heads or branches usually borne directly in the axils of leaves (Fig. 14), rarely supra-axillary; nutlet not ribbed at summit, about the same texture above and below the middle (Figs. 13, 15)...5
   4. At least some of the heads or branches supra-axillary, borne at some distance above the leaf (Fig. 18); nutlet ribbed or angled at summit, surface rather shining and greenish above the middle, dull and often somewhat spotted with brown below the middle (Figs. 16, 17)..................................6
5. At least the middle leaves keeled; bracts ascending; fruiting heads 2.5–3.5 cm. in diameter; mature fruits with shining surface; stipe 2.5–4 mm. long (Fig. 13); body of fruit 5–7 mm. long, 2.5–3 mm. thick; beak 4.5–6 mm. long; stigma 2–4 mm. long; anthers 1–1.6 mm. long. **S. andrócladum** (Engelm.) Morong. Muddy shores and shallow water; Que. & Vt. to e. Mass. & D.C.; Wis., Ill. & Mo. (*S. lucidum*, G, B; not *S. americanum* var. *androcladum*, G, nor *S. androcladum*, B.)
5. Leaves flat or obscurely keeled; bracts spreading; fruiting heads 1.5–2.5 cm. in diameter; mature fruits with dull surface; stipe 2–3 mm. long (Fig. 15); body of fruit 4.5–5.5 mm. long, about 2 mm. thick; beak 1.5–5 mm. long; stigma 1–2 mm. long; anthers 0.8–1.2 mm. long. **S. americànum** Nutt. Fig. 14. Muddy or peaty shores and shallow water; Nfd. to Minn., s. to Fla. & Mo. This species closely resembles *S. chlorocarpum*, which usually has some heads supra-axillary, whereas they are usually axillary in *S. americanum* But this character is not always reliable. Young fruits of both species often shrivel below the constriction.

S. americanum: 14. Upper part of plant, $\times \frac{1}{2}$; the heads are all axillary, head *a* fitting snugly in the axil of leaf *a*, head *b* in the axil of leaf *b*, etc. 15. Nutlet, $\times$ 4. **S. angusti-folium:** 16. Nutlet, $\times$ 4. **S. chlorocarpum:** 17. Nutlet, $\times$ 4. 18. Plant, $\times \frac{1}{2}$; the heads are supra-axillary, head *a* appearing at some distance above leaf *a*, head *b* well above leaf *b*, etc. **S. minimum:** 19. Nutlet, $\times$ 4.

Above the constriction the surface of *S. chlorocarpum* is shining, and that of *S. americanum* is dull. *S. androcladum* also has fruits shining above the constriction, but they are larger than those of *S. chlorocarpum*, and also shining below the constriction.

6. Plants usually erect; fruit ribbed at the summit between the angles (Fig. 17), with a beak about as long as the body...................................**7**

6. Plants usually with floating leaves; fruit scarcely ribbed between the less conspicuous angles (Fig. 16), with a beak shorter than the body............**8**

7. Pistillate heads (1–) 2–4, not touching each other, 1.5–2.7 cm. in diameter, the lowest 1–6.5 dm. above the base of the plant; staminate part of inflorescence 2–10 cm. long, of 4–9 heads. **S. chlorocárpum** Rydb. Fig. 18. Muddy shores and shallow water; Nfd. & Que. to Ont., s. to N.J., N.Y., Ind. & Iowa. (*S. diversifolium*, G, B.)

7. Pistillate heads 1–3, at least the upper touching each other, 1.2–2.2 cm. in diameter, the lowest 1–18 cm. above the base of the plant; staminate part of inflorescence 1–4 (–5) cm. long, of 2–5 heads. **S. chlorocarpum** var. **acaúle** (Beeby) Fernald. Nfd. to N.D., s. to Va. & W. Va.; commoner northward than *S. chlorocarpum.* (*S. diversifolium*, G; *S. acaule*, B.)

8. Leaves rounded on the back, 1.5–4 (–5) mm. wide, the middle and upper ones with enlarged slightly inflated sheathing bases; nerves on under surfaces of leaves 0.2–0.8 mm. apart; pistillate heads 1.2–2 cm. in diameter; stigmas 0.6–1.5 mm. long. **S. angustifòlium** Michx. Floating-leaf Bur Reed; Narrow Ribbon-leaf. Fig. 16. Shallow or deep water, sometimes on muddy shores; Nfd. to Alaska, s. to Conn., N.J., Pa., Mich., Colo. & Calif.

8. Leaves flat and ribbon-like, 5–12 mm. wide, scarcely inflated at base; nerves 0.8–2 mm. apart; pistillate heads 2–2.5 cm. in diameter; stigmas 1–1.8 mm. long. **S. multipedunculàtum** (Morong) Rydb. Labrador to Alaska, s. to N.H., Vt., Colo. & Calif. (*S. simplex*, G.)

9. Pistillate heads all borne directly in the axils of leaves; fruit with a beak 0.5–1.5 mm. long (Fig. 19). **S. mínimum** Fries. Springy spots, pools, ponds, etc.; Nfd. to Alaska, s. to Conn., N.J., Pa., Mich., Wis., Utah & Ore.

9. One or more pistillate heads borne a short distance above the leaf (supra-axillary); fruits with or without a beak...**10**

10. Fruiting heads 1–3, distinct, 5–12 mm. in diameter; fruits 3.5–4.5 mm. long, rounded to a beakless summit. **S. hyperbòreum** Laestad. Arctic regions, s. to N.S., Que., Man. & Alaska.

10. Fruiting heads 3–5, the upper densely crowded, 1–1.5 cm. in diameter; fruits 5–6.5 mm. long, tapered at both ends, with a slenderly conical beak. **S. glomeràtum** Laestad. Northern Europe· known in N. Am. only from Duluth, Minn., and Saguenay Co., Que.

References: Fernald, Rhodora, **27**, 190–193 (1925)—*S. multipedunculatum;* Lakela, Rhodora, **43**, 83–85 (1941)—*S. glomeratum* in N. Am.; Fernald, Rhodora, **24**, 26–34 (1922)—all other species. The preceding key is largely derived from that of Prof. Fernald, who has kindly supplied the additional portion dealing with *S. glomeratum.*

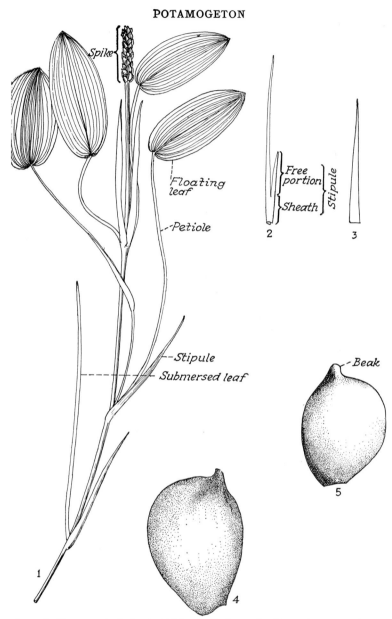

P. natans: 1. Upper part of plant, $\times \frac{1}{2}$. 4. Fruit, $\times 8$. **P. pectinatus:** 2. Leaf, $\times 2$.
3. Leaf tip, $\times 2$. 5. Fruit, $\times 8$.

PONDWEED FAMILY NAJADÀCEAE

1. Leaves alternate...2
1. Leaves opposite..3
  2. Flowers and fruits in spikes (Fig. 1) or heads (Fig. 77). **Potamogeton.**
  2. Flowers and fruits more or less stalked (Fig. 99). **Ruppia.**
3. Seeds tapered to each end (Fig. 90) and enclosed in a papery envelope (Fig. 85) **Najas.**
3. Seeds enclosed in a fruit coat which is narrowed to a stipe at one end and a slender style at the other and is toothed down one side (Fig. 84). **Zannichellia.**

PONDWEED **Potamogèton**

Leaves often of two kinds, the floating of firm texture, and the submersed thin and membranous. The flowers are mostly borne in spikes which rise above the water's surface for wind pollination; the numerous pencil-like erect emersed spikes are conspicuous in early summer. They usually go beneath the surface as the fruit matures. In some species there are also few-flowered globular spikes.

Identification of species of *Potamogeton* is notoriously difficult. The following key is arranged so that the 3 species of most interest in the field of wildlife management come first. The species with thread-like or ribbon-like leaves require careful study, but are well understood and may be named with some certainty, thanks to two recent studies of these groups. The broad-leaved Pondweeds are at present not so well understood.

1. Floating leaves heart-shaped at base (Fig. 1).............................2
1. Floating leaves tapered or rounded at base, or absent......................3
  2. Submersed leaves 0.8-2 mm. wide; floating leaves 4-9 cm. long, 2.5-6 cm. wide, with petioles 1-2.5 mm. thick; stipules 4.5-11 cm. long, strongly keeled; mature fruits 3.5-5 mm. long, 2.5-3.5 mm. wide, with keels scarcely developed (Fig. 4). **P. nàtans** L. Floating Brownleaf; Floating-leaf Pondweed. Fig. 1. The petioles are upright, with an abrupt joint at the summit which allows the broad brownish-green blade to float on the surface.
  2. Submersed leaves 0.25-1 mm. wide (Fig. 24), or else broad and flat (Fig. 26); floating leaves 1.5-5.5 cm. long, 1-3 cm. wide, with petioles 0.2-1 mm. thick; stipules 1-5.5 cm. long, with keels prominent only at base (or if plant is larger and coarser, the submersed leaves are flat and broad as in Fig. 26); fruit 2.5-3.5 mm. long, 2-2.5 mm. wide, with prominent keels (Fig. 25)........**23**
3. Submersed leaves thread-like (Fig. 6), or ribbon-like with parallel sides (Figs. 39, 40)..**4**
3. Submersed leaves broader, the sides curved and not parallel (Figs. 9-12).....**6**
  4. Stipules joined to the base of the leaf, making a sheath 1 cm. or more long (Figs. 2, 27) on the larger ones; spike of flowers like a string of beads (Fig. 6).....**5**
  4. Stipules not joined to the leaves (Fig. 39), or joined to them in a sheath less than 5 mm. long (Figs. 75, 76); spike usually continuous (Figs. 10, 77)....**25**
5. Leaf tips blunt or rounded (Fig. 29); fruit without a beak (Fig. 28).........**24**
5. Leaf tips with long tapering points (Fig. 3); fruit with a short beak (Fig. 5). **P. pectinàtus** L. Sago Pondweed. Hard or brackish water of lakes and slow-

POTAMOGETON

P. pectinatus: 6. Plant, $\times \frac{1}{2}$. **P. amplifolius:** 7. Upper part of plant, $\times \frac{1}{2}$.

flowing streams, widespread. Variable in length and thickness of leaves, and often coarser throughout than as represented in Fig. 6. This may usually be recognized by its much-branched stems and numerous thread-like leaves spreading in fan-like fashion. Tubers are frequently found. (Includes *P. interruptus,* G, B.)

6. Floating leaves with 30–50 nerves, or rarely absent; submersed leaves curved, 0.8–2 dm. long and 2.5–7 cm. broad. **P. amplifòlius** Tuckerm. Large-leaf Pondweed; Muskie Weed; Bass Weed. Fig. 7. Usually in hard or medium water; common.

6. Floating leaves with less than 30 nerves, or floating leaves absent; submersed leaves smaller...**7**

7. Leaf margins finely toothed (Fig. 9). **P. críspus** L. In hard or brackish water, often where polluted; naturalized from Europe.

7. Leaf margins not toothed..**8**

    8. Base of leaf clasping the stem (Figs. 10, 11)...........................**9**

    8. Base of leaf not clasping the stem....................................**11**

9. Stipules firm, persistent, 1.5–8 cm. long (Fig. 10); tip of leaf boat-shaped (Fig. 8) **P. praelòngus** Wulf. Whitestem Pondweed; Muskie Weed. N.S. to B.C., s. to N.J., Iowa & Calif.; Mex.

9. Stipules inconspicuous, soon reduced to shreds (Fig. 11), 2 cm. or less long; tip of leaf flat..**10**

    10. Leaves with 3–7 prominent nerves and with wavy margins. **P. Richardsònii** (Benn.) Rydb. Clasping-leaf Pondweed; Bass Weed. Fig. 11. Que. to Mackenzie & B.C., s. to N.E., N.Y., Ind. & perhaps Neb. (Includes *P. perfoliatus,* G, B; true *P. perfoliatus* is more northern.)

    10. Leaves with 1 prominent nerve (rarely more), and with flat margins. **P. bupleuroìdes** Fernald. Mostly in brackish water, Nfd. to Fla.; rarely inland to w. N.Y. & Mich.

11. Submersed leaves with an abrupt awl-shaped tip 2–4 mm. long (Fig. 12). The classification of the following 3 species is not yet well understood...........**12**

11. Submersed leaves with a shorter tip or none.............................**14**

    12. Fruit with 1 keel (Fig. 18); leaves all submersed and membranous. **P. lùcens** L. Local; N.S. to Calif., s. to Fla. & Mex.

    12. Fruit 3-keeled (Fig. 16); thicker floating leaves often present...........**13**

13. Upper submersed leaves definitely petioled. **P. illinoénsis** Morong. Ill., Iowa & Minn.

13. Upper submersed leaves obscurely or not at all petioled. **P. angustifòlius** Berchtold & Presl. Fig. 12. Que. & Mass. to Calif., s. to Fla., Tex. & Wyo.

    14. Submersed leaves petioled (Fig. 13); mature spike 4–5.5 cm. long........**15**

    14. Submersed leaves with obscure short petioles (Fig. 17) or none (Fig. 19); mature spike 1.5–3 cm. long....................................**16**

15. Floating leaves pointed at tip. **P. americànus** C. & S. Fig. 13. Widespread and common.

15. Floating leaves rounded at tip. **P. americanus** var. **novaeboracénsis** (Morong) Benn. Local.

    16. Upper leaves tinged with red; fruit keeled, 1.7–3 mm. long..............**22**

    16. Upper leaves not tinged with red; fruit scarcely keeled, 2.5 mm. long. **P. gramíneus** L. Variable Pondweed. The size, shape and arrangement of leaves vary greatly; some of the better marked forms are described below...**17**

17. Submersed leaves absent; emersed leaves crowded (Fig. 14). **P. gramineus** var. **graminifòlius** forma **terréstris** (Schlecht.) Carpenter. A form on muddy drying shores.

17. Submersed leaves present..**18**

P. praelongus: 8. Leaf tip, × 3. 10. Upper part of plant, × ½. P. crispus: 9. Plant, × ½. P. Richardsonii: 11. Upper part of plant, × ½. P. angustifolius: 12. Upper part of plant, × ½.

POTAMOGETON

P. americanus: 13. Upper part of plant, × ½. **P. gramineus va·. graminifolius f.
terrestris:** 14. Plant, × ½. **P. gramineus** var. **spathulaeformis:** 15. Branch, × ½.
P. illinoensis: 16. Fruit, × 8. **P. gramineus** var. **graminifolius f. maximus:** 17. Upper
part of plant, × ½. **P. lucens:** 18. Fruit, × 8.

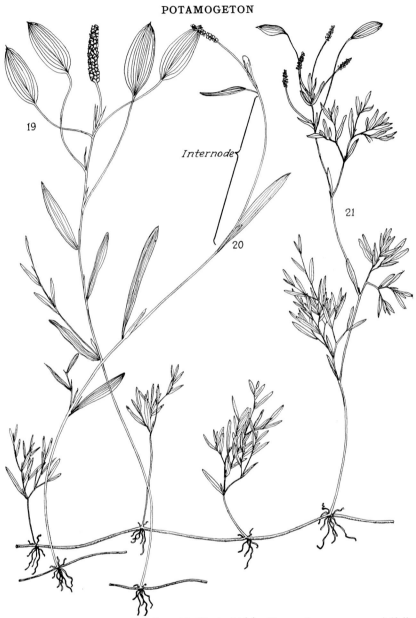

Internode

P. gramineus var. graminifolius: 19. Plant, × ½. P. gramineus var. graminifolius f. longipedunculatus: 20. Plant, × ½. P. gramineus var. graminifolius f. myriophyllus: 21. Plant, × ½.

P. tenuifolius: 22. Upper part of plant, × ½. P. tenuifolius var. subellipticus: 23.
Upper part of plant, × ½.

18. Submersed leaves acute at tip..**19**
18. Submersed leaves obtuse at tip (Fig. 15). **P. gramineus** var. **spathulaefòrmis**
Robbins. Nfd. to Conn., w. to Minn.; rare. (× *P. spathaeformis*, G; *P. varians*, B.)
19. Submersed leaves 1–6 cm. long. **P. gramineus** var. **graminifolius** forma máximus
(Morong) House. Fig. 17. A deep-water form.
19. Submersed leaves 1–3 cm. long...**20**
20. Submersed leaves scattered on long stems...........................**21**
20. Submersed leaves small and very numerous on short branches (Fig. 21).
P. gramineus var. **graminifolius** forma **myriophýllus** (Robbins) House. A
common form in shallow water.
21. Internodes much longer than the leaves, sometimes 2 dm. long. **P. gramineus**
var. **graminifolius** forma **longipedunculàtus** (Merat) House. Fig. 20. A deep-
water form.
21. Internodes about as long as the leaves. **P. gramineus** var. **graminifòlius** Fries.
Fig. 19. Common and widespread. (*P. heterophyllus* G, B.)
22. Leaves 0.7–2.5 dm. long, 5–15 mm. wide, 7-nerved, acute. **P. tenuifòlius**
Raf. Fig. 22. Greenland to Mass., w. to Mich., Colo. & Calif. (*P. alpinus*,
G, B.)
22. Leaves shorter and broader, 4–8 cm. long and 8–20 mm. wide, 7–13-nerved,
blunt. **P. tenuifolius** var. **subellípticus** Fernald. Fig. 23. Nfd. to B.C.,
s. to Mo., Mich. & Wyo.

POTAMOGETON

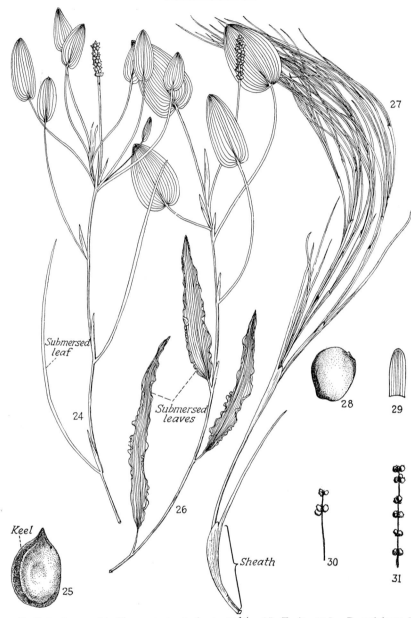

P. Oakesianus: 24. Upper part of plant, × ½. 25. Fruit, × 8. **P. pulcher:** 26. Upper part of plant, × ½. **P. vaginatus:** 27. Upper part of plant, × ½. 31. Inflorescence, × ½. **P. filiformis:** var. **borealis:** 28. Fruit, × 8. 29. Leaf tip, × 8. 30. Inflorescence, × ½.

23. Submersed leaves thread-like (Fig. 24). **P. Oakesiànus** Robbins. Local; Gulf of St. Lawrence to N.J., w. to N.Y.; central Wis. In its foliage this appears like a miniature *P. natans*, but its fruit is very distinct.
23. Submersed leaves broad and flat (Fig. 26). **P. púlcher** Tuckerm. Local; N.S. to Fla.; n.w. Wis. & e. Minn.; e. Mo.
    24. Stems slender; sheaths tight about the stem; blades thread-like; flowers in 2–5 whorls (Fig. 30). **P. filifòrmis** Pers., var. **boreàlis** (Raf.) St. John. In hard water; Nfd. to Alaska, s. to n. Maine, Pa., Wis. & Colo.
    24. Stems 2–3 mm. thick; sheaths large, loose, inflated (Fig. 27); larger leaves with flat blades; flowers in 5–12 whorls (Fig. 31). **P. vaginàtus** Turcz. Local, in hard or brackish water; Labrador to Alberta; s. to N.Y., Wis. & N.D.
25. Submersed leaves thick and coarse, few in number (Fig. 24). Some individuals of *P. Oakesianus* and of *P. natans* may have floating leaves rounded or even tapered at base. See **2.**
25. Submersed leaves thin, flat and ribbon-like (Fig. 39), or thread-like and very numerous (Figs. 63, 77)...**26**
    26. Base of leaves extended into short rounded lobes (Fig. 33); leaves extending stiffly in opposite directions so that the whole plant is flat (Fig. 32)......**27**
    26. Base of leaves not extended into lobes..............................**28**
27. Leaves minutely toothed on the margins (Fig. 33). **P. Robbínsii** Oakes. Robbins' Pondweed. Fig. 32. N.B. to Del., w. to n. Ind. & Wis.; B.C. to Wyo. & Ore.
27. Leaves not toothed. **P. Robbinsii** forma **cultellàtus** Fassett. Local.
    28. Submersed leaves with a conspicuous reticulate portion on each side of the midrib (Fig. 34), easily seen by holding the leaf to the light..............**29**
    28. Submersed leaves with central reticulate portion small and inconspicuous (Fig. 43) or absent. These narrow-leaved Pondweeds are superficially very similar and are differentiated on characters which, although definite, are often minute and require close study. The Water Star Grass may be confused with members of this group....................................**30**
29. Submersed leaves 5–10 mm. broad, 7–13-nerved; blades of floating leaves 3–8 cm. long, 1.5–3.5 cm. wide, 19–41-nerved; fruit 3–4.5 mm. long, with a keel 0.6–1.2 mm. broad (Fig. 35). **P. epihỳdrus** Raf. Leafy Pondweed. Lakes, pools and streams; Que. to N.J. & W.Va., w. to Minn., Iowa, & s. Ill.; Wash. to Idaho & Calif. (*P. epihydrus* var. *cayugensis*, G.)
29. Submersed leaves (1–) 2–8 mm. broad, (3–) 5–7-nerved; blades of floating leaves 2–7.5 cm. long, 0.4–2.5 cm. wide, (7–) 11–33-nerved; fruits 2.5–3.5 mm. long, with a keel 0.2–1 mm. broad (Fig. 36). **P. epihydrus** var. **Nuttállii** (C. & S.) Fernald. S. Labrador to Del. & Ga., w. to s. Alaska & Calif. (*P. epihydrus*, G.)
    30. Stipules all free from the leaf base (Figs. 43, 55); spikes uniform, on peduncles 0.3–9 (–24) cm. long; fruits plump or slightly flattened, beaked, the form of the embryo usually not visible (Fig. 42); winter buds (Fig. 41) frequently present....................................**31**
    30. Stipules of submersed leaves often fused with the base of the leaf (Fig. 75); flowers or fruits arranged (when floating leaves are present) in two or three ways, those in the axils of the broader floating leaves in spikes, and those in the axils of the narrow submersed leaves in small heads (Fig. 77); fruit strongly flattened, not beaked, the coiled form of the embryo clearly visible (Fig. 73)....................................**52**
31. Leaves all thread-like or ribbon-like and submersed........................**32**
31. Leaves of two sorts, the submersed narrow or thread-like, the floating 2–8 mm. wide, some plants of a colony with, and others without, floating leaves.......**51**

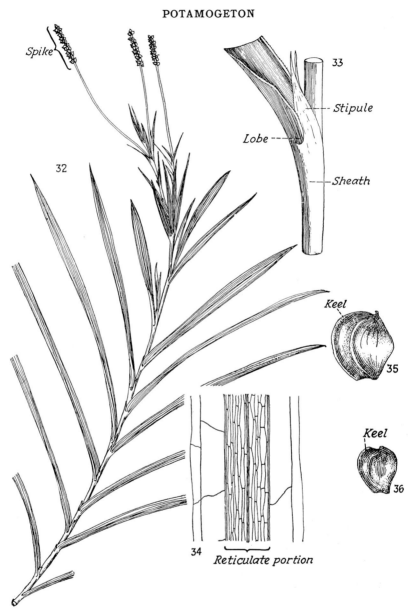

P. **Robbinsii**: 32. Upper part of plant, × ½.   33. Base of leaf, × 3.   **P. epihydrus**: 34. Portion of leaf, × 5.   35. Fruit, × 5.   **P. epihydrus** var. **Nuttallii**: 36. Fruit, × 5.

POTAMOGETON

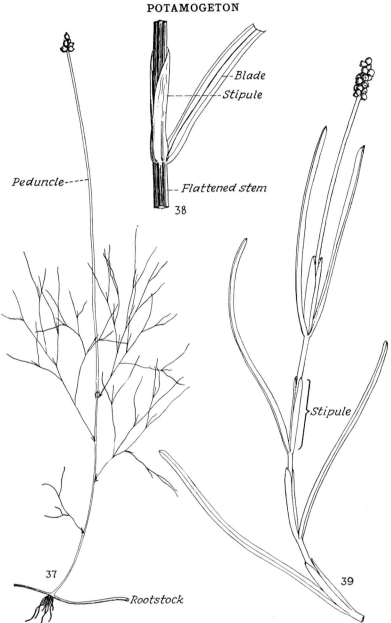

P. confervoides: 37. Plant, × 1.   P. zosteriformis: 38. Leaf base, × 2.   39. Upper part
of plant, × 1.

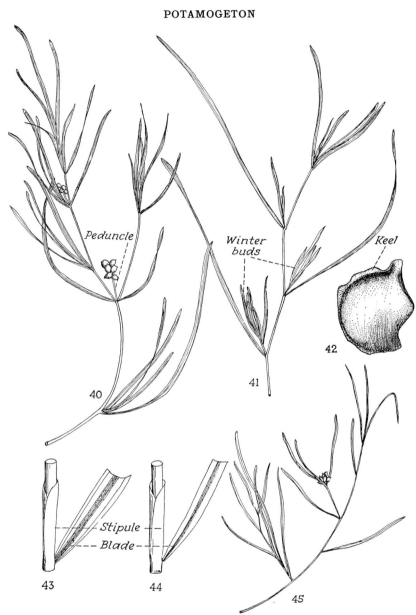

Peduncle

Winter buds

Keel

42

40

41

Stipule

Blade

43

44

45

P. foliosus: 40. Branch, × 1.   41. Branch with winter buds, × 1.   42. Fruit, × 8.   43. Leaf base, × 5.   **P. foliosus** var. **macellus:** 44. Leaf base. × 5.   45. Branch, × 1.

32. Rootstock long and creeping; peduncle straight and erect, 15–24 cm. long, continuing the main stem and exceeding the leaves (Fig. 37); leaves thread-like, 0.1–0.5 mm. wide, without glands at the base; stipules essentially nerveless. **P. confervoìdes** Reichenb. Local; Nfd.; N.S.; alpine pools in Maine, N.H., Vt., N.Y. & Pa.; ponds from Mass. to N.J.; Wis.

32. Rootstock mostly short or wanting; peduncles mostly in the upper forks of the stem (Fig. 40), 0.3–9 cm. long; leaves broader, or, if very narrow, generally with 2 glands at their base (Fig. 47); stipules nerved.................**33**

33. Leaves with 9–35 nerves; stem flattened....................................**34**

33. Leaves 1–9 nerves; spikes with 5 or fewer whorls of flowers or fruits.........**35**

  34. Principal leaves 2–5 mm. wide, up to 35-nerved, without basal glands and abruptly narrowed to the tip. **P. zosterifòrmis** Fernald. Flat-stemmed Pondweed. Figs. 38, 39. Que. to Va., w. to Ill., B.C. & Calif. (*P. zosterifolius*, G; *P. compressus*, B.)

  34. Principal leaves 1.5–2 mm. wide, 5–9-nerved, often with a pair of glands at base, very gradually tapering to a bristle tip. **P. longiligulatus, 42**

35. Stipules united into a tube which sheaths the stem (Fig. 43). Fresh plants will show this character; with a little practice a bit of material can be pinched in two with the nails just above the base of the leaf, and the stem partly withdrawn from the sheath, which is then examined with a lens. Dried material must be softened by boiling and spread out with needles......................................**36**

35. Stipules not united to sheath the stem (Fig. 53)..........................**42**

  36. Leaves without basal glands (Fig. 43); fruit slightly flattened, with a thin toothed keel (Fig. 42); fruiting spike 2–5 mm. long (Fig. 40)............**37**

  36. Leaves with a pair of basal glands (Fig. 47); fruit plump, rounded on the back (Fig. 46) or with a broad obscure keel; fruiting spike 6–15 mm. long (Fig. 51) **38**

37. Stems simple to loosely branched; leaves deep green to bronze, the primary ones 1.4–2.7 mm. wide, 3–5-nerved, the midrib with 1–3 rows of loosely cellular tissue on each side near the base (Fig. 43); winter buds on very short branches (Fig. 41). **P. foliòsus** Raf. Leafy Pondweed. Fig. 40. Hard or brackish water; s. Ont. & e. Pa. to W.I. & C.A., w. to the Pacific. (*P. foliosus* var. *niagarensis*, G.)

37. Stems bushy-branched; leaves bright-green, the primary ones 0.3–1.5 mm. wide, 1–3-nerved, the midrib without loosely cellular tissue or with a single row on each side (Fig. 44); winter buds on long branches. **P. foliosus** var. **macéllus** Fernald. Fig. 45. Que. to Fla., w. to Mackenzie & Calif. (*P. foliosus*, G.)

  38. Stipules delicately veined, membranous (Fig. 47), greenish..............**39**

  38. Stipules fibrous (Fig. 48), whitish......................................**40**

39. Main leaves 1–3 mm. wide. **P. panormitànus** Biv. Figs. 46, 47. Gulf of St. Lawrence to Va., w. to Alberta & Calif., s. to Ark., Tex. & Mex.

39. Main leaves 0.3–1 mm. wide. **P. panormitanus** var. **mìnor** Biv. Range about the same.

  40. Leaves flat 5–7-nerved, 1.5–3.5 mm. wide, blunt or rounded at tip; peduncles flattened. **P. Frièsii** Rupr. Limy or brackish water; s. Labrador to Mackenzie, s. to Va., n. Ind., S.D. & B.C.

  40. Leaves often rolled, 3 (–5)-nerved, 0.5–2.5 mm. wide, blunt to sharp-tipped; peduncles thread-like, enlarged toward the tip.....................**41**

41. Leaves rigid, blunt or abruptly narrowed to the tip (Fig. 49). **P. strictifòlius** Benn. Limy water; w. N.E. to Sask., s. to n. Ind. & Neb.

41. Leaves firm but scarcely rigid, tapering to the tip (Fig. 50). **P. strictifolius** var. **rutiloìdes** Fernald. Fig. 51. N.w. Vt. to Mackenzie, s. to n. Ind., Neb. & Utah.

  42. Leaves 5–9-nerved; midrib not bordered by loosely cellular tissue. **P.**

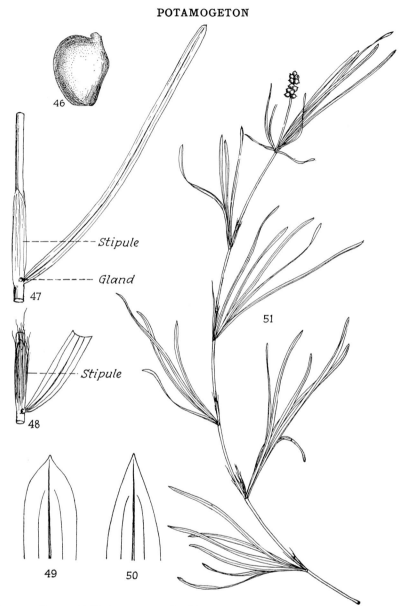

P. panormitanus: 46. Fruit, × 8.   47. Leaf base, × 5.   **P. strictifolius:** 48. Leaf base, × 5.
49. Leaf tip, × 10.   **P. strictifolius** var. **rutiloides:**   50. Leaf tip, × 10.   51. Branch, × 1.

POTAMOGETON

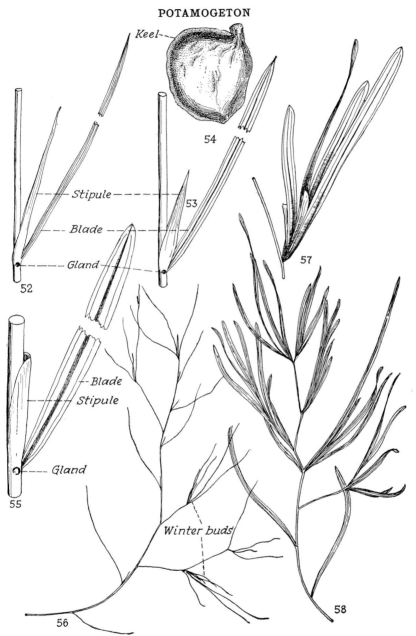

Keel

52

Stipule

Blade

Gland

54

53

57

Blade

Stipule

Gland

55

Winter buds

56

58

P. gemmiparus : 52. Leaf base, × 5. 56. Branch, × 1. **P. Hillii :** 53. Leaf base, × 5. 54. Fruit, × 8. 58. Branch, × 1. **P. obtusifolius :** 55. Leaf base, × 5. 57. Winter bud, × 1.

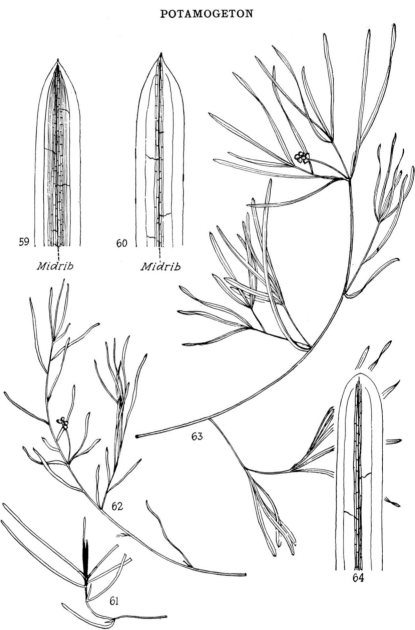

Midrib   *Midrib*

P. pusillus, var. **lacunatus**: 59. Leaf tip, × 10.   Var. **typicus**: 60. Leaf tip, × 10.   63. Branch, × 1.   Var. **mucronatus**: 61. Winter bud, × 1.   64. Leaf tip, × 10.   Var. **tenuissimus**: 62. Branch, × 1.

longiligulàtus Fernald. Limy water, local; Nfd.; Conn.; cent. N.Y.; s. Mich.; s.e. Minn.

42. Leaves 3-, rarely 5-, nerved; midrib bordered narrowly by loosely cellular tissue (Fig. 55)..**43**

43. Leaves tapering to bristle tips, and stipules long-pointed (Fig. 52)..........**44**

43. Leaves not bristle-tipped (Figs. 59–63), and stipules blunt or rounded (Fig. 55)..**45**

    44. Leaves bristle-like, 0.2–0.4 mm. wide; winter buds abundant (Fig. 56); fruits rare, plump, 2.2 mm. long, obscurely keeled. **P. gemmíparus** Robbins. Quiet water; Maine to Conn.

    44. Leaves ribbon-like 1–2.2 mm. wide; winter buds absent; fruits abundant, slightly flattened, 3–3.6 mm. long, prominently keeled (Fig. 54). **P. Híllii** Morong. Fig. 58. Shallow water, local; Vt. to n.w. Pa., Ohio & Mich.

45. Leaves reddish green, 2–4 mm. wide, rounded at tip (Fig. 55); fruits 3–4 mm. long, 2–3 mm. wide, with a thin keel; winter buds (Fig. 57) with bodies 1.5–4 cm. long and 2.5–7 mm. wide..**46**

45. Leaves green, 0.3–2.4 mm. wide, pointed (Fig. 59) or rounded (Fig. 64) at tip; fruits 2–2.8 mm. long, rounded on the back; winter buds (Fig. 61) with bodies 7–18 mm. long and 0.6–2.5 mm. broad. **P. pusíllus** L. Figs. 59–64. A variable species..**47**

    46. Stipules fibrous, soon splitting and fraying into persistent fibers; basal glands absent or only 0.2–0.4 mm. in diameter. **P. Portèri** Fernald. Cold streams; Lancaster County, Pa.

    46. Stipules membranous, not strongly fibrous; basal glands (Fig. 55) 0.6–1.2 mm. broad. **P. obtusifòlius** Mert. & Koch. Cold water; s. Labrador to N.J., w. to Mich. & Minn.; Vancouver Is.

47. Leaf tips sharp (Figs. 59, 60)..**48**

47. Leaf tips blunt or rounded (Fig. 64)....................................**50**

    48. Midrib bordered on each side by 1 or 2 rows of loose cellular tissue (Fig. 60) **49**

    48. Midrib bordered on each side by 3–5 rows of loose cellular tissue (Fig. 59). **P. pusillus** var. **lacunàtus** (Hagst.) Fernald. Cent. Maine to Man., s. to Md., Pa., n. Wis. & Minn.

49. Leaves 0.5–1.5 mm. wide. **P. pusillus** var. **týpicus** Fernald. Fig. 63. Que. to N.J. & Pa.; Tenn.; La.; n. Wis. & Minn.; Alaska to Calif.

49. Leaves 0.3–1 mm. wide. **P. pusillus** var. **tenuissímus** Mert. & Koch. Fig. 62. Nfd. & N.S. to N.J. & Pa., w. to Wis., S.D., and rarely to Okla.; Alaska to Calif.

    50. Principal leaves 3–7 cm. long. **P. pusillus** var. **mucronàtus** (Fieber) Graebner. Figs. 61, 64. Greenland; Nfd. to Del., w. to n. Ind., s. Wis., Neb., Vancouver & Alaska.

    50. Principal leaves 0.8–2.5 cm. long. **P. pusillus** var. **polyphýllus** Morong. Local; Que. to Mass. & N.Y.; Wis.; Neb.; Alberta.

51. Fruiting plants with floating leaves (Fig. 65); sterile plants with submersed leaves only, and with winter buds (Fig. 66); submersed leaves 0.1–0.5 mm. wide and tapering to a bristle tip (Fig. 67); fruit keeled, with a beak 0.3–0.5 mm. long (Fig. 69). **P. Vasèyi** Robbins. S. N.B. to e. Pa., w. to n. Ill., Wis. & e. Minn. The sterile plants, since they lack floating leaves, may be confused with *P. gemmiparus* and with *P. pusillus* var. *tenuissimus*. The two latter species have coarse winter buds on elongate ascending branches (Fig. 56, 62), whereas *P. Vaseyi* has tiny winter buds on short spreading branches (Fig. 66).

51. Fruiting plants with submersed leaves only; plants with floating leaves sometimes with flowers but not with fruits; both forms with winter buds; submersed leaves 0.4–1 mm. wide, merely pointed (Fig. 68); fruit not keeled and only slightly

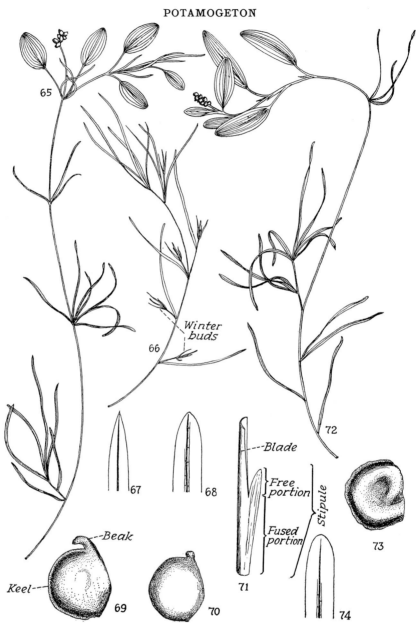

Winter buds

Blade

Free portion

Fused portion

Stipule

Beak

Keel

P. Vaseyi: 65. Branch, × 1.   66. Branch with winter buds, × 1.   67. Leaf tip, × 10. 69. Fruit, × 8.   **P. lateralis:** 68. Leaf tip, × 10.   70. Fruit, × 8.   **P. Spirillus:** 71. Leaf base, × 5.   72. Branch, × 1.   73. Fruit, × 8.   74. Leaf tip, × 10.

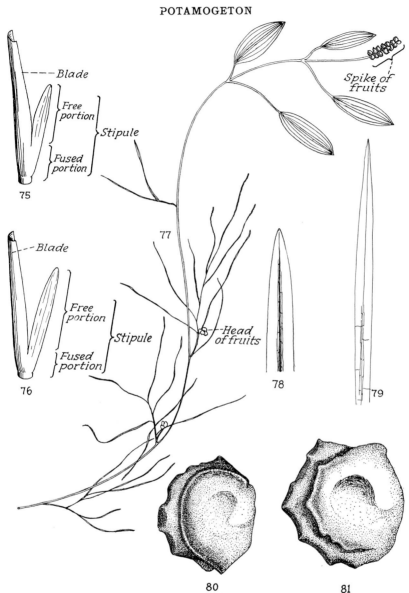

P. diversifolius: 75. Leaf base, × 5. 78. Leaf tip, × 10. **P. capillaceus**: 76. Leaf base, × 5. 77. Branch, × 1. 79. Leaf tip, × 10. 80. Fruit, × 8. **P. bicupulatus**: 81. Fruit, × 20.

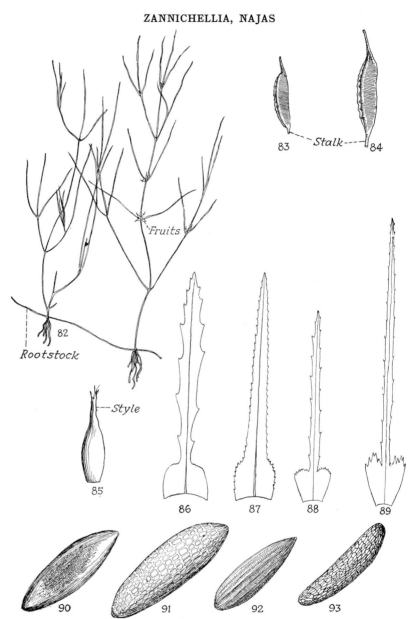

Z. palustris: 82. Part of plant, × 1. 83. Fruit, × 5. Z. palustris var. major: 84. Fruit, × 5. N. flexilis: 85. Fruit, × 4. 87. Leaf, × 5. 90. Seed, × 12. N. marina: 86. Leaf, × 5. N. minor: 88. Leaf, × 5. 92. Seed, × 12. N. gracillima: 89. Leaf, × 5. 93. Seed, × 12. N. guadalupensis: 91. Seed, × 12.

beaked (Fig. 70). **P. lateràlis** Morong. Local; Mass.; Conn.; N.Y.; Mich.; Minn.

52. Submersed leaves 0.5–2 mm. wide, not long tapering (Figs. 74, 78); floating leaves, when present, rounded at tip (Fig. 72), the larger 5–15-nerved; free portion of stipule not more than twice as long as the fused portion (Figs. 71, 75)..**53**

52. Submersed leaves 0.1–0.6 mm. wide, tapered to a long point (Fig. 79); floating leaves, when present, pointed (Fig. 77), 3–7-nerved; free portion of stipule several times as long as the fused portion (Fig. 76)......................**54**

53. Submersed leaves blunt or rounded at tip (Fig. 74); fused portion of stipule much longer than the free portion (Fig. 71); fruits 1.3–2.2 mm. wide, wings on the sides scarcely developed, beak lacking (Fig. 73). **P. Spiríllus** Tuckerman. Fig. 72. Nfd.; s. N.S. & N.B. to Del. & n.w. Ohio, w. to Minn. & Iowa; S.D. (*P. dimorphus,* G, B.)

53. Submersed leaves pointed (Fig. 78); fused portion of stipule about half as long as the free portion (Fig. 75); fruits 1–1.5 mm. wide, with low fine wings on the sides and a minute beak. **P. diversifòlius** Raf. Mex., Tex. & Ga., n. to s. N.Y., Pa., s. Ind., s. Minn., S.D., Mont. & Ore. (*P. dimorphus,* in part, G, B.)

    54. Fruits greenish, 1–1.5 mm. wide, the narrow keel entire or minutely toothed, the wings on the sides low and nearly toothless (Fig. 80). **P. capillàceus** Poir. Fig. 77. In sandy, muddy or peaty pools and streams; Maine to Fla. & Tex.; n. N.Y.; n. Ind.; n. & cent. Wis.

    54. Fruits light brownish, 1.6–2.2 mm. wide, the broad keel coarsely toothed, and wings on the sides toothed (Fig. 81). **P. bicupulàtus** Fernald. Mountain pools; s.e. Pa. & s. Tenn.

References: Hagström, Kungl. Svenska Vet. Handl., **55** No. 5 (1916)—notes on many species; St. John, Rhodora, **18**, 121–138 (1916); and **20**, 191–192 (1918)—*P. pectinatus* and its relatives; Fernald, Rhodora, **23**, 189–191 (1921)—*P. gramineus* and its varieties; **32**, 76–83 (1930); and **33**, 209–211 (1931)—*P. tenuifolius;* House, N.Y. State Mus. Bull. **254**, 53–54 (1924)—forms of *P. gramineus;* Fernald, Mem. Amer. Acad. Arts and Sci., **17** pt. 1, 1–183 (1932)—the narrow-leaved Pondweeds—the finest monograph of any group of aquatic plants in America, its introduction should be studied by anybody desiring an understanding of this genus; Fassett, Rhodora, **35**, 388–389 (1933)—*P. Robbinsii* forma *cultellatus;* Dole, Fl. Vt., 25 (1937)—*P. gramineus* var. *graminifolius* forma *terrestris.*

HORNED PONDWEED   **Zannichéllia**

This genus, as well as the next, differs from the Pondweeds in having all the leaves opposite. The fact that they are opposite may sometimes be obscured by the fact that there are bunches of smaller leaves in the axils of the larger ones (Fig. 97). The leaves are usually narrower, longer and less crowded in *Zannichellia* than in *Najas,* but the true distinguishing feature of *Zannichellia* lies in the flattish fruit which is usually toothed down one side (Figs. 83, 84).

Z. palústris L. Stems numerous and thread-like, from extensively creeping rootstocks (Fig. 82); fruits in bunches of 2–5, scarcely stalked, the body 2–2.5 mm. long (Fig. 83). Hard to brackish water; Sask. to Wis., Mo. & Tex., w. to the Pacific. **Z. palustris** var. **màjor** (Boenn.) Koch has the fruit with a longer stalk (Fig. 84), the body 2.5–3 mm. long, and is found in brackish water along the Atlantic Coast, rarely inland to N.Y. & Wis.

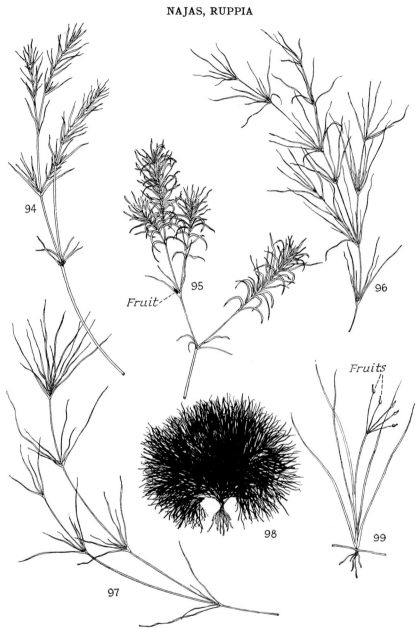

N. flexilis: 94–97, branches, × 1.   98. Plant, × 1.   R. maritima: 99. Part of plant, × 1

Bushy Pondweed   Nàjas

Stems slender and much branched; leaves narrow and ribbon-like, enlarged at base (Figs. 86–89); fruits borne singly in the leaf axils (Fig. 95), with papery walls, tapered to a short style (Fig. 85), and containing a single seed which is 2–5 mm. long, round in cross section, and tapered at each end (Figs. 90–93).

N. flexilis is by far the most common species. It is very variable, from long and slender (Fig. 97) to closely tufted (Fig. 98). The coarser forms may be confused with *Anacharis*, from which they are distinguished by the broadenings at the bases of the leaves.

1. Leaves coarsely toothed, the teeth visible to the naked eye (Fig. 86); backs of leaves often spiny; fruits 4–5 mm. long.   **N. marìna** L.   Salt springs and brackish water, local; Fla.; cent. N.Y.; w. Minn. & e. N.D.; Tex.; Ariz. to Utah & Calif.
1. Leaves with fine teeth, usually visible only under a lens; backs of leaves not spiny; fruits 2–3.5 (–4.5) mm. long...**2**
   2. Widenings of leaf bases tapered (Fig. 87)................................**3**
   2. Widenings of leaf bases lobe-like (Figs. 88, 89)...........................**5**
3. Style 1 mm. or more long (Fig. 85); seed very finely and obscurely marked with 30–40 rows of pits across the middle (Fig. 90), usually shining.................**4**
3. Style 0.5 mm. or less long; seed dull, coarsely and deeply pitted with 10–20 rows of pits across the middle (Fig. 91).   **N. guadalupénsis** (Spreng.) Morong.   Local; s. Mass., N.J. & Que. to s. Mich., s. Minn. & Ore.; southward to Peru, Bolivia & Argentina.
   4. Leaves 1.5–4 cm. long, 0.5–1 mm. wide at base above the lobes, gradually tapered to the tip (Fig. 87); seed shining.   **N. fléxilis** (Willd.) Rostk. & Schmidt. Common and widespread; very variable in proportions of leaves and length of stem (Figs. 94–98).
   4. Leaves 9–18 mm. long, 1.2–2 mm. wide, abruptly pointed; seed dull.   **N. olivàcea** Rosendahl & Butters.   Rare and local; N.Y.; Wis.; Minn.
5. Leaves often recurving, relatively coarsely toothed, the basal lobes fan-shaped and finely toothed (Fig. 88); pits on seeds broader than long, in regular vertical rows (Fig. 92).   **N. mìnor** Allioni.   About the Hudson River, N.Y., and probably of recent introduction.
5. Leaves not recurving, thread-like, finely toothed, the basal lobes wedge-shaped and coarsely jagged (Fig. 89); pits on seeds longer than broad, deep (Fig. 93). **N. gracíllima** (A. Br.) Morong.   Maine to Va. & cent. N.Y., and locally to Mich., Wis., Minn., Ind. & Mo.

Wigeon Grass   Rúppia

R. marítima L.   Fig. 99.   Leaves thread-like; fruits more or less asymmetrical, on stalks which may be several centimeters long or may be so short as to be almost lacking.   Mostly in salt water along the coast, very rarely inland.

References: Fernald, Rhodora, **23**, 110 (1921)—*Zannichellia palustris* var. *major;* Fernald, Rhodora, **25**, 105–109 (1923), and Clausen, Rhodora, **38**, 333–345 (1936)— *Najas* in N.A.; Rosendahl and Butters, Rhodora, **37**, 345–348 (1935)—*N. olivacea;* Fernald and Wiegand, Rhodora, **16**, 119–127 (1914)—varieties of *Ruppia maritima.*

} *Anther*

} *Filament*

1

Beak

2

3

4

S. cuneata: 1. Stamen, × 10.   2. Nutlet, × 10.   3, 4. Plants, × ½.   (Figs. 1–4 from J. G. Smith.)

WATER PLANTAIN FAMILY   **ALISMÀCEAE**

Very common plants, with rosettes of arrow-shaped, elliptical or ribbon-like leaves. Sometimes the plants are reduced to rosettes of narrow leaves (Fig. 50); *Sagittaria graminea, S. cristata* and *S. cuneata* often appear in this form and are frequently sterile. Sometimes long ribbon-like leaves are formed in deep water, which may be distinguished from Wild Celery only by the venation (see p. 7). The flowers are borne on a naked stem (the scape) and have 3 green sepals, 3 white or rarely pink petals, many stamens and a head of carpels which develop into nutlets. There are usually several underground stems, each ending in an edible tuber (Fig. 42), called "Duck Potato."

Owing to the great variability of the leaf shape within each species, the classification within this group is based almost entirely on flowers and fruits. Mature fruits are necessary for the determination of most species. Most commonly seen are *Sagittaria latifolia*, with leaves usually arrow-shaped but very variable in width, and *Alisma Plantago-aquatica*, with leaves not lobed.

1. Lowest flowers with only stamens, or with only carpels (Figs. 9, 30). **Sagittaria.**
1. Lowest flowers with both stamens and carpels...............................2
  2. Upper flowers with stamens only. **Lophotocarpus.**
  2. All flowers with both stamens and carpels................................**3**
3. Stamens 6–21; fruits in a dense head (Fig. 54). **Echinodorus.**
3. Stamens 6; fruits in a ring (Figs. 59, 64). **Alisma.**

ARROWHEAD, DUCK POTATO   **Sagittària**

Stamens and carpels in different flowers (rarely some flowers with both), the staminate flowers above, the pistillate below, or sometimes the two kinds of flowers on different plants; leaves of 4 types, the arrow-shaped (Fig. 3) and elliptical (Fig. 46) commonly emersed, and the tongue-like (Fig. 50) and ribbon-like (Fig. 51) submersed; two or more types may be found on the same plant.

  1. At least some of the leaves with a petiole and a flattened blade (Figs. 14, 32, 46).**2**
  1. Leaves not differentiated into petiole and blade (Figs. 50, 51). These plants sometimes have flowers and fruits, but are often sterile and nearly impossible to identify. The following key may help in some cases.......................**19**
    2. Blade with all veins branching from a point at its junction with the petiole (Fig. 3), or with but one central vein...................................**3**
    2. Blades with most of the side veins branching at intervals from the midrib (Fig. 44)...**17**
  3. Bracts at the base of each whorl of flowers separate (Figs. 8, 9, 16) or united for a very short distance at base; anthers about twice as long as wide, on slender filaments (Figs. 1, 12); leaves usually lobed at base **4**
  3. Bracts united (Fig. 35), the uppermost for more than half their length (Fig. 31); anthers not over 1.5 times as long as wide, on inflated filaments (Fig. 27); leaves not lobed (Fig. 30), or with short basal lobes (Fig. 28).......................**14**

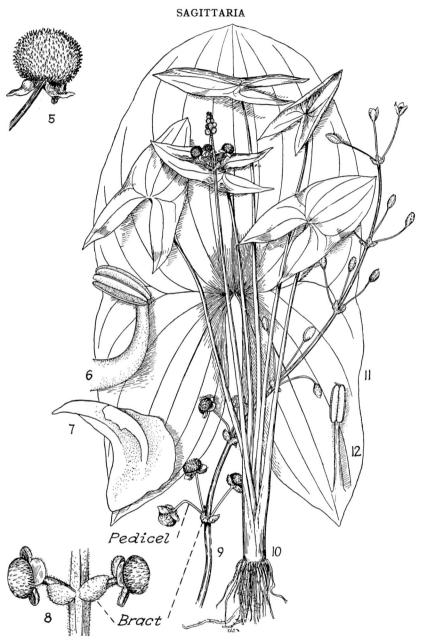

S. latifolia: 5. Fruiting head, × 1. **S. latifolia** var. **obtusa**: 6. Stamen, × 10. 7. Young nutlet, × 10. 9. Inflorescence, × 1. 10. Plant, × ⅜. 11. Leaf, × ½. 12. Stamen, × 10. **S. latifolia** var. **pubescens**: 8. Whorl of flowers, × 1. (Figs. 5–12 from J. G. Smith.)

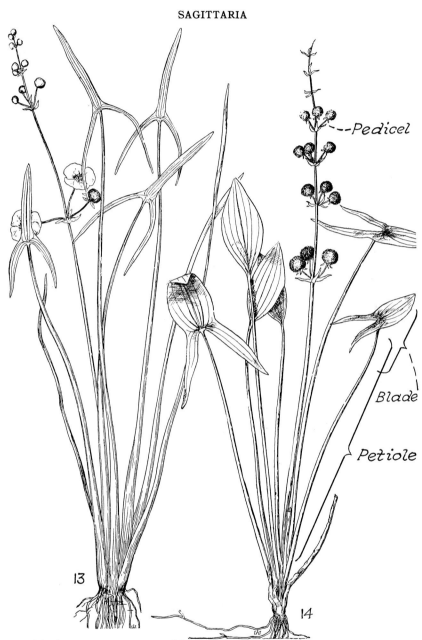

S. latifolia f. gracilis: 13. Plant, × ¼. S. latifolia f. diversifolia: 14. Plant, × ¼; the lobed leaves are the type of f. hastata. (Figs. 13, 14 from J. G. Smith.)

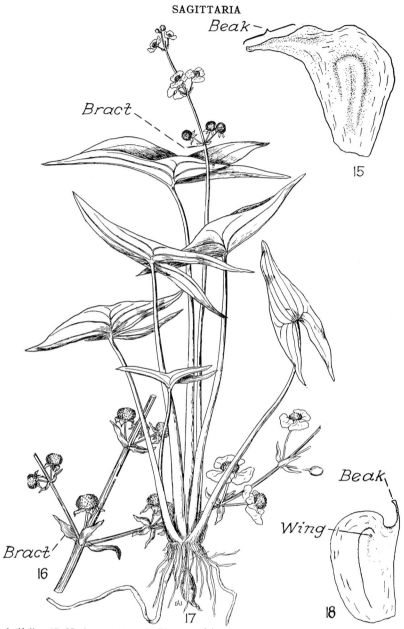

Beak

Bract

15

Bract

16

Beak

Wing

17

18

S. latifolia : 15. Nutlet, × 10.   17. Plant, × ¼.   S. brevirostra : 16. Inflorescence, × ½.
18. Nutlet, × 10.   (Figs. 15–18 from J. G. Smith.)

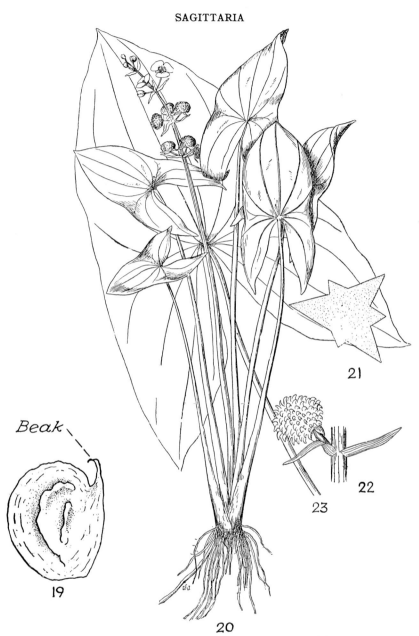

S. australis: 19. Nutlet, × 10. **S. brevirostra:** 20. Plant, × ⅓. 21. Cross section of scape, × 2. 22. Fruiting head and bracts, × 1. 23. Leaf, × ½. (Figs. 20–23 from J. G. Smith.)

S. Engelmanniana. 24. Nutlet, × 10.   25. Plant, × ½.   26. Leaf, × 1.   (Figs. 24–26 from J. G. Smith.)

4. Nutlets mostly 2–3 mm. long, without wings on the sides and with a short beak 0.5 mm. or less long (Fig. 2); leaves with blades mostly 2–18 cm. long. **S. cuneàta** Sheldon. Figs. 1–4. Que. to Mich., N.D. & B.C., s. to Maine, Conn., N.Y., Kans., Mex. & Calif. When submersed this often produces leaves with limp petioles and floating blades, as well as the types described in **20.** The emersed leaves are variable in width. (*S. arifolia*, G.)

4. Nutlets 3–3.5 mm. long, with 1 or 2 wings on each side (Fig. 18) and an erect or horizontal beak (Figs. 15, 18) 0.5–2 mm. long; leaves with blades mostly 1–3 dm. long...**5**

5. Beak of nutlet horizontal (Fig. 15); bracts 1 cm. or less long (Figs. 8, 17); scape round in cross section, scarcely angled. **S. latifòlia** Willd. Wapato. A variable species...**6**

5. Beak of nutlet erect (Figs. 18, 19); bracts 1.5–3 cm. long (Fig. 16); scape angled (Fig. 21) except in *S. Engelmanniana*.....................................**11**

   6. Plants not hairy...**7**

   6. Plants finely hairy (Fig. 8). **S. latifolia** var. **pubéscens** (Muhl.) J.G.Sm. N.J. & Pa. to N.C. (*S. pubescens*, B.)

7. Tip of leaf sharply angled (Figs. 13, 14); pedicels of pistillate flowers less than twice as long as the fruiting head (Figs. 14, 17)...................................**8**

7. Tip of leaf generally blunt or rounded (Fig. 11); pedicels of pistillate flowers more than twice as long as the fruiting heads (Fig. 9). **S. latifolia** var. **obtùsa** (Muhl.) Wiegand. Nearly as common as typical *S. latifolia* and often growing with it. (*S. latifolia* f. *obtusa*, G.)

   8. All leaves with basal lobes...**9**

   8. Some leaves without basal lobes, others with them (Fig. 14). **S. latifolia** forma **diversifòlia** (Engelm.) Robinson.

9. Leaves several centimeters wide...**10**

9. Leaves and their lobes very narrow, usually 1 cm. or less wide (Fig. 13). **S. latifolia** forma **grácilis** (Pursh) Robinson. This grades into the typical form.

   10. Basal lobes triangular (Fig. 17). **S. latifòlia** Willd. Very common; N.S. to B.C., s. to Fla., Mex. & Calif.; often accompanied by and grading into the other forms.

   10. Basal lobes with parallel sides (Fig. 14). **S. latifolia** forma **hastàta** (Pursh) Robinson.

11. Plants stout, 0.5–1 m. tall...**12**

11. Plants slender, 1.5–5 dm. tall...**13**

   12. Faces of nutlet with a single wing, and notch between the top of the nutlet and the beak rounded (Fig. 18). **S. breviróstra** Mack. & Bush. Figs. 16, 20–23. N.S. to Minn., s. to Tenn., Mo. & Kans. Commonly confused with *S. latifolia*, from which it is distinguished by its angled scape, long bracts and erect beak of the nutlet. Small individuals are distinguished from *S. cuneata* by the longer beak on the nutlet of *S. brevirostra*.

   12. Faces of nutlet with 2 or more wings (Fig. 19), and notch deep and narrow. **S. austràlis** (J.G.Sm.) Small. N.J. to s. Ind., s. to Ky. & Ala. (*S. longirostra*, G, B.)

13. Leaf blades ribbon-like, 5 mm. or less wide (Figs. 25, 26). **S. Engelmanniàna** J.G.Sm. Figs. 24–26. S. N.E. to Del. Superficially resembles *S. latifolia* f. *gracilis*, from which it may be distinguished by its longer bracts and erect beak of the nutlet.

13. Leaf blades oval or triangular. **S. Engelmanniana** forma **dilatàta** Fernald. With the preceding and much less common.

SAGITTARIA

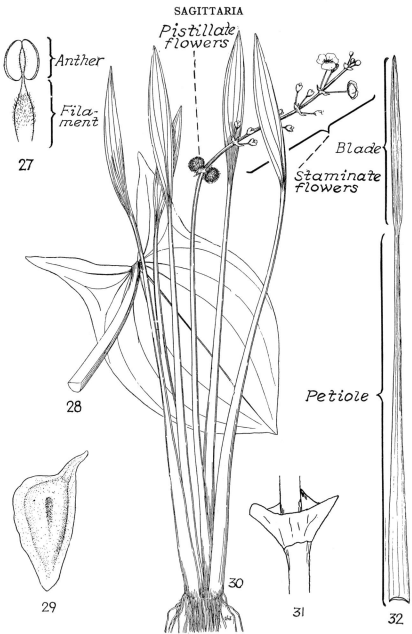

S. rigida: 27. Stamen, × 10. 29. Nutlet, × 10. 30. Plant, × ½. 31. Bracts, × 2. 32. Leaf, × ½. S. rigida f. elliptica: 28. Leaf, × ½. (Figs. 27–30 from J. G. Smith.)

S. graminea: 33. Leaf with well-developed blade, × 1.   34. Plant, × ½.   35. Bracts of inflorescence, × 2.   **S. cristata:** 36. Partly submersed plant, × ½; **S. graminea** may have the same appearance.   (Figs. 33–35 from **J. G.** Smith.)

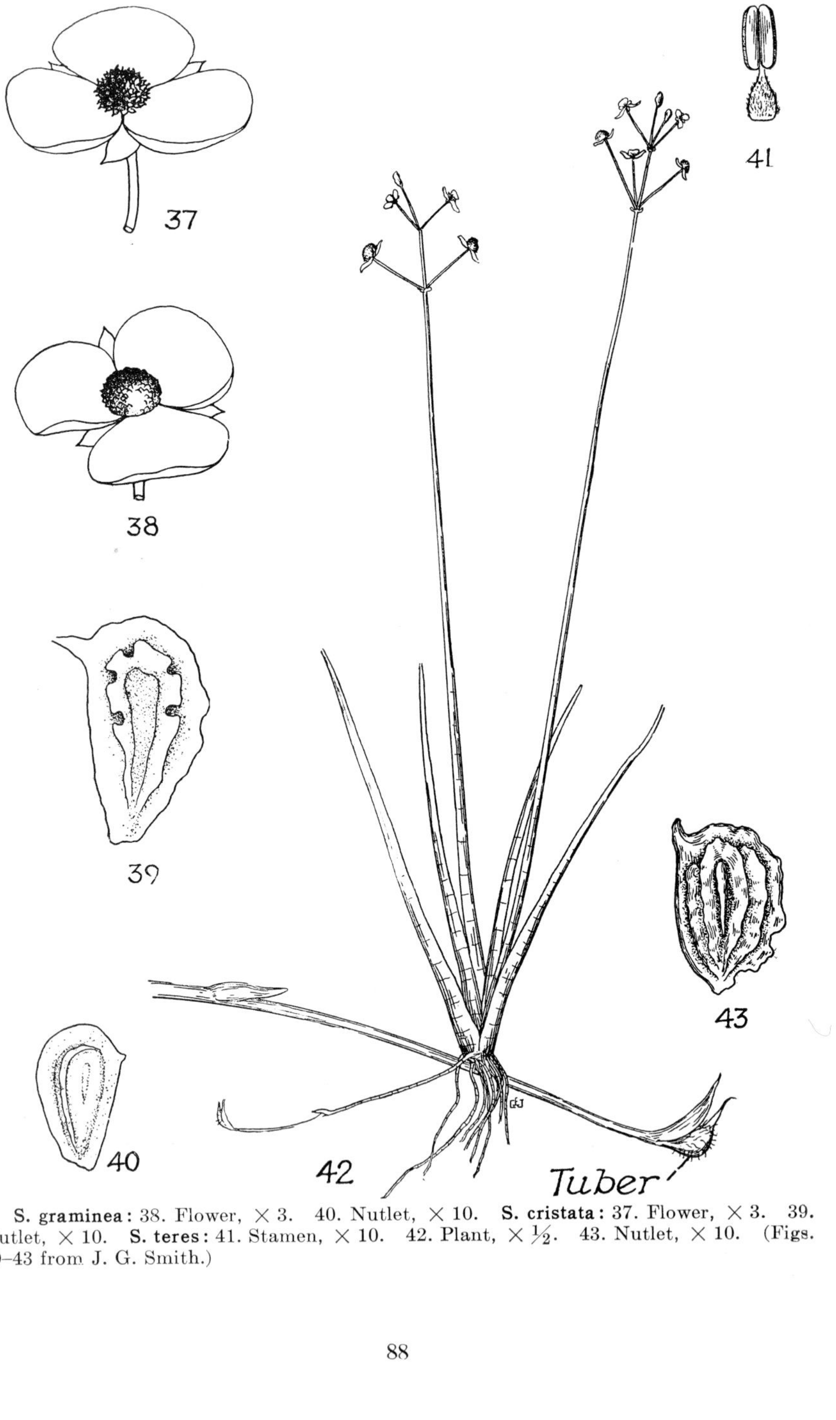

SAGITTARIA

S. graminea : 38. Flower, × 3. 40. Nutlet, × 10. S. cristata : 37. Flower, × 3. 39.
Nutlet, × 10. S. teres : 41. Stamen, × 10. 42. Plant, × ½. 43. Nutlet, × 10. (Figs.
39–43 from J. G. Smith.)

88

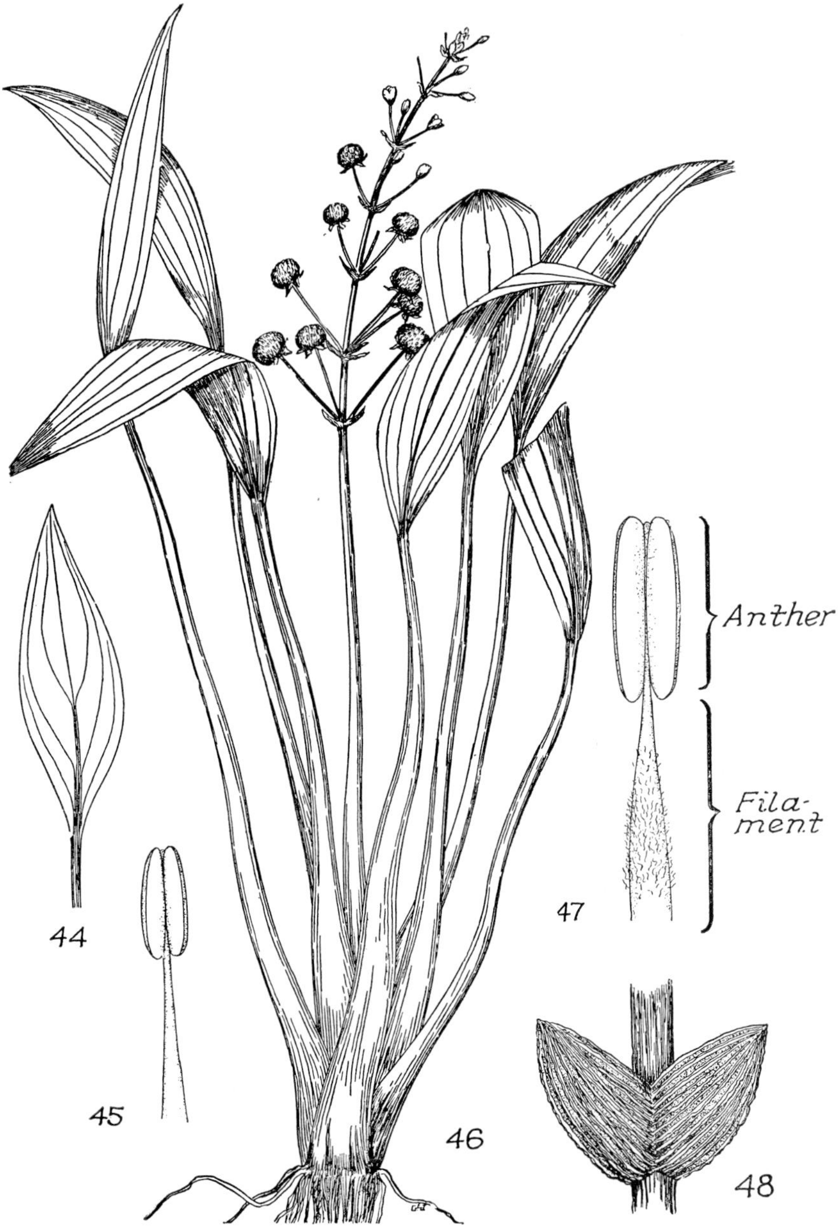

Anther

*Fila-
ment*

S. falcata: 44. Blade of leaf, $\times \frac{1}{6}$. 47. Stamen, $\times 10$. 48. Bracts, $\times 4$. **S. ambigua:**
45. Stamen, $\times 10$. 46. Plant, $\times \frac{1}{3}$. (Figs. 44–48 from J. G. Smith.)

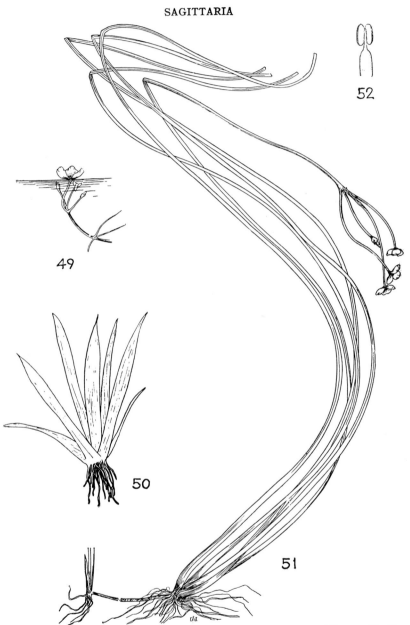

S. cristata: 50. Submersed rosette, × ½. **S. subulata** var. **gracillima:** 49. Tip of scape at water's surface, × 1. 51. Plant, × 1. 52. Stamen, × 10. (Figs. 49, 51, 52 from J. G. Smith.)

14. Pistillate flowers almost without pedicels (Fig. 30), rarely with short stout pedicels..**15**

14. Pistillate flowers with slender pedicels (Fig. 34)........................**16**

15. Leaf blades scarcely wider than the petiole, erect and stiff (Fig. 32). **S. rígida** Pursh. Stiff Wapato. Figs. 28–32. Usually in somewhat limy soil; s. Que. & s. Maine to Fla. & Neb. (*S. heterophylla* var. *rigida*, G.)

15. Leaf blades oval, sometimes with 1 or 2 short basal lobes (Fig. 28). **S. rigida** forma **ellíptica** (Engelm.) Fernald. With the last, and less common. (*S. heterophylla* and var. *elliptica*, G.)

    16. Beak of nutlet less than a fourth as long as the width of the nutlet (Fig. 40); sides of nutlet without wings; pistillate heads, both in flower and in fruit, appearing smooth (Fig. 38). **S. gramínea** Michx. Figs. 33–36. Nfd. to Ont., s. to Fla. & Tex. Very variable in foliage, often represented only by sterile rosettes (Fig. 50).

    16. Beak of nutlet about half as long as the width of the nutlet (Fig. 39); sides of nutlet winged; pistillate heads appearing minutely prickly owing to the projecting beaks of carpels or nutlets (Fig. 37). **S. cristàta** Engelm. Minn. to Bruce Peninsula, Ont., s. to Iowa & Wis. Not distinguishable from the last in absence of flowers and fruits. The sterile rosettes are often abundant in 1–2 m. of water (Fig. 50).

17. Bracts roughened with little elongate bumps (Fig. 48); pedicels of pistillate flowers 0.5–3.5 cm. long...**18**

17. Bracts smooth; pedicels of pistillate flowers 2.5–6.5 cm. long. **S. Weatherbiàna** Fernald. Va. to S.C.

    18. Filaments cobwebby (Fig. 47). **S. falcàta** Pursh. Coastal Wapato. Md. to Ky. & southward. (*S. lancifolia*, G.)

    18. Filaments without hairs (Fig. 45). **S. ambígua** J.G.Sm. Fig. 46. Mo., Kans. & Okla.

19. Leaves rigid enough to keep their shape when taken from the water (Fig. 50).**20**

19. Leaves long, flexible and ribbon-like (Fig. 51). These are forms in deep water and may be confused with Wild Celery and Bur Reed (see p. 7)............**21**

    20. Leaves round in cross section, thickest at base and tapering to the tip (Fig. 42). **S. tères** Wats. Dwarf Wapato. Mass. to Fla.

    20. Leaves flat, usually widest a little above the middle (Fig. 50). Probably all species may produce leaves of this type, and if fruits are present they may often be identified. Rosettes of this type are most often formed by *S. cuneata*, *S. graminea* and *S. cristata*, and are sometimes called Stubby Wapato...**3**

21. Leaves about 2 mm. wide. **S. subulàta** (L.) Buchenau, var. **gracíllima** (Wats.) J.G.Sm. Ribbon Wapato. Figs. 49, 51. Streams in e. Mass. & R.I. (*S. lorata*, B.)

21. Leaves 5–10 mm. wide. **S. rigida** forma **flùitans** (Engelm.) Fernald. This probably occurs throughout the range of *S. rigida*. Other species perhaps have a similar form.

Lophotocárpus

    L. calycìnus (Engelm.) J.G.Sm. Figs. 67, 68. This has the appearance of a *Sagittaria* but may be distinguished from the common species of that genus by its thickened pedicels and the manner in which the sepals are closely appressed to the heads of fruits. The leaves are variable and may be very narrow and without lobes, or as much as 3 dm. in width. Muddy shores and shallow water, mostly in the Mississippi valley, n. to S.D., Wis. & Ohio.

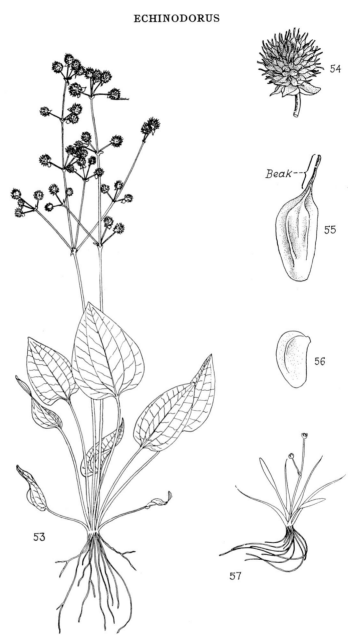

54

Beak--

55

56

53

57

E. cordifolius: 53. Plant, × ½.   54. Head of fruits, × 2.   55. Nutlet, × 16.   E. tenellus:
56. Nutlet, × 16.   57. Plant, × 1.

BURHEAD   Echinódorus

1. Small plants, seldom more than 1.5 dm. tall, with narrow leaves (Fig. 57); nutlets with a very small beak or none (Fig. 56); stamens 9. **E. tenéllus** (Martius) Buchenau. Mass. to Minn., s. to Fla. & Mex. (*Helianthium parvulum*, B.)
1. Large plants, with usually broad leaves (Fig. 53); nutlets with a beak (Fig. 55); stamens 12 or more..**2**
     2. Flowering stems erect; flowers 8–10 mm. wide; stamens 12. **E. cordifòlius** (L.) Griseb. Figs. 53–55. Ill. to Kans., s. to Calif. & Fla.
     2. Flowering stems prostrate, often rooting at the nodes; flowers 12–20 mm. wide; stamens about 21. **E. radìcans** (Nutt.) Engelm. N.C. to Ill. & Kan., s. to Fla. & Tex.

WATER PLANTAIN   Alísma

Leaves usually with broad flat blades which may be rounded or tapered at base; nutlets in a ring. Both species may grow submersed, with only ribbon-like blades present.

1. Petals white; rays ascending in fruit (Fig. 60); mature nutlet with 2 ridges and a groove down the back (Fig. 62)...**2**
1. Petals pink; rays more spreading in fruit (Figs. 63, 66); mature nutlet with 3 ridges and 2 grooves down the back (Fig. 65)..................................**3**
     2. Petals 3–6 mm. long, much exceeding the sepals (Fig. 58); nutlets 2.2–3 mm. long. **A. Plantàgo-aquática** L. Figs. 58–60, 62. Very common; Que. & N.S. to B.C., s. to N.Y., N.D. & Calif. (*A. brevipes*, B.)
     2. Petals 1–2 mm. long, slightly exceeding the sepals (Fig. 61); nutlets 1.5–2 mm. long. **A. Plantago-aquatica** var. **parviflòrum** (Pursh) Farwell. N.S. to Minn., s. to Fla. & Tex. (*A. subcordatum*, B.)
3. Fruiting stems longer than the leaves (Fig. 63). **A. gramíneum** Gmel. Local; s. Que. to Alberta, s. to n. N.Y., Wis., Neb., Utah & Ore. Also widely distributed in Europe, Asia and Africa.
3. Fruiting stems shorter than the leaves (Fig. 66). **A. gramineum** var. **Geyèri** (Torr.) Samuelsson. About the same range in N. A., but absent on other continents. (*Alisma Geyeri* G, B.)

References: Wiegand, Rhodora, **27**, 186 (1925)—*S. latifolia* var. *obtusa;* Wiegand and Eames, Cornell Univ. Agric. Exp. Sta. Mem., **92**, 53 (1926)—*A. Plantago-aquatica* var. *parviflorum;* Samuelsson, Arkif für Bot., **24A**, No. 7, 35–45 (1932)—*A. gramineum;* Fernald, Rhodora, **37**, 387–389 (1935)—*S. Weatherbiana;* Rhodora, **38**, 73–74 (1936)— forms of *S. rigida, S. Engelmanniana,* etc. Most of the drawings of *Sagittaria* and *Lophotocarpus* are from J. G. Smith, Ann. Rept. Mo. Bot. Gard., **6**, Pl. 1–29 (1895) and **11**, Pl. 53–58 (1900), and here are reproduced by permission of Director G. T. Moore of the Missouri Botanical Garden.

FLOWERING RUSH FAMILY   BUTOMÀCEAE

FLOWERING RUSH   Bùtomus

B. umbellàtus L. Fig. 69. Leaves ribbon-like, erect from a stout rootstock; flowers in an umbel terminating an erect scape about 1 m. tall, pink, with 3 sepals and 3 petals. Shores of the St. Lawrence River and some of its tributaries; Lake Champlain; Lake Erie. Naturalized from Europe. Very showy when in flower. **B. umbellatus** forma **vallisneriifòlius** Sagorski is a sterile form in deep water, with very long limp leaves the ends of which float at the surface of the water.

58

61 Sepal Petal

59

60

62

A. Plantago-aquatica: 58. Flower, $\times$ 2. 59. Head of nutlets, $\times$ 4. 60. Plant, $\times$ ½. 62. Nutlet, $\times$ 10. **A. Plantago-aquatica** var. **parviflorum:** 61. Flower, $\times$ 2.

A. gramineum: 63. Plant, $\times \frac{1}{2}$. 64. Head of nutlets, $\times 4$. 65. Nutlet, $\times 10$. **A. gramineum** var. **Geyeri**: 66. Plant, $\times \frac{1}{2}$.

L. calycinus: 67. Nutlet, $\times$ 5. 68. Plant, $\times \frac{1}{2}$. **B. umbellatus:** 69. Plant, $\times \frac{1}{5}$. (Figs. 67, 68 from J. G. Smith.)

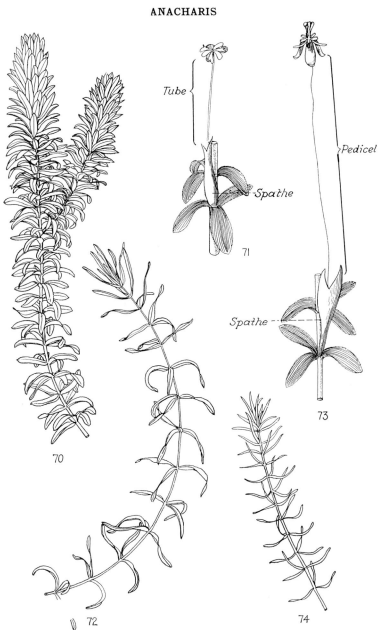

Tube

Pedicel

Spathe

Spathe

71

73

70

72

74

A. canadensis: 70. Pistillate plant, × 1.   71. Pistillate flower, × 2.   72. Staminate plant, × 1.   73. Staminate flower, × 2.   **A. occidentalis:** 74. Branch, × 1.   (Figs. 71, 73 redrawn from Victorin.)

VALLISNERIA, LIMNOBIUM

V. americana: 75. Staminate flower, × 6.   77. Staminate plant, × ½.   78. Pistillate plant with ripening fruit, × ½.   79. Pistillate plant with flowers at the surface ready for fertilization, × ½.   **L. Spongia**: 76. Floating plant, × ½.

FROGBIT FAMILY   HYDROCHARITÀCEAE

1. Leaves opposite or whorled, on a slender branching stem (Fig. 70).   **Anacharis.**
1. Leaves in rosettes..2
   2. Leaves ribbon-like (Fig. 77).   **Vallisneria.**
   2. Leaves divided into petiole and blade (Fig. 76).   **Limnobium.**

WATERWEED   Anácharis

Plants submersed, the stems branching, forming large masses near the bottom, sometimes purple-tinged; leaves whorled or rarely opposite, their bases embracing the stem, their margins with microscopic teeth; flowers from a spathe (Fig. 71), the pistillate with a long thread-like tube which reaches the surface.

1. Leaves 1.2–4.5 mm. wide, averaging about 2 mm. (Fig. 70); spathe of staminate flower 11–13 mm. long, the apex 2-lobed (Fig. 73); staminate flower long-pediceled, often remaining attached (Fig. 73), its sepals 3.5–5 mm. long; sepals of pistillate flower 2.3–2.7 mm. long. **A. canadénsis** (Michx.) Planchon. Usually in calcareous water; Que. & Maine to Sask. & Wyo., s. to N.Y., Ky. & Ill. (*Elodea canadensis*, G; *Philotria canadensis*, B.) The pistillate plant (Fig. 70) has the leaves densely crowded toward the tip; the staminate plant (Fig. 72) often has them less crowded and has been treated as *Elodea Planchonii* by St. John, but Wiegand and Eames consider these to be phases of one species.

1. Leaves 0.7–1.8 mm. wide, averaging about 1.3 mm. (Fig. 74); spathe of staminate flower 2 mm. long, sharp-pointed; staminate flower not pediceled, breaking from the spathe and floating on the surface, its sepals 2–2.5 mm. long; sepals of pistillate flowers 1.2–1.8 mm. long. **A. occidentàlis** (Pursh) Victorin. Usually in non-calcareous water, but occasionally found with *A. canadensis;* Maine to Minn. & Ore., s. to D.C., Mo. & Neb. (*Philotria angustifolia*, B.)

A. densa (Planchon) Victorin is the large species commonly grown in aquariums under the name *"Elodea canadensis* var. *gigantea"* and is rarely found as an escape.

References: St. John, Rhodora, **22,** 17–29 (1920)—key and descriptions; Wiegand and Eames, Cornell Univ. Agric. Exp. Sta. Mem., **92,** 55 (1926)—key and discussion; Marie-Victorin, Contrib. Lab. Bot. Univ. Montréal, No. 18 (1931)—descriptions of plants, nomenclature of the genus and remarks on the Waterweed in Europe; Weatherby, Rhodora, **34,** 114–116 (1932)—nomenclature of the genus.

WILD CELERY, TAPE GRASS   Vallisnèria

V. americàna Michx. Stems buried in the mud, sending up tufts of ribbon-like leaves (Fig. 77) in the water; staminate flowers in a spathe (Fig. 77) which breaks and allows them to come to the surface, where they float (Fig. 75); pistillate flowers on a long peduncle, the sepals and stigmas projecting above the surface (Fig. 79), the peduncle after fertilization becoming curved or wavy to pull the fruit below the surface (Fig. 78). N.S. to S.D., s. to the Gulf of Mexico. (*V. spiralis*, G, B.)

Rydberg, N. Amer. Fl., **17,** 68–69 (1909), treats our plants as consisting of 2 species, *V. spiralis* and *V. americana,* but Fernald, Rhodora, **20,** 108–110 (1918), shows that there is probably but one species, *V. americana,* and that in any case *V. spiralis* is a species of southern Europe, not known in North America. Wylie, Bot. Gaz., **63,** 135–145 (1917) describes the method of pollination in the American plant, and Svedelius, Svensk. Bot. Tidsk., **26,** 1–12 (1932), lists many differences between our plant and the European.

FROGBIT   Limnòbium

L. Spóngia (Bosc) Richard. Fig. 76. Plants floating in still water or rooting in soft mud; leaves in a tuft, long-petioled, the blades somewhat heart-shaped; flowers on stout peduncles, white; peduncles recurving as the fruit matures. W. N.Y.; N.J. (?); Del. to the Gulf of Mexico, n. in the Mississippi Valley to s. Ill. & Mo.

GRASS FAMILY   GRAMÍNEAE

Many kinds of grasses grow in damp places and may at times be found in the water. Those described below are the ones most likely to be found in the water and are arranged in order of their importance to the aquatic biologist.

The sedges and rushes resemble grasses superficially; see pp. 122, 124 and 125.

Some grasses, especially Wild Rice and Manna Grass, are occasionally found submersed and sterile with floating leaves like those of Wild Celery. These may be recognized as grasses by the ligule at the junction of blade and sheath (see Fig. 1, p. 9), and Fig. 9, p. 11). Manna Grass may be told from Wild Rice by its rootstocks, which *Zizania* lacks; but in many cases these aquatic forms can be identified, if at all, only by a study of the normally developed plants in the vicinity.

The most useful comprehensive work on the grasses is the Manual of the Grasses of the United States, by A. S. Hitchcock, to be had from the Superintendent of Documents, Washington, D.C., for $1.75. The names used below agree with Hitchcock in most cases; where they do not, a reference is given to the authority. Most of the illustrations are from Hitchcock, and a few are original.

1. Stamens and pistils in different spikelets...................................2
1. Stamens and pistils borne in the same spikelet; spikelets essentially uniform....3
   2. Pistillate spikelets borne above; staminate spikelets, which are red or yellow, borne below (Fig. 2). **Zizania**, p. 102.
   2. Pistillate and staminate spikelets intermixed (Fig. 60). **Zizaniopsis**, p. 121.
3. Plants 2–4 m. tall, with spreading pennant-like leaves; inflorescence feathery (Fig. 22). **Phragmites**, p. 107.
3. Plants usually smaller; inflorescence not feathery..........................4
   4. Spikelet covered with little spines (Fig. 24). **Echinochloa**, p. 107.
   4. Spikelet without spines...5
5. Spikelets all on one side of a close spike (Figs. 71, 75).....................16
5. Spikelets not all on one side of a close spike................................6
   6. Spikelets in a close cylindrical spike (Figs. 49, 50).......................15
   6. Spikelets in a loose open arrangement (Fig. 59) or in irregular masses (Fig. 46).7
7. Spikelets made up of several florets (Figs. 20, 64, 70).......................8
7. Spikelets 1-flowered (Figs. 39, 58, 61)...................................12
   8. Lemmas deeply creased or corrugated (Fig. 20). **Glyceria**, p. 105.
   8. Lemmas not corrugated...9
9. Tip of leaf boat-shaped (Fig. 67); lemma cottony at base (Fig. 69). **Poa**, p. 121.
9. Tip of leaf flat or tapered; lemma not cottony at base......................10
   10. Plants seldom more than 3 dm. tall, sometimes prostrate at base and rooting at the lower nodes (Figs. 42, 57).......................................11
   10. Plants 1–1.5 m. tall, not prostrate at base (Fig. 66). **Fluminea**, p. 121.

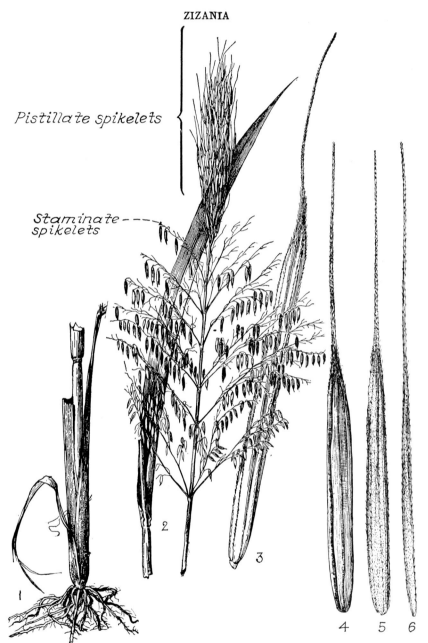

ZIZANIA

Pistillate spikelets

Staminate spikelets

1

2

3

4   5   6

Z. aquatica var. **interior:** 1. Base of plant, × ½.   2. Leaf, and panicle, × 2.   3. Pistillate spikelet filled with grain,   × 5.   4. Pistillate spikelet, empty,   × 5.   **Z. aquatica:** 5. Pistillate spikelet filled with grain, × 5.   6. Pistillate spikelet, empty, × 5.   (Figs. 1–3 from Hitchcock.)

11. Lemmas pointed (Fig. 40); common plants. **Eragrostis,** p. 113.
11. Lemmas cut off square at tip (Fig. 56); rare plants. **Catabrosa,** p. 121.
 12. Spikelets with fringed margins, overlapping in a single row (Figs. 32, 33).
  Leersia, p. 111.
 12. Spikelets without fringed margins, not overlapping in a single row........**13**
13. Spikelets in close irregular masses (Fig. 46). **Phalaris,** p. 113.
13. Spikelets in loose open arrangement (Fig. 59)............................**14**
 14. Lemma surrounded by a tuft of long hairs from its base (Figs. 37, 39). **Cala-**
  magrostis, p. 111.
 14. Lemma without long hairs (Fig. 58). **Agrostis,** p. 115.
15. Glumes and lemmas with long awns from their tips (Fig. 50). **Elymus,** p. 115.
15. Glumes blunt (Figs. 48, 54); lemmas with delicate awns from the back (Figs.
 47, 53). **Alopecurus,** p. 115.
 16. Spikelet long and narrow (Fig. 73). **Spartina,** p. 121.
 16. Spikelet nearly as wide as long (Fig. 74). **Beckmannia,** p. 121.

WILD RICE Zizània

Plants annual, with short roots which are easily pulled up (Fig. 1); spikelets in a much-branched panicle, the upper pistillate and the lower staminate (Fig. 2); staminate spikelets red or yellow; pistillate spikelet with a lemma which has a long bristle-like awn (Figs. 3–6). Plants of shallow water, sometimes on muddy shores.

1. Pistillate lemma thin and papery, dull, minutely roughened all over (Fig. 5); some lemmas not filling with grain, these appearing thread-like (Fig. 6). **Z. aquática** L. Brackish water along the coast, s. Maine to Fla.; inland to n. Ind. & e. Wis. This is the Wild Rice of the Eastern states, often a giant plant reaching 3 m. in height, with leaves 1–5 cm. wide and a ligule 1–2 cm. long. (*Z. palustris,* G.)
1. Pistillate lemma firm and tough, straw-like, smooth between the nerves (Fig. 3); lemmas which are not filled with grains having a distinct flattened body (Fig. 4).**2**
 2. Plant 0.7–1.5 m. tall; leaves 4–12 mm. wide; ligules 3–5 mm. long; lower pistillate branches with 2–6 spikelets; lower or middle staminate branches with 1–15 spikelets. **Z. aquatica** var. **angustifòlia** Hitchc. S. N.B. to n. Mass., w. to n. Minn. & n. Ind. This is the northern Wild Rice, a much smaller plant than the eastern. (*Z. aquatica,* G.)
 2. Plant 0.9–3 m. tall; leaves 1–4 cm. wide; ligules 1–1.5 cm. long; lower pistillate branches with 11–29 spikelets; lower or middle staminate branches with 30–60 spikelets. **Z. aquatica** var. **intèrior** Fassett. Figs. 1–4. Ind. & s. Minn. to Neb. & Tex. This is the large Wild Rice of the Middle West. Ordinarily distinct from the two other varieties, it intergrades with var. *angustifolia* where the ranges of the two overlap in n. Ind. and n.w. Wis.

Reference: Fassett, Rhodora, **26,** 153–160 (1924).

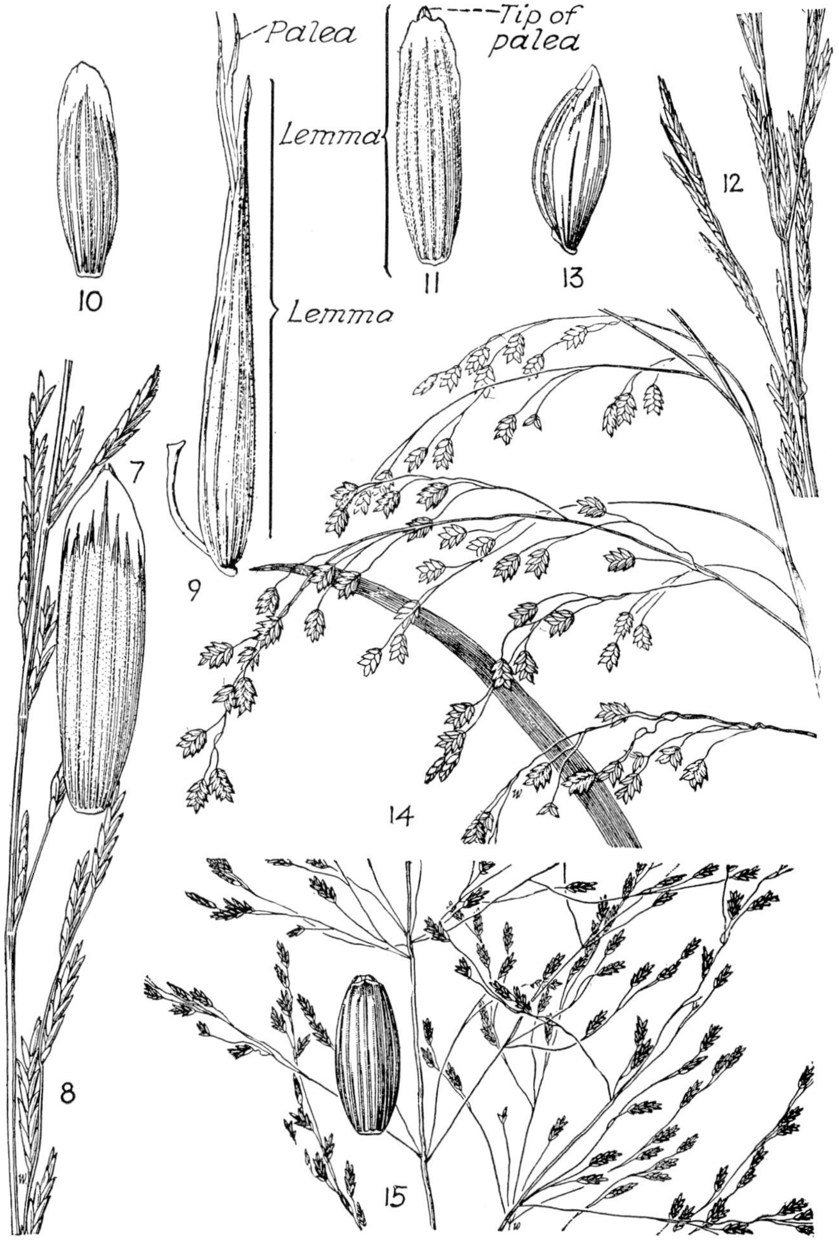

G. fluitans: 7. Lemma, × 10. 8. Part of panicle, × 1. **G. acutiflora:** 9. Lemma and palea, × 7.5. **G. borealis:** 10. Lemma, × 10. **G. septentrionalis:** 11. Lemma and palea, × 10. 12. Part of panicle, × 1. **G. canadensis:** 13. Floret, side view, × 7.5. 14. Part of panicle, × 3/4. **G. grandis:** 15. Part of panicle, × 4/5, and lemma, × 7.5. (Figs. 7–15 from Hitchcock.)

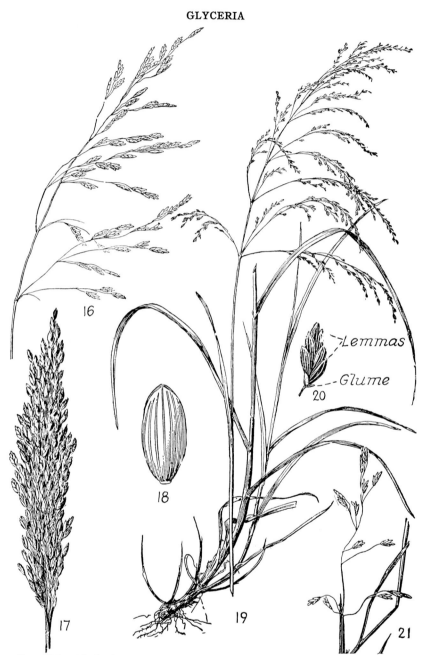

G. **pallida**: 16. Panicle, × 1. G. **obtusa**: 17. Panicle, × 1. 18. Lemma, × 7.5. G. **striata**: 19. Plant, × ⅜. 20. Spikelet, × 9. G. **neogaea**: 21. Panicle × ¾. (Figs. 16–21 from Hitchcock.)

MANNA GRASS   Glycèria

Spikelets several-flowered (Fig. 20), the glumes shorter than the first lemma; lemmas strongly nerved or ridged (Figs. 10, 18). This genus is unique among the grasses in having frequently a closed sheath. (*Panicularia*, B.)

1. Spikelets linear, usually 1 cm. or more long (Fig. 8)..........................2
1. Spikelets cvate, usually 5 mm. or less long (Fig. 20).........................5
  2. Lemma sharp-pointed, shorter than the palea (Fig. 9). **G. acutiflòra** Torr. Wet soil and shallow water; N.H. to Del., w. to Mich., s. to s. Ind., Tenn. & Mo.
  2. Lemma very blunt (Fig. 10)...3
3. Leaves 2–4 mm. wide; lemmas 3–4 mm. long, not roughened between the nerves (Fig. 10). **G. boreàlis** (Nash) Batchelder. Wet soil and shallow water; Nfd. to Alaska, s. to Conn., Ind., Iowa, S.D., N.M. & Calif.
3. Leaves 4–8 mm. wide; lemmas 4–6 mm. long, microscopically roughened between the veins (Figs. 7, 11)..4
  4. Lemma 4 mm. long (Fig. 11). **G. septentrionàlis** Hitchc. Fig. 12. Wet places and shallow water; Que. to Minn., s. to S.C. & Tex.
  4. Lemma 5–6 mm. long (Fig. 7). **G. flùitans** (L.) R. Br. Fig. 8. Shallow water; Nfd. to Que. & N.Y.; S.D.
5. Spikelets crowded closely together (Fig. 17). **G. obtùsa** (Muhl.) Trin. Wet places; N.S. to N.C.
5. Spikelets not crowded...6
  6. Spikelets 3–4 mm. wide; lemmas obscurely nerved (Fig. 13). **G. canadénsis** (Michx.) Trin. Fig. 14. Rattlesnake Manna Grass. Wet places, rarely in shallow water, and very common; Nfd. to Minn., s. to Md. & Ill.
  6. Spikelets 2.5 mm. or less wide; lemmas strongly nerved (Fig. 20)............7
7. Stems weak, somewhat trailing, often on the surface of the water.............8
7. Stems erect..9
  8. Plants 3–10 dm. high; leaves 2–8 mm. wide; spikelets 4–7-flowered, 6–7 mm. long; anther 1 mm. long. **G. pállida** (Torr.) Trin. Fig. 16. Shallow water; Maine to Wis., s. to N.C.
  8. Plants 2–4 dm. high; leaves 1–3 mm. wide; spikelets 3–5-flowered, 4–5 mm. long; anther 0.2–0.5 mm. long. **G. neogaèa** Steud. Fig. 21. Shallow water; Nfd. to Minn., s. to Conn. (*G. pallida* var. *Fernaldii*, G.)
9. Leaves 2–4 (–8) mm. wide; first glume 0.5 mm. long. **G. striàta** (Lam.) Hitchc. Meadow Grass. Figs. 19, 20. Wet soil; Nfd. to B.C., s. to Fla., Tex., Ariz. & n. Calif. (*G. nervata*, G; *P. nervata*, B.)
9. Leaves 6–12 mm. wide; first glume 1.5 mm. long. **G. grándis** Wats. Reed Meadow Grass. Fig. 15. Wet places, rarely in the water; P.E.I. to Alaska, s. to Tenn., Ohio, Neb., N.M. & Ore.

P. maxímus var. **Berlandieri:** 22. Rootstock, panicle, and part of stem, $\times \frac{1}{3}$, and floret, $\times 3$. 23. Spikelet, $\times 3$. (Figs. 22, 23, from Hitchcock.)

REED GRASS   Phragmìtes

P. máximus (Forsk.) Chiov., var. **Berlandièri** (Fourn.) Moldenke. Fig. 22. Plants reaching 4 m. tall, from stout rootstocks, sometimes with long stems creeping over the moist soil; leaves flat, 1–5 cm. wide, often somewhat spreading; spikelet with long silky hairs (Fig. 23), so that the whole flowering portion of the plant is a silky mass. Occurs mostly in brackish places across the continent. This is sometimes confused with Wild Rice, but, if the spikelets are not present, it may be distinguished by its stout rootstocks which make it difficult to pull up. (*P. communis*, G; *P. Phragmites*, B.)

References: Fernald, Rhodora, **34**, 211 (1932)—differences between *Phragmites* in Europe and in America; Merrill, Trans. Amer. Philos. Soc. new ser., **24**, 79–80 (1935), and Moldenke, Torreya, **36**, 93 (1936)—nomenclature.

WILD MILLET   Echinóchloa

Coarse grasses, 0.5–1, rarely –2, m. tall; spikelets covered with stiff bristles, and often with a long purple or green awn.

1. Body of spikelet about half as broad as long (Fig. 24), its upper glume not awned; sheaths not hairy...2
1. Body of spikelet about a third as broad as long (Fig. 29), its upper glume awned; lower sheaths usually with spreading bristly hairs (Fig. 27). **E. Waltèri** (Pursh) Nash. Often making a rank growth on mucky shores or in brackish marshes; along the coast from N.H. to Fla. & Tex.; inland to n. Ind., s. Minn. & s. Man. **E. Walteri** forma **laevigàta** Wieg. lacks the hairs on the sheaths. **E. Walteri** forma **brevisèta** Fern. & Grisc. has the awns very short.
   2. Spikelets 3.3–4.5 mm. long, 1.8–2.2 mm. wide (Fig. 24). **E. púngens** (Poir.) Rydb. Fig. 25. Wet places; Maine to Fla., w. to s. Minn., Kans., Okla. & N.M. (*E. crusgalli*, G, B, in part; the true *E. crusgalli* is a native of Europe, found in N.A. as a weed of roadsides, barnyards, etc.)
   2. Spikelets 2.5–3.4 mm. long, 1.4–1.8 mm. wide............................3
3. Spikelets with many spreading bristles over the back (Fig. 30). **E. pungens** var. **microstàchya** (Wieg.) Fern. & Grisc. Moist sand and wet shores; Maine to Ont., s. Minn. & southward.
3. Spikelets nearly free from bristles on the back (Fig. 28). **E. pungens** var. **occidentàlis** (Wieg.) Fern. & Grisc. Sometimes in more peaty soil; Maine to R.I.; n. Wis.; Ill. & Mo. to Wash. & N.M.

References: Wiegand, Rhodora, **23**, 49–65 (1921)—keys and descriptions; Fernald and Griscom, Rhodora, **37**, 136–137 (1935)—nomenclature.

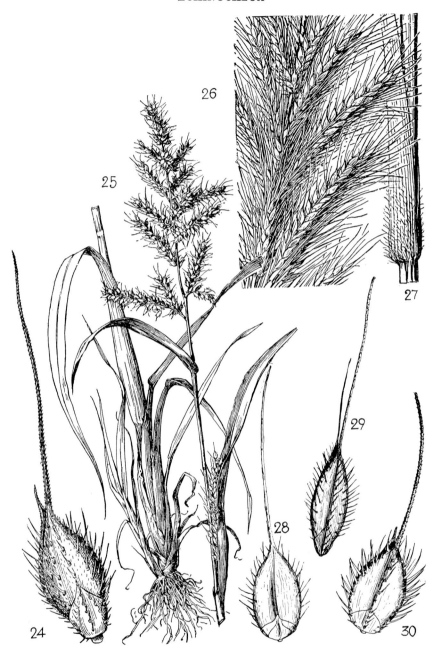

E. pungens: 24. Spikelet, × 10.   25. Base of plant, and panicle, × ½.   **E. Walteri:** 26. Part of panicle, × 9/10.   27. Sheath, × 9/10.   29. Spikelet, × 10.   **E. pungens** var. **occidentalis:** 28. Spikelet, × 10.   **E. pungens** var. **microstachya:** 30. Spikelet, × 10. (Figs. 24–27 from Hitchcock.)

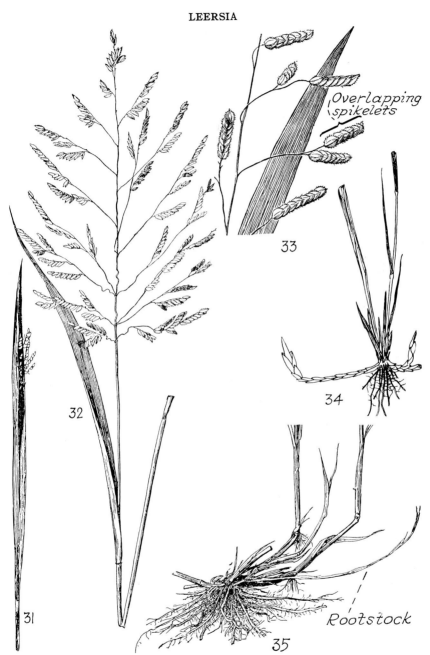

Overlapping spikelets

33

34

32

31

Rootstock

35

L. oryzoides f. **inclusa**: 31. Upper sheath, with spikelets barely exposed, $\times$ ½. **L. oryzoides**: 32. Upper part of plant, $\times$ ½. 35. Base of plant, $\times$ ½. **L. lenticularis**: 33. Part of panicle and of leaf, $\times$ ⅚. 34. Base of plant, $\times$ ½. (Figs. 32, 33, 35 from Hitchcock.)

36

37

38

39

C. inexpansa: 36. Panicle, ½. 37. Spikelet, × 10. C. canadensis: 38. Entire plant, × ½. 39. Spikelet, × 10. (Figs. 36–39 from Hitchcock.)

CUT-GRASS Leérsia

1. Rootstock stout, densely scaly (Fig. 34); spikelets 3–4 mm. wide (Fig. 33). **L. lenticulàris** Michx. Catchfly Grass. Wet shores and ditches; Ind. to s. Minn., s. to Fla. & Tex.
1. Rootstock slender (Fig. 35); spikelets 1.5–2 mm. wide......................2
   2. Leaves and sheaths very rough...3
   2. Leaves and sheaths nearly or quite smooth. **L. oryzoìdes** forma **glàbra** Eaton. Usually a submersed form.
3. Panicle exposed (Fig. 32). **L. oryzoìdes** (L.) Swartz. Rice Cut-grass. Common in wet places, often making a zone around ponds; Que. & Maine to e. Wash., s. to Fla., Tex., Ariz. & Calif. Often so abundant as to make walking difficult, the rough leaves tearing cloth and flesh.
3. Panicle enclosed in the sheath (Fig. 31). **L. oryzoides** forma **inclùsa** (Wiesb.) Dörfl. A late summer phase.

BLUEJOINT Calamagróstis

Tall slender grasses, from extensively creeping rootstocks, making swales.

1. Inflorescence open (Fig. 38). **C. canadénsis** (Michx.) Beauv. Bluejoint. Common in moist places; Greenland to Alaska, s. to Md., N.C., Mo., Kans., Colo., Ariz. & Calif.
1. Inflorescence compact (Fig. 36)...2
   2. Ligule, at summit of the sheath, 4–6 mm. long. **C. inexpánsa** Gray. Figs. 36, 37. Northern Reed Grass. Greenland to Alaska, s. to Maine, N.Y., Ill., Mo., N.M. & Calif.
   2. Ligule 1–3 mm. long. **C. neglécta** (Ehrh.) Gaertn. Greenland to Alaska, s. to Maine, N.Y., Mich., Wis., Colo. & Ore.

Several varieties of these species have been described; they may be found in Stebbins, Rhodora, **32**, 35–57 (1930).

Lemma

40

41

42

E. **Frankii**: 40. Part of panicle, × 1, and florets, × 10. **E. pectinacea**: 41. Part of panicle, × 1. **E. hypnoides**: 42. Plant, × ½. (Figs. 40–42 from Hitchcock.)

LOVE GRASS   Eragróstis

Small slender grasses, rarely more than 3 dm. high; spikelets many-flowered (Fig. 41), the glumes shorter than the lemmas. The spikelets are similar to those of *Glyceria*, but the lemmas are not so strongly nerved as in that genus, and are only 3-nerved.

1. Stems rooting at the nodes and making a mat on the mud (Fig. 42)............2
1. Stems erect or ascending except sometimes at base, not rooting at the nodes....3
   2. Anthers 0.2 mm. long; stamens and pistils in the same spikelet. **E. hypnoides** (Lam.) BSP. Fig. 42. Muddy shores, rarely in shallow water; Que. to Wash., s. to W.I. & S.A.
   2. Anthers 2 mm. long; stamens and pistils in different plants. **E. réptans** (Michx.) Nees. S.D., Ill. & Ky., s. to Tex. & La.
3. Spikelets long and narrow, many-flowered (Fig. 41). **E. pectinàcea** (Michx.) Nees. Mostly in waste places and along roadsides, sometimes in sandy soil of river bottoms and rarely in the water; Maine to N.D., s. to Fla. & e. Tex. (*E. Purshii*, not *E. pectinacea*, G, B.)
3. Spikelets oval, not more than 7-flowered (Fig. 40)..........................4
   4. Spikelets 2–5-flowered, 2–3 mm. long, shorter than their pedicels. **E. Fránkii** Mey. Fig. 40. Sand bars and muddy banks; N.H. to Minn., s. to the Gulf of Mexico.
   4. Spikelets 5–7-flowered, 3–4 mm. long, mostly longer than their pedicels. **E. Frankii** var. **brévipes** Fassett. S. Wis., e. Iowa & n.w. Ill.

REED CANARY GRASS   Phálaris

P. arundinàcea L. Figs. 43–46. Slender grass, reaching 1.5 m. in height, often in large clumps; leaves 5–15 mm. wide; spikelets in club-like masses (Fig. 46); lemma with 2 tufts of silky hairs at base. Damp meadows, both native and introduced from Europe; N.B. to Alaska, s. to N.C., Mo., Okla. & Calif.

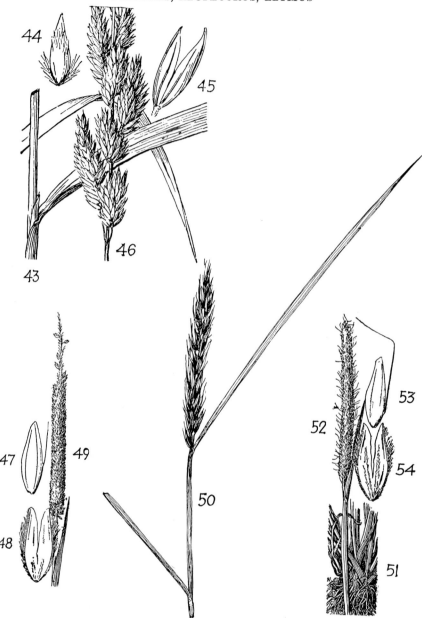

P. arundinacea : 43. Leaf and stem, × 1.   44. Floret, × 10.   45. Glumes, × 10.   46. Panicle,   × 1.   **A. aequalis :** 47. Lemma,   × 10.   48. Glumes,   × 10.   49. Spike,   × 1. **A. geniculatus** var. **ramosus :** 52. Summit of plant, × 1.   53. Lemma, × 10.   54. Glumes, × 10.   51. Base of plant, × 1.   **E. virginicus :** 50. Summit of plant, × ½.   (Figs. 43–49 51–54 from Hitchcock.)

FOXTAIL **Alopecùrus**

Low slender grasses (Fig. 51); spikelets in a close cylindrical spike; glumes fringed on the back (Fig. 54) lemma with a bent awn from its back (Fig. 53).

1. Awn from near the base of the lemma (Fig. 53), long enough to extend 2–3 mm. beyond the glumes (Fig. 52)...2
1. Awn from near the middle of the lemma (Fig. 47), so short as to extend 1 mm. or less beyond the glumes (Fig. 49), or not at all. **A. aequális** Sobol. Muddy borders of ponds, sometimes trailing on the surface in shallow water, also river bottoms, pools, etc.; Maine to Alaska, s. to Pa., Ind., Kans., N.M. & Calif. (*A. geniculatus* var. *aristulatus*, G; *A. aristulatus*, B.)
   2. Spikelet 2.5–3 mm. long. **A. geniculàtus** L. Water Foxtail. Wet places and shallow water, perhaps introduced; Nfd. to Va.; Sask. to Ariz. & Calif.
   2. Spikelet 2–2.4 mm. long. **A. geniculatus** var. **ramòsus** (Poir.) St. John. Figs. 51–54. Mostly on the coastal plain, Mass. to the Gulf of Mexico; n. in the Mississippi Valley to s. Wis.

References: St. John, Rhodora, **19**, 165–167 (1917)—*A. geniculatus* var. *ramosus*; Fernald, Rhodora, **27**, 196–199 (1925)—*A. aequalis.*

WILD RYE **Élymus**

E. virgínicus L. Fig. 50. Stems in dense tufts, erect, rigid, 6–12 dm. tall; spikelets in a dense spike on a somewhat zigzag rachis, the glumes extended into long rigid awns. River banks, etc.; Nfd. to Alberta, s. to Fla. & Ariz. Several varieties have been described; see Fernald, Rhodora, **35**, 196–198 (1933).

BENT GRASS **Agróstis**

Plants slender, erect or creeping; spikelets in open pyramidal panicles (Fig. 59), 1-flowered, the palea very short and filmy in texture (Fig. 58).

1. Plants erect. **A. stonífera** L., var. **màjor** (Gaud.) Farwell. Redtop. Fig. 59. Pastures, hayfields, etc., occasionally on sandy shores and rarely in the water. (*A. alba*, G, B, Hitchc.)
1. Plants creeping, rooting at the nodes, the flowering stems more or less erect. **A. stolonifera** var. **compácta** Hartm. Creeping Bent. Coastal marshes, and occasionally inland on wet shores. (*A. alba* var. *maritima*, G; *A. maritima*, B; *A. palustris*, Hitchc.)

Reference: Malte, Commercial Bent Grasses (*Agrostis*) in Canada, 1928, reprinted from Ann. Rept. for 1926, National Museum of Canada.

C. aquatica: 55. Floret, × 5. 56. Spikelet, × 5. 57. Plant, × ½. A. stolonifera var. major: 58. Spikelet, × 5. 59. Plant, × 1. (Figs. 57–59 from Hitchcock.)

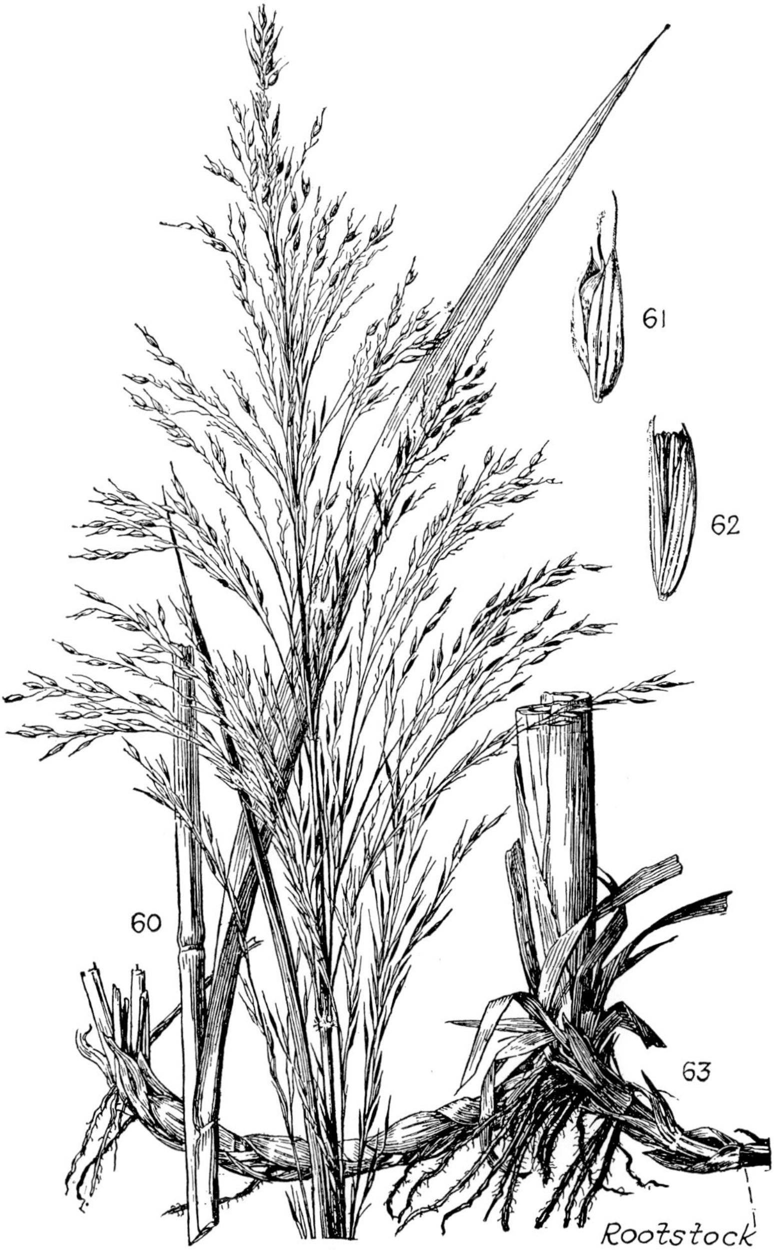

Z. milacea : 60. Panicle and part of stem & leaf, $\times \frac{1}{2}$. 61. Pistillate spikelet, $\times 5$. 62. Staminate spikelet, $\times 5$. 63. Base of plant, $\times \frac{1}{2}$. (Figs. 60–63 from Hitchcock.)

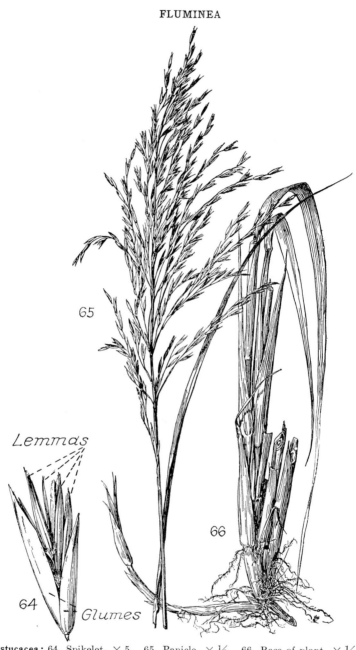

65

Lemmas

64

Glumes

F. festucacea: 64. Spikelet, × 5. 65. Panicle, × ½. 66. Base of plant, × ½. (Figs. 64–66 from Hitchcock.)

P. pratensis: 67. Leaf tip, × 3. 68. Plant, × ½. 69. Lemma, × 10. 70. Spikelet, × 5. (Figs. 68–70 from Hitchcock.)

S. pectinata: 71. Inflorescence, $\times \frac{1}{2}$. 72. Base of plant, $\times \frac{1}{2}$. 73. Spikelet, $\times 5$.
B. Syzigachne: 74. Spikelet, $\times 5$. 75. Plant, $\times \frac{1}{2}$. 76. Upper part of plant, $\times \frac{1}{2}$.
(Figs. 71–76 from Hitchcock.)

BROOK GRASS Catabròsa

C. aquática (L.) Beauv. Figs. 55–57. Stems creeping at base, rooting at the nodes; flowering stems erect; spikelets in a branched pyramidal panicle. Mountain meadows and in springs; Nfd. to Alberta, s. to w. Wis., Neb., Ariz. & Nev.

SOUTHERN WILD RICE, CUT-GRASS Zizaniópsis

Z. milàcea (Michx.) Doell & Aschers. Figs. 60–63. Plants 1–3 m. tall, from a stout scaly rootstock (Fig. 63); leaves 1–2 cm. wide, smooth except for the very rough margins; spikelets in a large narrow panicle, the pistillate (Fig. 61) toward the tips of the branches, the staminate (Fig. 62) below. Swamps and shallow water; Va. to Ohio & southward.

WHITETOP Flumínea

F. festucàcea (Willd.) Hitchc. Plants stout, 1–1.5 m. tall, from creeping rootstocks (Fig. 66); spikelets in a much-branched panicle (Fig. 65), several-flowered, the glumes much longer than the lemmas (Fig. 64). Shallow water and marshes; Man. to B.C., s. to Iowa, Neb. & Ore. (*Scolochloa festucacea*, G, B.)

BLUEGRASS Pòa

A large genus with many species, so abundant that occasionally a plant may be found in the water. Figs. 67–70 illustrate the Kentucky Bluegrass, **P. praténsis** L.

CORD GRASS Spartìna

S. pectinàta Link. Stout grasses, 1–2 m. tall, from stout very scaly rootstocks (Fig. 72) leaves prolonged into tapering whip-like tips; spikelets in several strongly 1-sided close spikes (Fig. 71). Margins of salt marshes, low prairies, sandy river bottoms and lake shores; Nfd. & Que. to e. Wash. & Ore., s. to N.C., Ky., Ill., Ark., Tex. & N.M. (*S. Michauxiana*, G, B.)

SLOUGH GRASS Beckmánnia

B. Syzigáchne (Steud.) Fernald. Plants tufted, 3–10 dm. tall; spikelets round (Fig. 74), in scattered compact 1-sided spikes (Fig. 76). Marshes and ditches; Man. to Alaska, s. to N.Y., Ohio, Ill., Kan., N.M. & Calif. (*B. erucaeformis*, G, B.)

SEDGE FAMILY   **CYPERÀCEAE**

Superficially resembling the grasses and differing from them as follows. Sedges: stems triangular and solid (except in *Dulichium*) with leaves in 3 ranks and with closed sheaths (Fig. 5); flowers borne in the axils of overlapping scales (Fig. 8), each with a single pistil having 2–3 stigmas, and with 1–3 stamens, the filaments of which are fixed to the base of the anthers (Fig. 6); fruit a lens-shaped or 3-angled nutlet, sometimes with bristles at the base (Fig. 7). Grasses: stems round in cross section and hollow, with leaves in two ranks and with open sheaths the edges of which often overlap (Fig. 1); flowers borne in spikelets, each consisting usually of 2 glumes (Figs. 3, 4), and one or more florets, each with a lemma, palea and a flower, the last consisting of a pistil with 2 stigmas, and usually 3 or 6 stamens the filaments of which are fixed to the side of the anthers (Fig. 2); fruit a mealy grain.

For the determination of most genera and species of sedges the mature nutlet is necessary. As far as is possible, other characters are also used in the key below, to allow determination of at least some immature material.

1. Stamens and pistils in different flowers (in different spikes in most aquatic species); nutlet within a sac called the perigynium (Figs. 162, 167, 171, 184, 195, 209). **Carex,** p. 157.
1. Stamens and pistils in the same flower; nutlet not enclosed in a sac............2
    2. Stem tipped by a single spikelet (Figs. 50, 59, 106).....................3
    2. Stem with one or more leaves extending beyond the spikelet (Fig. 100) or spikelets (Figs. 25, 28, 91, 115, 138, 146).....................................4
3. Bristles if present not long and silky (Fig. 34), hidden behind the scales, but commonly absent. **Eleocharis,** p. 129.
3. Bristles long and silky (Fig. 107), extending beyond the scales (Fig. 106). **Scirpus hudsonianus,** p. 143.
    4. Stem hollow, round in cross section (Fig. 28). **Dulichium,** p. 129.
    4. Stem solid or nearly so, 3-angled, the angles sometimes rounded............5
5. Scales in 2 ranks, so that the spikelet is flattened (Figs. 9, 20, 27). **Cyperus,** p. 123.
5. Scales borne all around the spikelet, so that it is essentially cylindrical or round in cross section (Figs. 94, 115, 125, 138, 144, 151).............................6
    6. Nutlet with a tubercle formed from the persistent base of the style (Figs. 89, 140, 143, 153)...8
    6. Nutlet without a tubercle...7
7. Spikelets with all but the lowest scale bearing flowers and eventually enclosing nutlets. **Scirpus,** p. 143, **Hemicarpha,** p. 151, and **Fimbristylis,** p. 141. Since the two latter genera are separated from *Scirpus* on minute characters, they are for convenience included in the key to that genus.
7. Spikelets with only the uppermost scale bearing flower and fruit. **Cladium,** p. 155.
    8. Each spikelet with several empty scales below the 1 or 2 that enclose nutlets; tubercle of different texture from the body of the nutlet (Figs. 140, 147). **Rynchospora,** p. 151.

8. Each spikelet with one empty scale below the several that enclose nutlets; tubercle of same texture as the body of the nutlet (Figs. 88, 131)..........9
9. Scales hairy, each tipped by a recurved bristle (Fig. 133). **Fuirena,** p. 149.
9. Scales not hairy, not bristle-tipped (Fig. 90). **Psilocarya,** p. 141.

Cypèrus

Spikelets flattened, the scales in 2 rows on opposite sides of the rachilla; rachilla often with thin wings somewhat enclosing the nutlet (Fig. 24) or running to the scale next above; spikelets often in heads, with leafy bracts at base. Some species may be exceedingly variable in size, from a few centimeters to a meter in height.

1. Plant with a tuft of roots and no stolons or enlarged base...................5
1. Plant with stolons (Fig. 11) or an enlarged hard base (Fig. 12). Members of this group are called Nut Grass..2
    2. Scales pointing forward, each reaching to a little below the middle of the one next above on the same side (Fig. 9)...................................3
    2. Scales spreading, each reaching to above the middle of the one next above on the same side (Fig. 13)..4
3. Base of plant with slender scaly stolons which often end in edible tubers (Fig. 11). **C. esculéntus** L. Chufa. Fig. 9. Moist soil; N.B. to Minn. & Neb., s. to Fla. & Tex.; Calif. to Alaska.
3. Base of plant with a corm (Fig. 12) but no stolons. **C. strigòsus** L. Fig. 10. Moist soil; Maine & Ont. to Minn., s. to Fla. & Tex. Very variable in size of plant, length of spikelet, density of inflorescence, etc.; see also **13.**
    4. Spikelets with parallel sides (Fig. 13), or often replaced by bulblets (Fig. 14). **C. dentàtus** Torr. Fig. 14. Mostly on sandy shores; N.S. & Maine to N.Y., s. to W.Va. & S.C.
    4. Spikelets nearly round, closely crowded (Fig. 15). **C. pseudovégetus** Steud. N.J. to the Gulf of Mexico, n. in the Mississippi Valley to Mo. & Kans.
5. Style 2-cleft and nutlets 2-sided (Fig. 21)................................6
5. Style 3-cleft and nutlet 3-sided (Fig. 22)................................9
    6. Nutlet about as broad as long. **C. flavéscens** L. N.Y. to Mich. & Ill., s. to C.A. & W.I.
    6. Nutlet longer than broad..7
7. Scales dark- or purplish-brown, the backs curved more than the edges (Fig. 20)..8
7. Scales light-brown or yellowish, the edges curved more than the backs (Fig. 16). **C. filicìnus** Vahl, var. **microdóntus** (Torr.) Fernald. Lake shores; s.e. Mass. to the Gulf of Mexico. (*C. microdontus,* G, B.)
    8. Styles 2-cleft nearly to the base (Fig. 18), extending beyond the usually dull scales (Fig. 17). **C. diándrus** Torr. Sandy and mucky shores; N.B. to Minn., s. to S.C. & Kans.
    8. Styles 2-cleft to about the middle (Fig. 21), hidden by the usually shining scales (Fig. 20). **C. rivulàris** Kunth. Low grounds, shores, etc.; Maine to Minn., s. to N.C., Mo. & Kans.
9. Scales tapering to a recurved tip (Fig. 19). **C. infléxus** Muhl. Rather local; N.B. to B.C., s. to Fla., Tex., Mex. & Calif. (*C. aristatus,* G.)
9. Scales without recurved tips..10
    10. Spikelets in close heads about 1 cm. broad (Fig. 23). **C. acuminàtus** Torr. & Hook. Low ground; Ill. to S.D., Ore., Calif. & Tex.
    10. Spikelets in loose heads..11

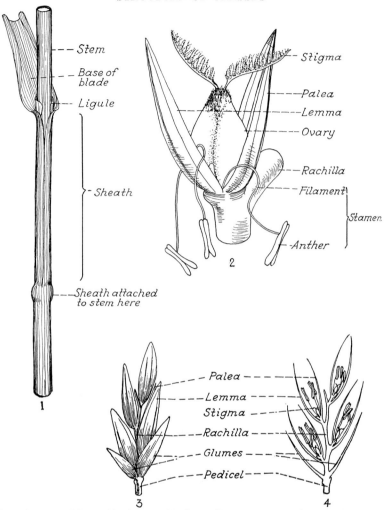

1. Part of stem and base of leaf, generalized.   2. Floret, generalized.   3 Spikelet, superficial view, generalized.   4. Spikelet, dissected, diagrammatic.

STRUCTURE OF SEDGES

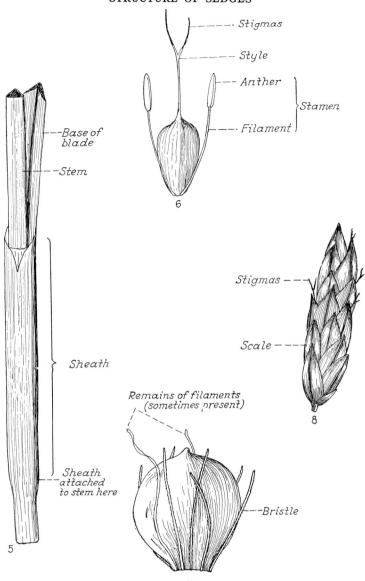

5. Part of stem and base of leaf, generalized.   6. Flower, generalized.   7. Nutlet, generalized.   8. Spikelet, generalized.

125

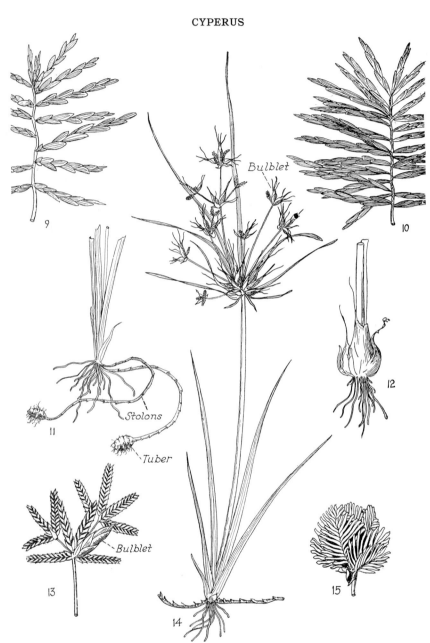

C. esculentus: 9. Part of inflorescence, × 2.   11. Base of plant, × ½.   C. strigosus: 10. Part of inflorescence, × 2.   12. Base of plant, × ½.   C. dentatus: 13. Part of inflorescence, × 2.   14. Plant, × 1.   C. pseudovegetus: 15. Part of inflorescence, × 4.

C. filicinus var. microdontus: 16. Part of inflorescence, × 2. C. diandrus: 17. Part of inflorescence, × 2. 18. Nutlet, × 10. C. inflexus: 19. Inflorescence, × 2. 22. Nutlet, × 10. C. rivularis: 20. Part of inflorescence, × 2. 21. Nutlet, × 20. C. acuminatus: 23. Part of inflorescence, × 2. C. erythrorhizos: 24. Spikelet, × 10. 25. Inflorescence, × 1. C. ferruginescens: 26. Spikelet, × 5. C. Engelmanni: 27. Spikelet, × 5.

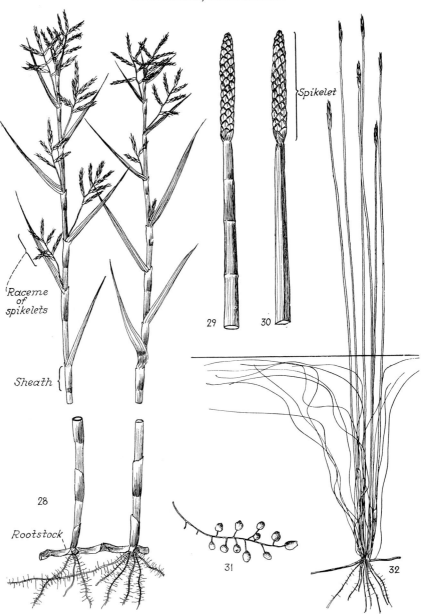

D. arundinaceum: 28. Plant, × ½. E. equisetoides: 29. Summit of culm, × 1.
E. quadrangulata: 30. Summit of culm, × 1. 31. Underground branch with tubers,
× ½. E. Robbinsii: 32. Plant, with thread-like stems below water, × ½.

11. Scales about 1 mm. long; wings of rachilla white, extending to and fused with base of scale next above (Fig. 24). **C. erythrorhìzos** Muhl. Fig. 25. Muddy shores; Mass. to Minn., s. to Fla., Tex., Kans. & Calif.
11. Scales about 2 mm. long; wing of rachilla not extended upward, hidden by scales (Fig. 27)..**12**
 12. Each scale reaching the base of the next scale above on the same side (Fig. 26) **13**
 12. Each scale ending below the base of the next scale above on the same side (Fig. 27). **C. Engelmánni** Steud. Mass. to Wis., s. to N.J. & Mo.
13. Scales opaque, membranous, 2 mm. long, reddish-brown; nutlets reddish or golden brown, 1–1.5 mm. long. **C. ferruginéscens** Boeckl. Fig. 26. W. N.E. & Md. to the Pacific Coast. (*C. ferax*, in part, G; *C. speciosus*, B.) This may be confused with *C. strigosus*, which has scales 4 mm. long.
13. Scales firm, somewhat shining, 2–3.5 mm. long, brownish or yellowish; nutlets 1.5–2 mm. long, dull gray to black, coarsely pebbled. **C. fèrax** Richard. Mostly in brackish or saline soil; tropical America to Mass. and to Calif.

References: Fernald, Rhodora, **19**, 153 (1917)—*C. filicinus* var. *microdontus;* Fernald and Griscom, Rhodora, **37**, 148–150 (1935)—*C. ferruginescens.*

Three-way Sedge **Dulíchium**

D. arundinàceum (L.) Britton. Fig. 28. Stems hollow, cylindrical, sometimes nearly 1 m. tall, from a rootstock; leaves many, with a conspicuous sheath; spikelets in racemes in the axils of the upper leaves. Wet shores and shallow water; Nfd. to Wash., s. to Fla. & Tex.

Spike Rush **Eleócharis**

Leaves without blades, represented by sheaths at the base of the stem (Fig. 50); spikelets solitary, terminating the stems; base of style persistent and enlarged, forming a tubercle on the top of the nutlet (Figs. 47, 52, 63). The shape and size of the tubercle are of great importance in the classification within this genus, so that identification of species is almost impossible without mature nutlets.

1. Stem nearly as thick as the spikelet (Figs. 30, 32); scales persistent...........**2**
1. Stem much more slender than the spikelet (Figs. 50, 67); scales easily removed when nutlets are mature..**4**
 2. Stem jointed (Fig. 29). **E. equisetoìdes** (Ell.) Torr. Jointed Spike Rush. Shallow water; Mass. to Fla., inland about the Great Lakes to Wis.; Mo. (*E. interstincta*, G, B.)
 2. Stems not jointed...**3**
3. Stems 1.5–5 mm. thick, sharply 4-angled (Fig. 30); tubers frequently present (Fig. 31). **E. quadrangulàta** (Michx.) R. & S. Square-stem Spike Rush. Shallow water of ponds; e. Mass. to Ala. & Tex., inland about the Great Lakes to e. Wis.
3. Stems 1–2 mm. thick, bluntly 3-angled (Fig. 32); tubers absent, but limp thread-like stems frequently produced. **E. Robbínsii** Oakes. Triangle Spike Rush. Shallow water and wet shores; N.S. to Fla., inland about the Great Lakes; w. Wis.
 4. Plants from a rootstock 2 mm. or more thick (Figs. 49, 50)...............**5**
 4. Plants without rootstocks (Fig. 59) or with thread-like rootstocks (Figs. 54, 58)..**19**

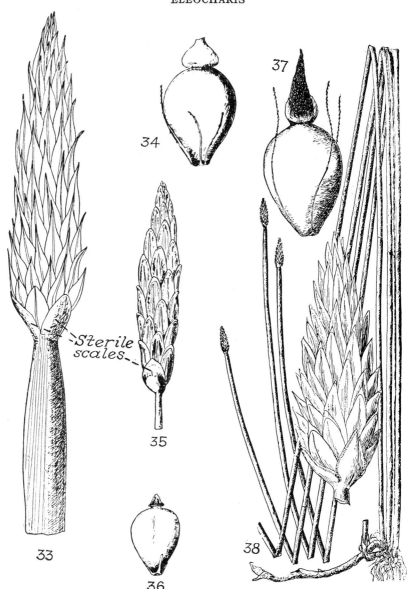

Sterile scales

E. **Smallii:** 33. Spikelet, × 5. 34. Nutlet, × 20. **E. calva:** 35. Spikelet, × 5. 36. Nutlet, × 20. **E. palustris** var. **major:** 37. Nutlet, × 20. 38. Plant, × ½, and spikelet, × 5. (Figs. 33–38 from Fernald & Brackett.)

5. Style 2-cleft; nutlet 2-angled (Fig. 34). This is the group treated in Gray's Manual as *E. palustris* and in Britton and Brown as *E. palustris* and *E. Smallii* . . **6**

5. Style 3-cleft; nutlet 3-angled (Figs. 45–48) **11**

    6. Sterile scales at base of spikelet 2–3 (Fig. 33) **8**

    6. Sterile scale solitary, encircling the base of the spikelet (Fig. 35) **7**

7. Spikelets 9–17 mm. long, usually with 40 or more scales. **E. cálva** Torr. Figs. 35, 36. Wet shores and bogs; Que. to Alberta, s. to Fla., Okla. & Mex.

7. Spikelets 4–9 mm. long, few-flowered. **E. ámbigens** Fernald. Pond-margins and marshes; e. Mass. to Va.; Fla.; La.

    8. Tubercle higher than broad (Fig. 37) **9**

    8. Tubercle as broad as high or broader (Fig. 34) **10**

9. Stems 0.5–2 mm. in diameter; nutlet 1.2–1.7 mm. long. **E. palústris** (L.) R. & S. Creeping Spike Rush. Muddy and sandy shores and shallow water; Nfd. to B.C., s. to Maine, n. Mich., N.D., Wyo., Idaho & Ore.

9. Stems 1.5–5 mm. in diameter; nutlet 1.4–2.1 mm. long. **E. palustris** var. **màjor** Sonder. Figs. 37, 38. Labrador to B.C., s. to Pa., Mich., Ill., S.D., Wyo., Idaho & Calif.

    10. Stems wiry; scales sharp-pointed (Fig. 33). **E. Smállii** Britton. N.S. to Wis. & Neb., s. to Del., Ill. & Mo.

    10. Stems flattened; scales blunt. **E. mamillàta** Lindb. Ill. to B.C., s. to La., Tex., Mex. & Calif.

11. Nutlet 0.7–1.5 mm. long; upper sheath cut square across the top (Fig. 40) **12**

11. Nutlet 1.7–2 mm. long; upper sheath cut at an angle across the top (Fig. 39). **E. fállax** Weatherby. Perhaps a hybrid, found but once, at Dinah's Pond, Yarmouth, Mass.

    12. Nutlet with 3 keel-like angles (Fig. 45). **E. tricostàta** Torr. S.e. Mass. to Fla.

    12. Nutlet without keel-like angles **13**

13. Stems flattened; scales sharp-pointed (Fig. 41) **14**

13. Stems angled or wiry **15**

    14. Scales dark chestnut-brown. **E. compréssa** Sull. W. Que. to Sask., s. to Ga., Tex. & Colo. This intergrades with *E. tenuis* in the east. (*E. acuminata*, G, B.)

    14. Scales blackened. **E. compressa** var. **atràta** Svenson. Gulf of St. Lawrence to the Great Lakes.

15. Tip of upper sheath whitened; nutlet 0.7–1 mm. long. **E. nítida** Fernald. Fig. 49. Nfd. to n. N.H.

15. Tip of upper sheath dark-margined; nutlet 0.9–1.5 mm. long **16**

    16. Pits on surface of nutlet shallow (Fig. 46); stem 6–8-angled (Fig. 42). **E. ellíptica** Kunth. Fig. 50. Nfd. to B.C., s. to N.J., Tenn., Ind., Wis. & Mont.

    16. Pits deep (Figs. 47, 48); stem 4–5-angled **17**

17. Stem with 4 wing-like angles (Fig. 43), 3–9 dm. tall. **E. ténuis** var. **pseudóptera** (Weatherby) Svenson. L.I. & N.J. to Va. & e. Tenn.

17. Stem not winged, rarely more than 3 dm. tall **18**

    18. Tubercle sharply pointed, often a fifth as high as the body of the nutlet (Fig. 47); stem 4–5-angled. **E. tenuis** (Willd.) Schultes. Slender Spike Rush. Fig. 44. N.S. to S.C., w. to w. Pa.

    18. Tubercle flattened (Fig. 48); stem 5-angled. **E. ténuis** var. **verrucòsa** Svenson. Ind. to Okla., s. to La. & e. Tex.

ELEOCHARIS

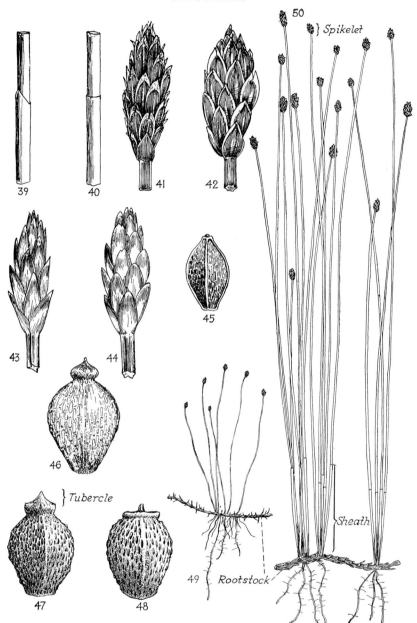

E. *fallax:* 39. Sheath, × 5. **E. tricostata:** 40. Sheath, × 5. 45. Nutlet, × 20.
E. compressa: 41. Spikelet, × 5. **E. elliptica:** 42. Spikelet, × 5. 46. Nutlet, × 20.
50. Plant, × ½. **E. tenuis** var. **pseudoptera:** 43. Spikelet, × 5. **E. tenuis:** 44. Spikelet,
× 5. 47. Nutlet, × 20. **E. tenuis** var. **verrucosa:** 48. Nutlet, × 20. **E. nitida:** 49.
Plant, × ½.

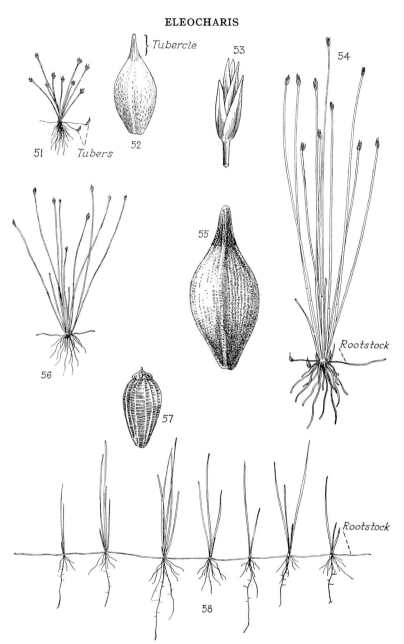

E. parvula : 51. Plant, × ½. 52. Nutlet, × 20. **E. pauciflora** var. **Fernaldii**: 53. Spikelet, × 5. 54. Plant, × ½. 55. Nutlet, × 20. **E. acicularis**: 56. Fruiting plant, × ½. 57. Nutlet, × 20. 58. Part of mat of sterile plants, × ½.

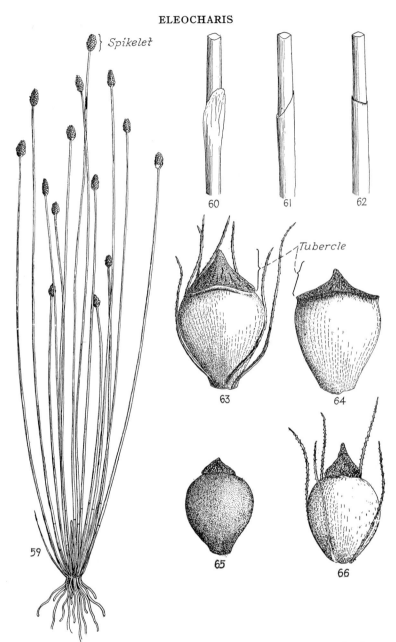

Spikelet

Tubercle

E. obtusa: 59. Plant, × ½. 62. Sheath, × 5. 63. Nutlet, × 20. E. flavescens: 60. Sheath, × 5. E. caribaea: 61. Sheath, × 5. E. Engelmanni var. detonsa: 64. Nutlet, × 20. E. diandra: 65. Nutlet, × 20. E. ovata: 66. Nutlet, × 20.

19. Spikelets 3–9-flowered (Figs. 51, 53, 56), rarely as much as 15-flowered; usually dwarf plants not over 5 cm. high..**20**

19. Spikelets many-flowered..**23**

    20. Bunches of stems connected by slender rootstocks (Figs. 54, 58), all stems about the same length..**21**

    20. Bunches of stems not connected by rootstocks; some stems usually very short (Fig. 84). **E. intermedia, 38.**

21. Scales dark-brown, often with a green midrib...........................**22**

21. Scales light-brown or greenish. **E. pàrvula** (R. & S.) Link. Mostly along the coast and often on salt marshes, Nfd. to Cuba; rarely inland in Wis. & Mo. Usually stouter than the next, although sometimes slender and looking like it, but *E. parvula* has tubers (Fig. 51), rarely found in *E. acicularis*. The nutlets of this (Fig. 52) and of *E. pauciflora* (Fig. 55) differ from those of most species of *Eleocharis* in having tubercle not sharply differentiated from the body; for this reason they are sometimes put in the genus *Scirpus*. (*Scirpus nanus*, G, B.)

    22. Stems often forming a turf; tubercle distinct from the body of the nutlet (Fig. 57). **E. aciculàris** R. & S. Needle Rush. Fig. 56. Very common on wet shores and in shallow water; Greenland to Alaska, s. to Fla., Ky., Okla., Wyo. & Idaho. Frequently sterile (Fig. 58). In the water this may send up quantities of elongate hair-like stems—this is **E. acicularis** forma **inundata** Svenson. **E. radicans** (Poir.) Kunth., from Okla. & Calif. to Mex., has been collected in s. Mich.; it has light-green scales and 2 stamens with anthers 0.3–0.4 mm. long, while *E. acicularis* has green scales with brown markings, and 3 stamens with anthers 1 mm. long.

    22. Stems solitary or a few together (Fig. 54 represents a large clump); tubercle confluent with the body of the nutlet (Fig. 55). **E. pauciflòra** (Lightf.) Link, var. **Fernáldii** Svenson. Figs. 53–55. Marl bogs and shores of the Great Lakes; James Bay & Nfd. to n. N.E., w. to Ind., Wyo. & Calif.

23. Style 2-cleft; nutlet 2-angled (Figs. 65, 66)...........................**24**

23. Style 3-cleft; nutlet 3-angled (Figs. 80, 82)...........................**35**

    24. Sheaths white and loose toward the tip (Fig. 60).....................**25**

    24. Sheaths close and often dark-margined at the tip (Figs. 61, 62)..........**26**

25. Nutlets 0.8–1 mm. long, reddish-brown when young and deep purplish-black when mature. **E. flavéscens** (Poir.) Urban. Del. & Va. to the Gulf of Mexico and S.A. (*E. ochreata*, G; *E. flaccida*, B.)

25. Nutlets 1 mm. or more long, olive-green. **E. olivàcea** Torr. Wet sand and peat; N.S. to Md., locally inland to Minn.

    26. Plants with rootstocks; base of tubercle much narrowed (Figs. 34, 36, 37). The more slender members of the **E. palustris** group may run here; see **6.**

    26. Plants without rootstocks (Fig. 59); tubercle broadest at base (Figs. 63–66) **27**

27. Mature nutlet black; summit of sheath cut at a slant, so that the triangular tip is longer than the width of the sheath (Fig. 61)...........................**28**

27. Mature nutlet pale-brown; summit of sheath cut nearly square, so that the triangular tip is shorter than the width of the sheath (Fig. 62)..............**29**

    28. Scales pale-brown. **E. caribaèa** (Rottb.) Blake. Md. to Fla., W.I. & S.A. (*E. capitata*, G, B.)

    28. Scales purple-brown. **E. caribaea** var. **díspar** (Hill) Blake. Rare, in wet sand; s. Ont., s. Mich. & n. Ind. (*E. capitata* var. *dispar*, G.)

29. Tubercle nearly or quite as broad as the body of the nutlet (Figs. 63, 64)......**30**

29. Tubercle less than two-thirds as broad as the body of the nutlet (Figs. 65, 66)..**34**

ELEOCHARIS

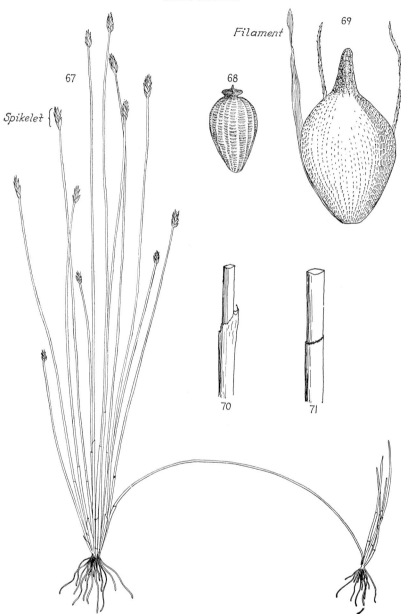

Filament

Spikelet {

E. rostellata: 67. Plant, × ½. 69. Nutlet, × 20. 71. Sheath, × 5. **E. Wolffii**: 68.
Nutlet, × 20. **E. intermedia**: 70. Sheath, × 5.

136

30. Tubercle a third to half as high as the body of the nutlet (Fig. 63); spikelet egg-shaped...**31**

30. Tubercle flattened, not more than a fourth as high as the body of the nutlet (Fig. 64); spikelet usually cylindrical...............................**32**

31. Nutlet with 6 or 7 long bristles (Fig. 63). **E. obtùsa** (Willd.) Schultes. Fig. 59. Common in muddy places; e. N.S. to n. Wis. & Neb., s. to the Gulf of Mexico; Calif. to B.C.; Hawaii. Very variable in size.

31. Nutlet without bristles. **E. obtusa** var. **Peásei** Svenson. Sandy pond shore, local; Que. to e. Maine & N.H.; Lake Nipissing, Ont.

    32. Spikelets 5–16 mm. long...**33**

    32. Spikelets 2 cm. long. **E. Engelmánni** var. **robústa** Fernald. Local; Mo.

33. Nutlet with bristles. **E. Engelmánni** Steud. Rather uncommon, s. Maine to Va., s. Mich. & Tex., w. locally to Wash. & s. Calif.

33. Bristles absent. **E. Engelmanni** var. **detónsa** Gray. Fig. 64. Local; Mass.; Pa.; Mich. & Ind. to Wis.; Ariz.

    34. Tubercle broader than high (Fig. 65); bristles absent. **E. diándra** Wright. Oneida Lake, N.Y.; Lake Champlain; tidal shores from Maine to Pa.

    34. Tubercle higher than broad (Fig. 66); bristles present. **E. ovàta** (Roth) R. & S. Local; Nfd. to N.J.; Mich. to Minn.; Wash.

35. Summit of upper sheaths loose, of very delicate texture, cut at an oblique angle (Fig. 70)..**36**

35. Summit of upper sheaths close, firm, green- or dark-margined, cut nearly square across the top (Fig. 71)...**39**

    36. Nutlet with longitudinal ridges (Fig. 68). **E. Wólffii** Gray. E. N.Y.; Ind. to Kans. & La.; Assiniboia.

    36. Nutlet without longitudinal ridges...............................**37**

37. Tubercle about as broad as high (Fig. 78). **E. microcàrpa** Torr., var. **filicúlmis** Torr. Figs. 76–78. Conn. to Tenn. & the Gulf of Mexico; n. Ind.; Okla. (*E. Torreyana*, G, B.)

37. Tubercle several times as high as broad (Fig. 82)....................**38**

    38. Bristles present. **E. intermèdia** (Muhl.) Schultes. Figs. 82–84. Que. & Minn. to s. Ind., e. Tenn. & s. N.Y.

    38. Bristles absent. **E. intermedia** var. **Haberèri** Fernald. Oneida Lake, N.Y.

39. Tubercle nearly or quite as large as the body of the nutlet (Fig. 75). **E. tuberculòsa** (Michx.) R. & S. Figs. 72–75. N.S. & N.H. to Tenn., Ark. & the Gulf of Mexico. The bristles are upwardly barbed: in **E. tuberculosa** forma **retrórsa** Svenson they are downwardly barbed (Fig. 74). and in forma **pubnicoénsis** (Fernald) Svenson they are smooth.

39. Tubercle much smaller than the body of the nutlet......................**40**

    40. Sterile stems often arching, rooting at the tip (Fig. 67); tubercle not sharply differentiated from the body of the nutlet (Fig. 69). **E. rostellàta** Torr. Beaked Spike Rush; Walking Sedge. Salt marshes, N.S. to W.I.; rarely inland about the Great Lakes; s. B.C. to Okla. & Tex., w. to the Pacific Coast.

    40. Sterile stems not rooting at tip, the fertile sometimes doing so in *E. melanocarpa;* tubercle sharply differentiated from the body of the nutlet..........**41**

41. Tubercle covering the whole summit of the nutlet (Fig. 80). **E. melanocàrpa** Torr. Figs. 79–81. Mass. to Tex.; n. Ind.

41. Tubercle much narrower than the summit of the nutlet..................**42**

    42. Tubercle higher than broad (Fig. 86). **E. tórtilis** (Link) Schultes. Figs. 85–87. L.I. to Tex. (*E. simplex*, B.)

    42. Tubercle broader than high. See **E. nitida, 15.**

References: Fernald and Brackett, Rhodora, **31**, 57–77 (1929)—*E. palustris* and its relatives (p. 130 is from this paper). Svenson, a series of papers in Rhodora, constituting a monograph of the genus, as follows: **31**, 121–135, 152–163, 167–191, 199–219, 224–242 (1929); **34**, 193–203, 215–227 (1932); **36**, 377–389 (1934); **39**, 210–231, 236–273 (p. 138 is from this paper) (1937); **41**, 1–19, 43–77, 90–110 (1939).

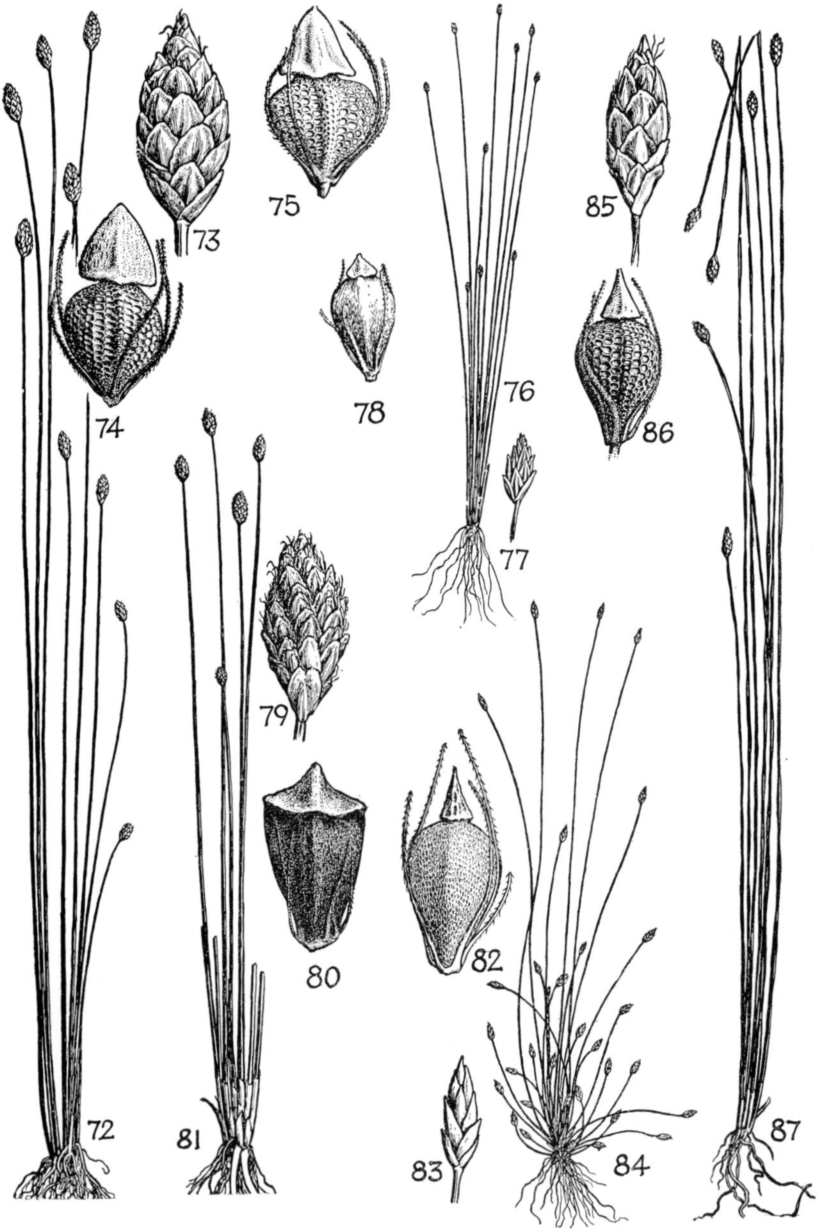

E. **tuberculosa:** 72. Plant, $\times \frac{1}{2}$. 73. Spikelet, $\times 2.5$. 75. Nutlet, $\times 20$. **E. tuberculosa** f. **retrorsa:** 74. Nutlet, $\times 20$. **E. microcarpa** var. **filiculmis:** 76. Plant, $\times \frac{1}{2}$. 77. Spikelet, $\times 2.5$. 78. Nutlet, $\times 20$. **E. melanocarpa:** 79. Spikelet, $\times 2.5$. 80. Nutlet, $\times 20$. 81. Plant, $\times \frac{1}{2}$. **E. intermedia:** 82. Nutlet, $\times 20$. 83. Spikelet, $\times 2.5$. 84. Plant, $\times \frac{1}{2}$. **E. tortilis:** 85. Spikelet, $\times 2.5$. 86. Nutlet, $\times 20$. 87. Plant, $\times \frac{1}{2}$. (Figs. 72–87 from Svenson.)

P. scirpoides: 88. Nutlet, × 20, before the deciduous portion of the style has fallen. 90. Plant, × ½. P. nitens: 89. Nutlet, × 20, after the deciduous portion of the style has fallen. F. Baldwiniana: 91. Plant, × ½.

F. Baldwiniana: 92. Spikelet, × 5. 93. Nutlet, × 20. **F. puberula:** 94. Spikelet, × 5. 95. Nutlet, × 20, before the style has fallen. 96. Nutlet, × 20, after the style has fallen. **F. autumnalis** var. **mucronulata:** 97. Inflorescence, × 2. **F. autumnalis:** 98. Inflorescence, × 2. 99. Nutlet, × 20.

BALD RUSH   Psilocárya

Plants variable in size, sometimes reaching 7 dm. in height; stems leafy, nearly every leaf with a flowering branch in its axil.

1. Nutlet finely wrinkled or roughened, with a tubercle higher than broad (Fig. 88). **P. scirpoìdes** Torr. Fig. 90. Wet sandy shores and boggy ground; Mass. & R.I.; n. Ind.; Wis.
1. Nutlet strongly cross wrinkled, with a tubercle broader than high (Fig. 89). **P. nìtens** (Vahl) Wood. L.I. to the Gulf of Mexico; n. Ind.

Fimbristỹlis

Base of style swollen (Fig. 95), thus differing from that of *Scirpus*, but dropping and not forming a tubercle (Fig. 96), thus differing from that of *Rynchospora*.

1. Style 2-cleft and nutlet 2-angled (Fig. 95); leaves usually more or less hairy.....**2**
1. Style 3-cleft and nutlet 3-angled (Fig. 99); leaves not hairy...................**4**
    2. Plants 0.5–7 dm. high; spikelets usually on rays (rarely congested); nutlets about 1 mm. long.. ....**3**
    2. Plants 0.3–2 dm. high; spikelets in a close head; nutlet 0.5 mm. long. **F. Váhlii** (Lam.) Link. Damp sand; N.C. to Fla., Tex. & Mo.
3. Scales usually finely hairy; nutlets marked with a fine network (Fig. 96). **F. pubérula** (Michx.) Vahl. Fig. 94. Salt marshes from N.Y. to the Gulf of Mexico, thence northward to Ont., Mich., Ill. & Neb. The spikelets are usually on rays and narrowed at tip; in **F. puberula** forma **eucýcla** Fernald they are rounded at tip, and in forma **pycnostàchya** Fernald they are all crowded into a compact head. (*F. castanea*, G; *F. interior* and *F. puberula*, B.)
3. Scales not hairy; nutlet with 6–8 ribs on each side, and fine lines between them (Fig. 93). **F. Baldwinìàna** (Schultes) Torr. Figs. 91–93. Low ground; Fla. & the Gulf of Mexico, n. to Ill. & Mo. (*F. laxa*, G.)
    4. Spikelets about a third as thick as long, on usually unbranched rays (Fig. 98). **F. autumnàlis** (L.) R. & S. Fig. 99. Low ground, sandy shores, etc.; Maine to Minn., s. to Tenn. & La. (*F. Frankii*, G; *F. geminata*, B.)
    4. Spikelets about a fifth as thick as long, on mostly branching rays (Fig. 97). **F. autumnalis** var. **mucronulàta** (Michx.) Fernald. Pa. to Ill., s. to S.A. (*F. autumnalis*, G, B.) Many individuals are intermediate.

Reference: Fernald, Rhodora, **37**, 396–399 (1935), where most of these species and varieties are discussed.

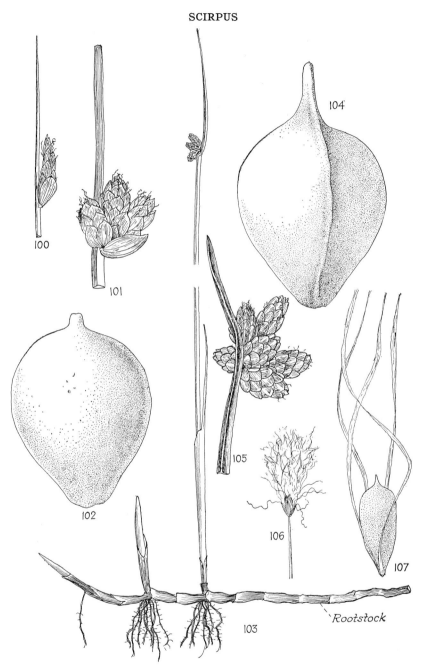

S. subterminalis: 100. Inflorescence, × 2. S. americanus: 101. Inflorescence, × 2. 102. Nutlet, × 20. 103. Plant, × ½. S. Torreyi: 104. Nutlet, × 20. S. Olneyi: 105. Inflorescence, × 2. S. hudsonianus: 106. Inflorescence, × 1. 107. Nutlet, × 20.

BULRUSH   Scírpus

Stems with sheaths at base, sometimes leafy and sometimes naked; spikelets usually several, with leaves (the involucre) from the base of the spikelets or of the branches; nutlet often with bristles at base.

1. Involucre of a single leaf, appearing as if a continuation of the stem (Fig. 100), or else very short..2
1. Involucre of 2 or more leaves (Figs. 115, 120, 125)..........................16
   2. Bristles long and silky (Figs. 106, 107). **S. hudsoniànus** (Michx.) Fernald. Boggy and limy meadows; Nfd. to Hudson Bay & B.C., s. to Conn., N.Y. & Minn. (*Eriophorum alpinum*, B.)
   2. Bristles not long and silky...3
3. Spikelets not stalked (Fig. 105); stems rarely more than 1 m. tall.............4
3. Spikelets on stalks or long rays (Fig. 110), or sometimes spikelets crowded; stems 1–2 m. tall...13
   4. Spikelet solitary (Fig. 100); stems weak, usually supported by the water, often with tufts of hair-like leaves (Fig. 7, p. 2). **S. subterminàlis** Torr. Water Bulrush. Shallow water, ponds and streams; Nfd. to B.C., s. to S.C., Pa., Mich., Wis. & Idaho; Mo.
   4. Spikelets usually 2 or more; stems erect................................5
5. Stems sharply 3-angled, 2–5 mm. thick (Fig. 101), from a rootstock (Fig. 103)..6
5. Stems bluntly 3-angled, 1–1.5 mm. thick, from tufted roots (Fig. 112)........8
   6. Sides of stem concave (Fig. 105); involucral leaf 1–3 cm. long. **S. Olnèyi** Gray. Olney's Three-square. Salt marshes, N.S. to the Gulf of Mexico and on the Pacific Coast; rarely inland in Mich. & Ark.
   6. Sides of stem nearly flat (Fig. 101); involucral leaf 4–15 cm. long.........7
7. Scales of spikelet red-brown; nutlet with an abrupt short point (Fig. 102). **S. americànus** Pers. Three-square. Fig. 103. Abundant in shallow water of sandy lakes, often in somewhat brackish water eastward.
7. Scales yellow-brown; nutlet with an abrupt long point (Fig. 104). **S. Torrèyi** Olney. Torrey's Three-square. Pond margins; Maine to Minn. & Man., s. to R.I., Pa. & Mo.
   8. Scales about ½ mm. long. See **Hemicarpha**, p. 151.
   8. Scales about 2 mm. long...9
9. Nutlet strongly cross wrinkled (Fig. 108). **S. Hállii** Gray. Very local on wet shores; Mass.; Ill. to the Gulf of Mexico.
9. Nutlet smooth or very lightly wrinkled..................................10
   10. Nutlet convex on both sides...11
   10. Nutlet flat on one side and convex on the other (Fig. 111)...............12
11. Bristles present. **S. débilis** Pursh. Figs. 109, 112. Maine to Minn., s. to Ga., Ala. & Neb.
11. Bristles absent. **S. debilis** var. **Williámsii** Fernald. Local.
   12. Bristles absent. **S. Smíthii** Gray. Maine to Del., w. to Wis.
   12. Bristles present. **S. Smithii** var. **setòsus** Fernald. Local.
13. Stems triangular; basal sheath with a blade. **S. etuberculàtus** (Steud.) Ktze. Swamp Bulrush. Md. to the Gulf of Mexico.
13. Stems nearly round in cross section; sheath without a blade................14
   14. Nutlet 3-angled, with 2–4 bristles. (Fig. 113). **S. heterochaètus** Chase. Slender Bulrush. Shallow water and shores; Mass. & Vt. to Ore., s. to Ill. & Neb.
   14. Nutlet 2-angled...15

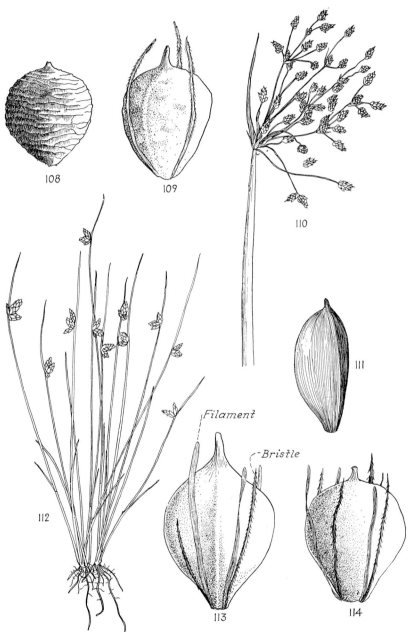

Filament

Bristle

108   109   110   111   112   113   114

S. Hallii: 108. Nutlet, × 20.   **S. debilis**: 112. Plant, × ½.   109. Nutlet in face view,
× 20.   **S. validus**: 110. Inflorescence, × ½.   114. Nutlet, × 20.   **S. Smithii**: 111.
Nutlet in side view, × 20.   **S. heterochaetus**: 113. Nutlet, × 20.

144

SCIRPUS

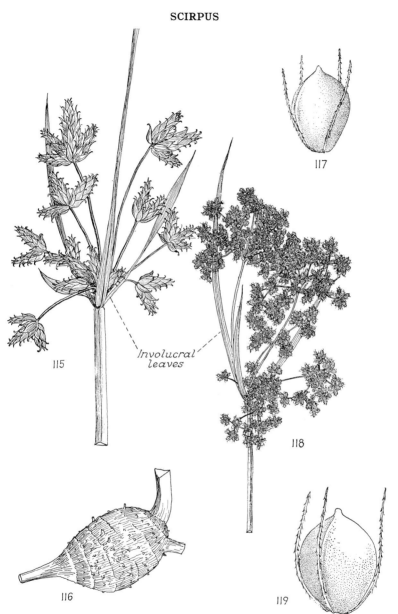

S. fluviatilis: 115. Inflorescence, × ½. 116. Tuber, × ½. **S. microcarpus** var. **rubrotinctus:** 117. Nutlet, × 20. **S. sylvaticus:** 118. Inflorescence, × ½. 119. Nutlet, × 20.

145

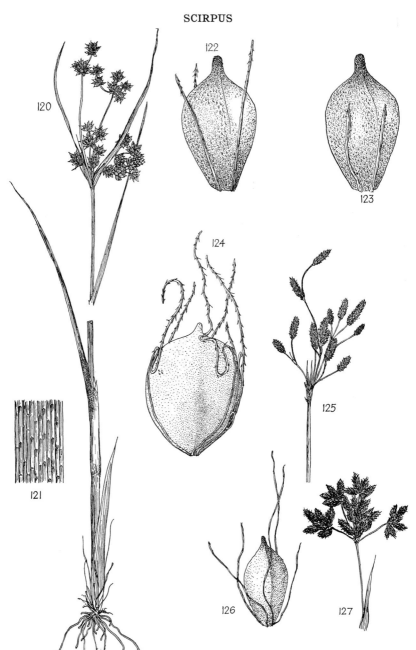

S. **atrovirens**: 120. Plant, $\times \frac{1}{2}$. 121. Surface marking of lower sheath, $\times$ 2. 122. Nutlet, $\times$ 20. **S. atrovirens** var. **georgianus**: 123. Nutlet, $\times$ 20. **S. polyphyllus**: 124. Nutlet, $\times$ 20. **S. lineatus**: 125. Inflorescence, $\times \frac{1}{2}$. **S. Peckii**: 126. Nutlet, $\times$ 20. 127. Inflorescence, $\times \frac{1}{2}$.

15. Stem 0.8–2.5 cm. thick at base, light-green, soft and spongy; nutlet 2 mm. long.
    S. válidus Vahl. Softstem Bulrush; Great Bulrush. Figs. 110, 114. Shallow
    water, widespread. A common plant, often growing in several feet of water.
15. Stem 3–10 mm. thick at base, dark olive-green, firm in texture; nutlet 2.5–3 mm.
    long. **S. acùtus** Muhl. Hardstem Bulrush, Big Bulrush. Commonly in hard
    water, widespread. Usually distinct from the last, but intermediates sometimes
    occur. (*S. occidentalis*, G, B.)
    16. Largest involucral leaves 2–10 mm. wide at base (Figs. 115, 118, 128); style
        not swollen at base...**17**
    16. Largest involucral leaves 1 mm. or less wide at base (Fig. 97); style swollen
        at base (Figs. 95, 99). **Fimbristylis,** p. 141.
17. Spikelets 5–10 mm. thick (Fig. 115); stems sharply triangular, 7–15 mm. thick
    toward the base, from stout tuber-bearing rootstocks (Fig. 116). **S. fluviátilis**
    (Torr.) Gray. River Bulrush. Muddy shores and shallow water; N.B. to s.
    Man., s. to N.J. & Kans.
17. Spikelets 1–5 mm. thick; stems mostly bluntly triangular and 2–5 mm. thick.
    All the following species normally have a much branched inflorescence several
    decimeters wide, but this is sometimes contracted into a close head. Occasionally
    some of the spikelets are replaced by bulb-like leafy shoots..................**18**
    18. Upright stems solitary from rootstocks; bristles about the nutlet barbed
        downward (Figs. 117, 124) when seen with a strong lens, or rarely absent.**19**
    18. Upright stems in large clumps, without rootstocks; bristles smooth (Fig.
        126)..**25**
19. At least the lower sheaths red; bristles barbed nearly to the base (Fig. 117)...**20**
19. All sheaths green; bristles barbed only above the middle (Fig. 122).........**22**
    20. Nutlet 2-angled, with 4 bristles (Fig. 117). **S. microcàrpus** Presl, var.
        rubrotínctus (Fernald) Jones. Nfd. to Man., s. to Conn., N.Y. & Minn.
    20. Nutlet 3-angled, with 3 or 6 bristles (Fig. 119)........................**21**
21. Spikelets 3–5 mm. long. **S. sylváticus** L. Figs. 118, 119. S. Maine to Ga., w.
    to Mich.
21. Spikelets 6–14 mm. long. **S. sylvaticus** var. **Bisséllii** Fernald. Rare and local.
    22. Bristles twice as long as the nutlet (Fig. 124)..........................**24**
    22. Bristles scarcely longer than the nutlet, or absent......................**23**
23. Bristles about equaling the nutlet (Fig. 122); lower sheaths with deep rectangular
    pits (Fig. 121). **S. atróvirens** Muhl. Fig. 120. Que. to Sask., s. to Ga. & Mo.
23. Bristles shorter than the nutlet (Fig. 123) or wanting; sheaths not deeply pitted.
    S. atrovirens var. **georgiànus** (Harper) Fernald. N.S. to Wis., s. to Ga. & Ark.
    24. Spikelets 2.5–3.5 mm. long. **S. polyphýllus** Vahl. W. N.E. to Ga., w. to
        Mich. & Ark.
    24. Spikelets 3–5 mm. long. **S. polyphyllus** var. **macróstachys** Boeckl. Conn.
        to N.Y.
25. Bristles scarcely longer than the scales (Figs. 125, 127)....................**26**
25. Bristles much longer than the scales (Figs. 129, 130), giving the inflorescence a
    woolly appearance (Fig. 128). The Wool Grass group. Species are usually
    distinct, but occasional individuals, particularly in the Middle West, are difficult
    to place..**28**
    26. Bristles scarcely equaling the nutlet. **S. divaricàtus** Ell. Va. to Mo. &
        southward.
    26. Bristles about twice as long as the nutlet (Fig. 126)....................**27**

S. cyperinus var. pelius: 128. Inflorescence, $\times \frac{1}{2}$. 129. Part of inflorescence, $\times 2$.
S. pedicellatus: 130. Part of inflorescence, $\times 2$. F. squarrosa: 131. Nutlet, $\times 12$.
F. pumila: 132. Nutlet, $\times 12$. F. simplex: 133. Upper part of plants, $\times 1$. 134.
Nutlet, $\times 12$.

27. Spikelets nearly all long stalked (Fig. 125); scales golden brown, with a broad green midrib. **S. lineàtus** Michx. Maine & N.H. to Ga., w. to Ore. & Tex.

27. Spikelets mostly without stalks and in close clusters at the tips of rays (Fig. 127); scales nearly black, and the midrib not green. **S. Péckii** Britton. Maine to Conn. & n. N.Y.

    28. Spikelets mostly without stalks, in clusters of 2–15 (Fig. 129)...........29

    28. Spikelets mostly stalked (Fig. 130)...................................31

29. Involucels reddish-brown (Figs. 128, 129)................................30

29. Involucels dull brown, with blackish bases. **S. cyperìnus** var. **pèlius** Fernald. Figs. 128, 129. Nfd. to Conn., and less commonly westward to Wis. & Man.

    30. Spikelets 3–6 mm. long. **S. cyperìnus** (L.) Kunth. N.S. to Va., Ark. & Wis.

    30. Spikelets 7–10 mm. long. **S. cyperinus** var. **Andréwsii** Fernald. Local; Maine to Conn.

31. Involucels (Fig. 130) brown or reddish..................................32

31. Involucels black..34

    32. Involucels (Fig. 130) reddish-brown. **S. Erióphorum** Michx. Gulf of Mexico, n. to Mo. & to Conn.

    32. Involucels dull brown...33

33. Spikelets 3–6 mm. long. **S. pedicellàtus** Fernald. Fig. 130. Que. to Conn., N.Y. & Minn.

33. Spikelets 7–10 mm. long. **S. pedicellatus** var. **púllus** Fernald. Local.

    34. Base of involucre not sticky; nutlets whitish or light yellowish; scales 1–2 mm. long. **S. atrocínctus** Fernald. Nfd. to Sask., s. to Conn., Pa. & Iowa.

    34. Base of involucre with a sticky gum; nutlets reddish-brown; scales 2–3 mm. long. **S. Lôngii** Fernald. Local; e. Mass. & N.J.

References: M. E. Jones, Univ. of Mont. Bull. No. 61, 20 (1910)—*S. microcarpus* var. *rubrotinctus;* Fernald, Rhodora, **13**, 6 (1911)—*S. Longii;* Fernald, Rhodora, **22**, 55 (1920)—*S. acutus;* Fernald, Rhodora, **23**, 134 (1921)—*S. atrovirens* var. *georgianus.*

UMBRELLA GRASS   **Fuirèna**

Tufted sedges with flat leaves which often have spreading hairs; spikelets in close clusters, mostly about 1 cm. long (Fig. 133), each with a leaf-like bract at its base; nutlet very small, triangular, long-stalked and long-beaked, with not only 3 bristles at its base, but also 3 flat perianth scales (Fig. 132).

1. Perianth scales cut square or indented at tip (Fig. 134). **F. símplex** Vahl. Fig. 133. Swamps and wet sand; Mo. & Kans. to Mex.

1. Perianth scales tapered at tip..2

    2. Annual, with stems 0.5–4.5 dm. tall, often depressed; perianth scales prolonged into a spine-like tip (Fig. 132); bristles longer than the nutlet, barbed. **F. pùmila** Torr. Bogs and wet shores; Mass. to Fla.; Mich. & Ind. (*F. squarrosa*, B, G.)

    2. Perennial, with a thick rootstock, and erect stems 0.25–1 m. tall; perianth scales merely pointed (Fig. 131); bristles shorter than the nutlet, mostly smooth. **F. squarròsa** Michx. Wet sand; N.J. to Ky. & Okla., s. to Tex., Fla. and the Gulf of Mexico. (*F. hispida*, B, G.)

Reference: Fernald, Rhodora, **40**, 396–398 (1938)—*F. pumila* and *F. squarrosa.*

H. micrantha: 135. Plant, × 1. 136. Spikelet, × 5. **H. Drummondi:** 137. Spikelet, × 5. **R. corniculata:** 140. Nutlet, × 5. 139. Part of inflorescence, × 1. **R. macrostachya:** 138. Inflorescence, × 1. **R. inundata:** 141. Part of inflorescence, × 1.

Hemicárpha

Separated from *Scirpus*, technically, by the fact that it has a minute translucent scale between the flower and the axis of the spikelet; distinguished, practically, by its very small size.

1. Scales with spreading tips (Fig. 136) 2
1. Scales with closely appressed tips (Fig. 137). **H. Drummóndi** Nees. W. Ont. & s. Wis. to Ind. & Tex.
    2. Scales with short abrupt tips. **H. micrántha** (Vahl) Pax. Figs. 135, 136. Sandy shores; N.H. to the Gulf of Mexico and thence westward to the Pacific and northward to Wis. & Minn.
    2. Scales tipped by a bristle about as long as the body of the scale. **H. micrantha** var. **aristulàta** Coville. Mo. to Wyo., s. to Tex.

Reference: Coville, Bull. Torr. Bot. Club, **21,** 36 (1894)—*H. micrantha* var. *aristulata.*

BEAK RUSH   **Rynchóspora**

Nutlet topped by a tubercle much as in *Eleocharis*, but differing from that genus in having the spikelets numerous and with leafy involucres.

1. Tubercle 1 cm. or more long (Fig. 140) 2
1. Tubercle 5 mm. or less long (Figs. 143, 147) 5
    2. Bristles longer than the body of the nutlet 3
    2. Bristles shorter than the body of the nutlet 4
3. Spikelets in close clusters of 10–50 together (Fig. 138); body of nutlet 5–6 mm. long; tubercle 1.9–2.2 cm. long. **R. macrostàchya** Torr. S. Me. to the Gulf of Mexico, n. to Ind.
3. Spikelets in small clusters of few spikelets each (Fig. 141); body of nutlet 4.2–4.8 mm. long; tubercle 1.0–1.8 cm. long. **R. inundàta** (Oakes) Fernald. E. Mass., L.I. & N.J.
    4. Nutlet 5–6 mm. long, 2.8–3.3 mm. broad, twice as wide as the base of the tubercle (Fig. 140). **R. corniculàta** (Lam.) Gray. Horned Rush. Fig. 139. Del. to the Gulf of Mexico, n. to Mo.
    4. Nutlet 4.4–5.3 mm. long, 2.4–2.6 mm. broad, the summit scarcely wider than the base of the tubercle. **R. corniculata** var. **intèrior** Fernald. Ala. to Tex., Ark. & Ind.
5. Nutlet wrinkled (Figs. 142, 143) 6
5. Nutlet smooth (Figs. 147, 149, 150, 153) 8
    6. Bristles long-fringed (Fig. 142). **R. oligántha** Gray. N.J. to Fla.
    6. Bristles merely barbed (Fig. 143) 7
7. Stems sharply triangular; leaves flat. **R. cymòsa** Ell. Figs. 143, 144. N.J. to Ill. & southward.
7. Stem nearly round in cross section; leaves inrolled. **R. Torreyàna** Gray. Fig. 145. N.H.; N.J. to Ga.
    8. Clusters of spikelets nearly globular, the lower in each cluster spreading or pointing downward (Figs. 146, 148) 9
    8. Clusters of spikelets oblong, all the spikelets ascending (Figs. 151, 152) 12

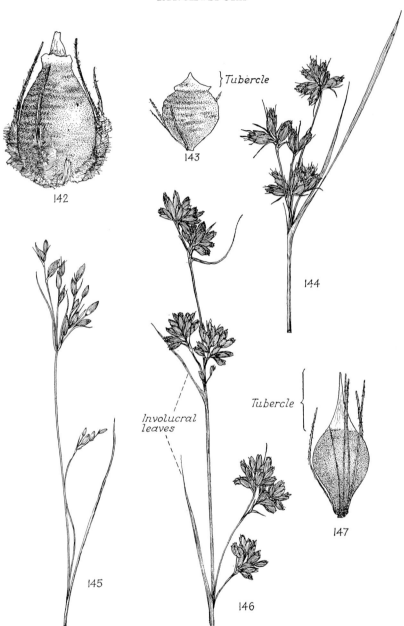

R. oligantha: 142. Nutlet, × 10. R. cymosa: 143. Nutlet, × 10. 144. Inflorescence, × 1. R. Torreyana: 145. Inflorescence, × 1. R. glomerata var. minor: 146. Inflorescence, × 1. 147. Nutlet, × 10.

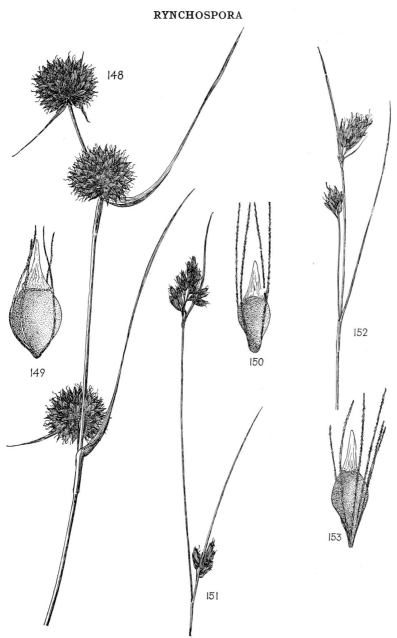

R. cephalantha : 148. Inflorescence, × 1. R. gracilenta : 149. Nutlet, × 10. R. fusca :
150. Nutlet, × 10. 151. Inflorescence, × 1. R. capillacea : 152. Inflorescence, × 1.
153. Nutlet, × 10.

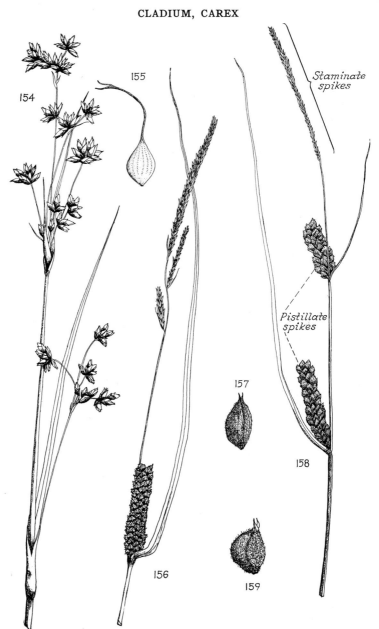

Staminate
spikes

Pistillate
spikes

154

155

157

156

158

159

Cladium mariscoides: 154. Upper part of plant, × 1. 155. Young nutlet, × 3. **Carex lanuginosa:** 156. Inflorescence, × 1. 159. Perigynium, × 5. **C. lasiocarpa:** 158. Inflorescence, × 1. 157. Perigynium, × 5.

154

9. Clusters loose (Fig. 146), usually many.....................................**10**
9. Clusters dense (Fig. 148), 1–6 in number................................**11**
   10. Stems stout, 0.5–2 m. tall; leaves 3–7 mm. wide; spikelets 5–6 mm. long; nutlets 1.7–2 mm. long. **R. glomeràta** (L.) Vahl. Del., Md., Tenn., Ark. & Okla., s. to the Gulf of Mexico. (*R. glomerata* var. *paniculata*, G.)
   10. Stems slender, 0.1–1.5 m. tall; leaves 0.5–4 mm. wide; spikelets 3–5 mm. long; nutlets 1.5–1.8 mm. long. **R. glomerata** var. **mìnor** Britton. Fig. 146. N.S. to the Gulf of Mexico, w. about the Great Lakes to Wis. (*R. glomerata*, G.) The bristles are downwardly barbed (Fig. 147); in **R. glomerata** var. **minor** forma **controvérsa** (Blake) Fernald they are upwardly barbed, and in **R. glomerata** var. **minor** forma **discùtiens** (Clarke) Fernald they are smooth.
11. Spikelets 5–7 mm. long, in 1–3 clusters; nutlets 2.2–2.6 mm. long, 1.4–1.6 mm. broad. **R. cephalántha** Gray. Fig. 148. L.I. to the Gulf of Mexico. (*R. axillaris*, G, B.)
11. Spikelets 3–4 mm. long, in 2–6 clusters; nutlets 1.5–2 mm. long, 0.8–1.2 mm. broad. **R. microcéphala** Britton. N.J. to the Gulf of Mexico. (*R. axillaris* var. *microcephala*, G.)
   12. Bristles upwardly barbed (Fig. 150)................................**13**
   12. Bristles downwardly barbed or smooth..............................**15**
13. Leaves inrolled, bristle-like. **R. fúsca** (L.) Ait. f. Figs. 150, 151. Nfd. to Ont., s. to Del. & Mich.
13. Leaves flat, 1–2.5 mm. wide...**14**
   14. Leaves 1 mm. wide; spikelets 3–4 mm. long; nutlets 1.2–1.5 mm. long. **R. gracilénta** Gray. Fig. 149. S. N.Y. to the Gulf of Mexico.
   14. Leaves 1.5–2.5 mm. wide; spikelets 4–5.5 mm. long; nutlets 1.6–2 mm. long. **R. gracilenta** var. **diversifòlia** Fernald. S. N.J. to the Gulf of Mexico.
15. Bristles downwardly barbed. **R. capillàcea** Torr. Figs. 152, 153. N.B. to Minn., s. to N.J. & Mo.
15. Bristles smooth. **R. capillacea** forma **levisèta** (Hill) Fernald. Local, and often with the typical form.

References: Fernald, Rhodora, **20**, 138–140 (1918)—species with long tubercle; Fernald, Rhodora, **37**, 252 (1935)—*R. capillacea* f. *leviseta*, and 399–405—*R. glomerata*, *R. cephalantha*, *R. microcephala*, and *R. gracilenta* var. *diversifolia*.

Twig Rush   Clàdium

C. mariscoìdes (Muhl.) Torr. Rather coarse plants, 0.4–1 m. high, from stout rootstocks; stems leafy, topped by a branched inflorescence, each branch terminated by a cluster of spikelets (Fig. 154). Sandy and boggy shores, often in shallow water; N.S. to Ont. & Minn., s. to Fla., Ky., Ind. & Iowa. (*Mariscus mariscoides*, B.)

C. **Pseudo-Cyperus:** 160. Perigynium, × 5. 161. Inflorescence, × 1. **C. comosa:**
162. Perigynium, × 5. 163. Inflorescence, × 1. **C. lacustris:** 164. Perigynium, × 5.
165. Inflorescence, × 1.

Càrex

A large and complex genus, characterized by having the nutlet enclosed in a sac, the perigynium, and by having the stamens and pistils in separate flowers. The staminate and pistillate flowers may be in separate spikes (Figs. 158, 173), in different parts of the same spike (Figs. 186, 192) or scattered and scarcely distinguishable in each spike (Fig. 198). To a large extent the species are based on the nature of the perigynium: this may be inflated (Fig. 171), closely filled by the nutlet (Fig. 196), or scale-like (Fig. 209); it may be beaked (Fig. 193) or beakless (Fig. 181). Mature perigynia are necessary for the identification of practically all species. The most recent comprehensive work is that of K. K. Mackenzie, N. Amer. Fl., **18**, 1–478 (1931–1935).

1. Staminate and pistillate flowers easily distinguishable in different spikes (Fig. 158), or in different parts of the same spike (Figs. 186, 192)..................**2**
1. Staminate and pistillate flowers intermixed in essentially uniform spikes (Figs. 198, 210)..**23**
  2. Perigynia 2-toothed, the teeth longer than thick (Figs. 156–176)...........**3**
  2. Perigynia not 2-toothed, or slightly notched at tip (Figs. 177–194).......**13**
3. Perigynia velvety with short spreading hairs (Figs. 157, 159)................**4**
3. Perigynia without hairs...**5**
  4. Leaves flat, 1.5–5 mm. wide; stem with 3 sharp angles. **C. lanuginòsa** Michx. Figs. 156, 159. Usually in shallow water; n. N.B. to B.C., s. to Tenn., Ark., N.M. & Calif.
  4. Leaves closely rolled, 2 mm. or less wide; stem with 3 rounded angles. **C. lasiocàrpa** Ehrh. Figs. 157, 158. Usually in shallow water; Nfd. to B.C., s. to N.J., Iowa, Idaho & Wash. (*C. filiformis*, G.)
5. Perigynia nearly all pointing slightly downward (Figs. 161, 163)..............**6**
5. Perigynia all pointing upward, except sometimes the lowermost (Figs. 165, 166, 169, 176)..**7**
  6. Teeth of perigynia 0.5–1 mm. long, parallel or slightly spreading (Fig. 160). **C. Pseùdo-Cypèrus** L. Fig. 161. Usually in shallow water; Nfd. to Sask., s. to Conn., N.Y. & Minn.
  6. Teeth of perigynia 1.5–2 mm. long, widely spreading (Fig. 162). **C. comòsa** Boott. Fig. 163. Usually in shallow water; Que. to Minn., s. to the Gulf of Mexico; Pacific Coast.
7. Perigynia abruptly narrowed to a slender beak (Figs. 172, 176)..............**10**
7. Perigynia gradually narrowed to the tip (Figs. 164, 167, 171)..................**8**
  8. Pistillate spikes with 50–150 perigynia (Fig. 165). **C. lacústris** Willd. Ripgut. Fig. 164. N.S. to Man., s. to D.C. & Iowa. Coarse sedges, with long scaly runners, forming extensive swales; base of plants strongly marked with red; leaves 5–15 mm. broad and 1 m. or more long. (*C. riparia*, G.)
  8. Pistillate spikes with 3–30 perigynia.....................................**9**

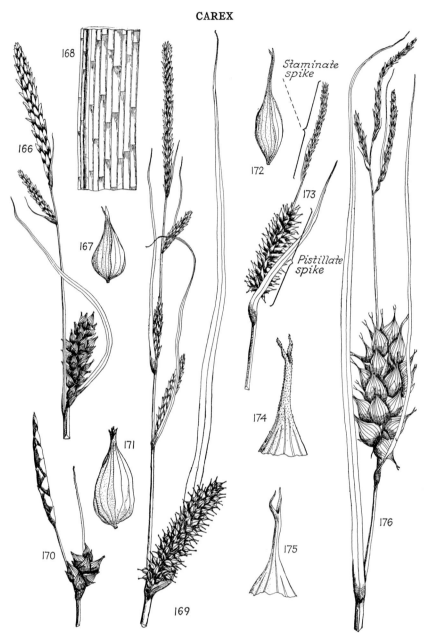

C. **Walteriana:** 166. Inflorescence, × 1. 167. Perigynium, × 5. **C. rostrata:** 168. Surface marking of leaves, × 3. 169. Inflorescence, × 1. **C. oligosperma:** 170. Inflorescence, × 1. 171. Perigynium, × 5. **C. hystericina:** 172. Perigynium, × 5. 173. Inflorescence, × 1. **C. bullata:** 174. Upper part of perigynium, × 5. **C. Tuckermani:** 175. Upper part of perigynium, × 5. 176. Inflorescence, × 1.

9. Pistillate spikes cylindrical (Fig. 166); leaves flat. **C. Walteriàna** Bailey. Fig. 167. Pine barren swamps; Mass. to Fla. (*C. striata* var. *brevis*, G.)
9. Pistillate spikes globose (Fig. 170); leaves closely inrolled. **C. oligospérma** Michx. Fig. 171. Mostly in Sphagnum bogs, rarely in shallow water; Labrador to Mackenzie, s. to Mass., Pa., Ind. & Wis.

    10. Base of plant thick and spongy; leaves with prominent cross markings between the veins (Fig. 168). **C. rostràta** Stokes. Fig. 169. Swamps and shallow water; Greenland to Alaska, s. to Del., Ind., N.M. & Calif.

    10. Base of plant not spongy; leaves with cross markings obscure or lacking. .**11**

11. Staminate spikes 2 or more (Fig. 176)......................................**12**
11. Staminate spike solitary (Fig. 173). **C. hystericìna** Muhl. Fig. 172. Swamps in calcareous regions; N.B. to Wash., s. to Va., Ky., Tex., Ariz. & Calif.

    12. Beak minutely roughened (Fig. 174). **C. bullàta** Schkuhr. Wet meadows in acid soil; N.S. to Ga.

    12. Beak smooth (Fig. 175). **C. Tuckermàni** Boott. Fig. 176. Wet woods and meadows in calcareous regions; N.B. to Minn., s. to N.J., Ind. & Iowa.

13. Perigynia ascending (Figs. 177, 182, 192)................................**14**
13. Perigynia spreading or pointing downward (Fig. 194).......................**22**

    14. Perigynia narrowed to a slender beak (Fig. 178). **C. scabràta** Schwein. Fig. 177. Wet woods; etc.; N.S. to Ont. & Mich., s. to S.C. & Tenn.

    14. Perigynia rounded (Fig. 185) or tapered to the tip (Fig. 189)............**15**

15. Nutlet 3-sided (Fig. 180); stigmas, present on young nutlet, 3 in number (Figs. 183, 184)..**16**
15. Nutlet 2-sided; stigmas 2 in number.......................................**18**

    16. Scales whitish, much shorter than the perigynia (Fig. 182). **C. prásina** Wahlenb. Wet woods; Que. to Mich., s. to D.C., Ky. & Ga.

    16. Scales brown, nearly or quite equaling the perigynia (Fig. 179)...........**17**

17. Spikes 1–2.5 cm. long. **C. limòsa** L. Figs. 179–181. Bogs; Labrador to Yukon, s. to Del., Iowa, Mont. & Calif.
17. Spikes 2.5–4 cm. long. **C. Barráttii** Schwein. & Torr. Swamps in pinelands; Conn. to N.C. (*C. littoralis*, G.)

    18. Scales of the pistillate spike pointed (Fig. 186)........................**19**
    18. Scales of the pistillate spike rounded at tip (Fig. 192).................**21**

19. Lower sheaths with fine brown fibers on the side opposite the leaf (Fig. 188). .**20**
19. Lower sheaths without fibers. **C. aquátilis** Wahl., var. **substrícta** Kükenth. Figs. 185, 186. In swamps; Nfd. to Wash., s. to N.J., Ind. & Neb. (*C. substricta* in N. Amer. Fl.)

    20. Stems in dense tussocks; sheaths smooth on the side opposite the leaf. **C. strícta** Lam. Niggerhead. Figs. 187–189. Maine to N.C. and the Gulf of Mexico, w. about the Great Lakes to Minn. The large tussocks are characteristic in meadows and at times stand well above the water in shallow pond margins.

    20. Stems forming large beds, but not tussocks; sheaths minutely roughened on the side opposite the leaf. **C. stricta** var. **stríctior** (Dewey) Carey. Crex. Meadows; N.S. to Minn., s. to N.C., Tenn. & Iowa; commoner westward than the preceding. (*C. strictior* in N. Amer. Fl.)

21. Perigynia without nerves, twisted at tip (Fig. 191); spike 2–7.5 cm. long. **C. tórta** Boott. Fig. 192. Along streams; N.S. to Minn., s. to N.C., Tenn. & Ark.
21. Perigynia nerved, flat (Fig. 196); spike 1.5–4.5 cm. long. **C. lenticulàris** Michx. Fig. 197. Wet shores; Labrador to Mackenzie, s. to Mass., Minn. & Idaho.

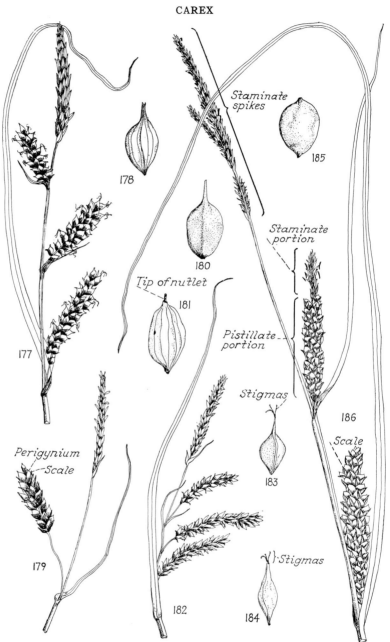

C. **scabrata**: 177. Inflorescence, $\times$ 1. 178. Perigynium, $\times$ 5. **C. limosa**: 179.
Inflorescence, $\times$ 1. 180. Nutlet, $\times$ 5. 181. Perigynium, $\times$ 5. **C. prasina**: 182.
Inflorescence, $\times$ 1. 183. Young nutlet, topped by 3 stigmas, $\times$ 5. 184. Perigynium,
enclosing nutlet, $\times$ 5. **C. aquatilis v. substricta**: 185. Perigynium, $\times$ 5. 186. Inflorescence, $\times$ 1.

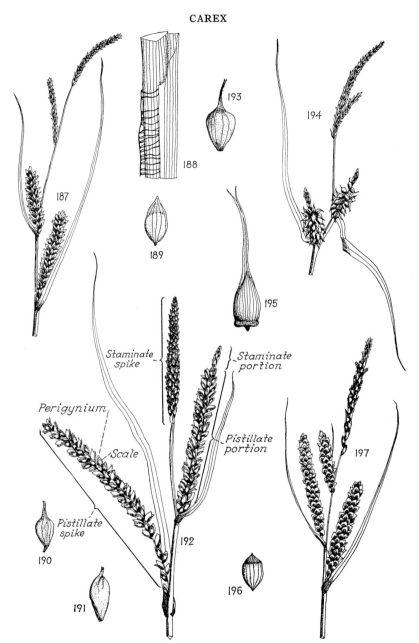

C. stricta: 187. Inflorescence, × 1. 188. Part of sheath & stem, × 4. 189. Perigynium, × 5. C. viridula: 190. Perigynium, × 5. C. torta: 191. Perigynium, × 5. 192. Inflorescence, × 1. C. cryptolepis: 193. Perigynium, × 5. 194. Inflorescence, × 1. C. crus-corvi: 195. Perigynium, × 5. C. lenticularis: 196. Perigynium, × 5. 197. Inflorescence, × 1.

161

C. stipata : 198. Inflorescence, × 1. 199. Part of sheath and stem, × 4. 205. Perigynium, × 5. **C. laevivaginata** : 200. Perigynium, × 5. 201. Part of sheath and stem, × 4. **C. vulpinoidea** : 202. Perigynium, × 5. **C. diandra** : 203. Perigynium, × 5. 204. Inflorescence, × 1. **C. straminea** : 206. Inflorescence, × 1. 207. Perigynium, × 5. **C. suberecta** : 208. Perigynium, × 5. **C. alata** : 209. Perigynium, × 5. 210. Inflorescence, × 1.

22. Beak about as long as the body of the perigynium (Fig. 193). **C. cryptólepis** Mackenzie. Fig. 194. Mostly on sandy shores and in wet meadows; Nfd. to Minn., s. to N.J. & Ind. (*C. flava*, G, in part.)

22. Beak about one-half as long as the body of the perigynium (Fig. 190). **C. virídula** Michx. Wet shores; Greenland to Alaska, s. to N.J., Ind., N.M. & Calif. (*C. Oederi* var. *pumila*, G; *C. chlorophila* Mackenzie.)

23. Beak of perigynium 2–3 times as long as the body (Fig. 195). **C. crus-córvi** Shuttlw. Swamps; Fla. to Tex., n. in the Mississippi Valley to Mich., Minn. & Neb.

23. Beak not longer than the body, or absent................................**24**

    24. Perigynia thin and scale-like (Figs. 207, 208, 209), closely overlapping (Figs. 206, 210)..**28**

    24. Perigynia not scale-like (Figs. 200, 202, 203), mostly spreading (Figs. 198, 204)..**25**

25. Perigynia 4 mm. or more long..**26**

25. Perigynia 3 mm. or less long...**27**

    26. Sheaths cross-wrinkled on the side opposite the blade, thin and papery at apex (Fig. 199). **C. stipàta** Muhl. Figs. 198, 199, 205. Wet meadows; Nfd. to Alaska, s. to N.C., Tenn., N.M. & Calif.

    26. Sheaths smooth, with a thickened margin at the mouth (Fig. 201). **C. laevivaginàta** (Kükenth.) Mackenzie. Fig. 200. Wet woods; Mass. to Minn., s. to Fla. & Mo.

27. Perigynia flat (Fig. 202). **C. vulpinoídea** Michx. Swamps; Nfd. to B.C., s. to Fla., Tex., Ariz. & Ore.

27. Perigynia plump (Fig. 203). **C. diándra** Schrank. Fig. 204. Wet meadows; Nfd. to Yukon, s. to N.J., Ind., Colo. & Calif.

    28. Spikes rounded at tip (Fig. 206); perigynia with nerves on the inner side (Fig. 207). **C. stramínea** Willd. Swamps and meadows; Mass. to D.C., w. to Mich. & Ind. (*C. hormathodes* var. *Richii*, G; *C. Richii* Mackenzie.)

    28. Spikes pointed at tip (Fig. 210); perigynia without nerves on the inner side (Fig. 208)..**29**

29. Perigynia mostly more than 2.8 mm. broad; scales of pistillate spikes with a bristle-like roughened tip. **C. alàta** Torr. Figs. 209, 210. Swamps; Mass. to Fla. & Tex., w. to n. Ind.

29. Perigynia less than 2.8 mm. broad; scales blunt. **C. suberécta** (Olney) Britton. Fig. 208. Moist meadows; Ont. to Va., Minn. & Iowa.

References: Mackenzie, N. Amer. Fl., **18**, 1–478 (1931–1935)—all species here discussed; Svenson, Rhodora, **40**, 329 (1938)—*C. stramínea*.

ARUM FAMILY   ARÀCEAE

Herbs growing on muddy shores or advancing into shallow water; juice very acrid or pungent; flowers small and inconspicuous, without petals, in a fleshy head, the spadix (Figs. 9–11), often with a leaf-like or petal-like spathe. In addition to the species here described are the Jack-in-the-pulpit and the Skunk Cabbage, inhabitants of moist places but not in the water.

1. Leaves lobed at base...2
1. Leaves not lobed at base..3
    2. Each basal lobe with a heavy central vein (Figs. 1–8). **Peltandra.**
    2. Veins of basal lobes not heavy (Fig. 11). **Calla.**
3. Leaves erect, grass-like (Fig. 9). **Acorus.**
3. Leaves spreading, tongue-like (Fig. 10). **Orontium.**

ARROW ARUM   Peltándra

P. virgínica (L.) Kunth. Figs. 1–8. Leaves 3-nerved, more or less arrow-shaped, on long petioles; spathe green, fleshy, inrolled. Muddy shores and shallow water; s. Maine to the Gulf of Mexico and n. to Mo.; w. rarely to Ind. Fig. 1 represents the common form; the following sporadic variations have been named: **P. virginica** forma **hastifòlia** Blake (Fig. 4); forma **latifòlia** (Raf.) Blake (Fig. 7); forma **brachyòta** Blake (Fig. 5); forma **angustifòlia** (Raf.) Blake (Fig. 2); forma **heterophýlla** (Raf.) Blake (Figs. 3, 6); forma **rotundàta** Blake (Fig. 8).

WATER ARUM   Cálla

C. palústris L. Leaves in a cluster at the end of a stout rootstock (Fig. 11), or scattered along it, long-petioled, the blade more or less heart-shaped; spathe white, flat. Cold bogs and mucky shores, the rootstocks trailing into shallow water; N.S. to Hudson Bay, s. to N.J., Pa., Wis., Iowa & Minn.

GOLDEN CLUB   Oróntium

O. aquáticum L. Fig. 10. Plants often floating; leaves in a rosette, tapered to the petiole (Fig. 5, p. 21); spadix naked, the spathe inconspicuous at the base of its stalk. Shallow water near the coast; s. Mass. to the Gulf of Mexico.

SWEET FLAG   Ácorus

A. Cálamus L. Fig. 9. Leaves erect, sword-like; spadix long, yellowish, appearing to come from the side of the stem; whole plant fragrant when bruised. Margins of swamps and in shallow water; N.S. to Ont. & Minn., s. to the Gulf of Mexico.

References: Tidestrom, Rhodora, **12**, 47–50 (1910); and Blake, Rhodora, **14**, 102–106 (1912)—forms of *Peltandra virginica*.

DUCKWEED FAMILY   LEMNÀCEAE

Plants floating, without leaves, consisting of globose or flattened fronds, with or without roots, very rarely flowering. Known as Duck's-meat, Water Lentils, Seed-moss, and Waterweed. The genera are easily recognized: *Spirodela* has several roots and is red beneath; *Lemna* has but one root to each joint and is green beneath; *Wolffia* consists of almost microscopic meal-like bodies without roots; and *Wolffiella* is made up of

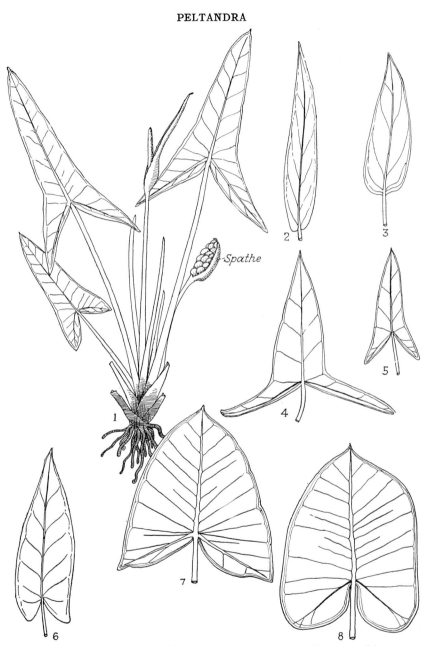

Spathe

P. virginica: 1. Plant, $\times \frac{1}{4}$.   2–8. Variations in shape of leaves, $\times \frac{1}{4}$.

A. **Calamus**: 9. Plant, $\times \frac{1}{6}$.　**O. aquaticum**: 10. Plant, $\times \frac{1}{20}$.　**C. palustris**: 11. Plant, $\times \frac{1}{4}$.

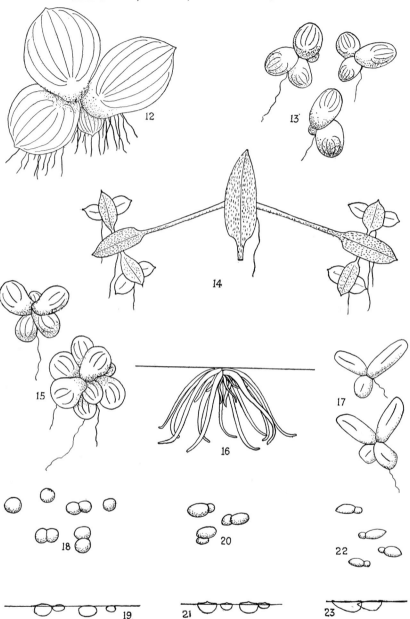

S. polyrhiza: 12. Plants, × 4. **L. minor**: 13. Plants, × 4. **L. trisulca**: 14. Plant, × 4.
L. perpusilla: 15. Plants, × 4. **L. valdiviana**: 17. Plants, × 4. **Wolffiella floridana**:
16. Side view, × 4, showing water line. **Wolffia columbiana**: 18. Plants, seen from above,
× 4. 19. Side view, × 4, showing water line. **W. papulifera**: 20. Plants, seen from above,
× 4. 21. Side view, × 4, showing water line. **W. punctata**: 22. Plants, seen from
above, × 4. 23. Side view, × 4, showing water line.

strap-shaped bodies without roots and occurring singly or radiating from a point. These include the smallest of the flowering plants. They occur in hard water, and as a rule several species are found intermixed.

1. Plants with roots...**2**
1. Plants without roots..**3**
    2. Plants red on the lower surface, each joint with several roots (Fig. 12). **Spirodela.**
    2. Plants green on the lower surface, each joint with one root. **Lemna.**
3. Plants globular, the size of grains of corn meal (Figs. 18–23). **Wolffia.**
3. Plants strap-shaped, 6–8 mm. long, solitary or cohering in star-like masses (Fig. 16). **Wolffiella.**

BIG DUCKWEED   Spirodèla

S. polyrhìza (L.) Schleid. FIG. 12. The largest of the Duckweeds, 3–10 mm. long, mostly 7-nerved. Common; N.S. to B.C., s. to Fla., Tex., Nev. & Calif.

DUCKWEED   Lémna

Plants 1–3-nerved, 2–5 mm. long (except in *L. trisulca*). The smaller species are difficult to distinguish.

1. Joints of plant long and narrow, stalked (Fig. 14); plants commonly sinking. **L. trisúlca** L. Star Duckweed. Common; N.S. to B.C., s. to N.C., Ala., Tex., N.M. & Calif.
1. Joints rounded, not stalked; plants floating until cold weather.................**2**
    2. Joints symmetrical or nearly so (Fig. 13). **L. mìnor** L. Lesser Duckweed. Common; nearly throughout N.A.
    2. Joints asymmetrical **3**
3. Joints with unequally rounded sides, 3-veined (Fig. 15). **L. perpusílla** Torr. Mass. to Fla. & westward, not common.
3. Joints with parallel sides, obscurely 1-veined (Fig. 17). **L. valdiviàna** Philippi. Mass. southward and westward, not common. (*L. cyclostasa*, B.)

WATERMEAL   Wólffia

Plants with almost globular joints 0.5–1.5 mm. long, floating at or near the surface of the water, often in great abundance. These are the smallest of the flowering plants.

1. Plants rounded on the back, not dotted (Figs. 18, 19). **W. columbiàna** Karst. Lying just below the surface of the water; e. Mass.; Conn. to Minn., s. to Fla. & La.
1. Plants flattened on the back (Figs. 21, 23), with minute brown dots............**2**
    2. Upper surface with a little pointed mound in the center (Fig. 21). **W. papulífera** Thompson. Fig. 20. Rare; Mo.
    2. Upper surface flat, upturned at the tip, or peanut-shaped (Fig. 23). **W. punctàta** Griseb. Fig. 22. Common; Ont. to Mich. & N.Y., to s. W.I. & Tex.

Wolffiélla

W. floridàna (Smith) Thompson. Fig. 16. Joints solitary or cohering, sometimes forming interwoven masses. E. Mass.; Va.; Mo. & Ga. to the Gulf of Mexico.

PIPEWORT FAMILY   ERIOCAULÂCEAE

PIPEWORT   Eriocaúlon

E. septangulàre With.   Fig. 24.   Leaves 2–8 cm. long, in a rosette, long and slender, taper-pointed, round in cross section, soft, somewhat translucent and showing many cross veins when held to the light; roots white, fibrous, slightly constricted between cross markings; flowers in button-like heads at the tips of slender stalks, which have an inflated bract at the base.   Sandy shores of soft water lakes, with stems erect and seldom more than 10 cm. tall, or in water up to 2 m., the rosettes forming green patches on the bottom and the stalks elongating so that the heads are at the surface.   Nfd. to N.J., w. about the Great Lakes to Minn.   (*E. articulatum*, G.)

Reference: Robinson and Fernald, Rhodora, **11**, 40 (1909)—*E. septangulare.*

YELLOW-EYED GRASS FAMILY   XYRIDÂCEAE

YELLOW-EYED GRASS   Xỳris

Leaves grass-like, stiff, borne at or near the base of the plant; flowers at the tip of a naked scape, yellow, in the axils of closely overlapping scales in a pine-cone-like head.   Plants of sandy or boggy shores, sometimes in shallow water.

1. Base of plant bulbous (Fig. 26); upper scales fringed with little hairs (Fig. 27) which are scarcely visible without a lens; sepals fringed (Fig. 33).   **X. tòrta** Sm. Wet sandy shores; s. N.H. to e. Minn. & n. Ind., s. to the Gulf of Mexico.   (*X flexuosa*, G, B.)

1. Plant not bulbous (Fig. 25); upper scales entire or minutely toothed, not fringed; sepals minutely toothed above the middle (Figs. 34, 36), or rarely entire (Fig. 35)..**2**

    2. Plants 1–9 dm. high, without creeping rootstocks; heads broadly ellipsoid (Figs. 28–30); scales with a central green portion..........................**3**

    2. Plants not over 1.5 dm. tall, tufted and forming a turf; heads narrowly ellipsoid (Fig. 31); scales dark-brown throughout, lacking a central green portion. **X. montàna** Ries.   Boggy land; Nfd. to Pa., w. to n. Wis.

3. Plants 4–9 dm. tall; leaves 2.5–6 dm. long and reaching 2 cm. in width; sepals with tips extending beyond the toothed margins of the scales (Fig. 28).   **X. Congdóni** Small.   Boggy shores and shallow water; s. Maine; e. Mass. to N.J.   (*X. Smalliana*, G.)

3. Plants variable in size, 1–5 dm. tall; leaves narrow; sepals entirely hidden by the entire-margined scales (Fig. 29) except in some rare plants of Wisconsin........**4**

    4. Scales not toothed, sometimes with the margin becoming torn with age; sepals not tipped by two points (Fig. 34); leaves not roughened, tapering to pointed tips which are straight or slightly turned to one side (Figs. 37–40).   **X. caroliniàna** Walt.   Fig. 29.   Wet sandy shores; Maine. to n. Ind., s., mostly near the coast, to the Gulf of Mexico.   **X. caroliniana** forma **phyllólepis** Fernald has the flowers replaced by leafy shoots.

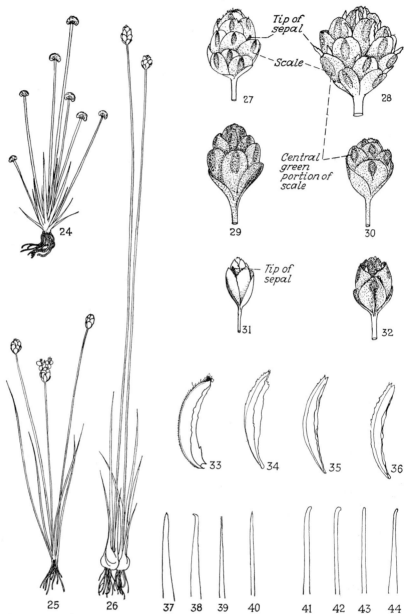

E. septangulare: 24. Plant, × ½.   X. caroliniana: 25. Plant, × ½.   29. Head, × 2.
34. Sepal, × 4.   37–40. Leaf tips, × 2.   X. torta: 26. Plant, × ½.   27. Head, × 2.
33. Sepal, × 4.   X. Congdoni: 28. Head, × 2.   X. montana: 31. Head, × 2.   X. papillosa:
30. Head, × 2.   36. Sepal, × 4.   41–44. Leaf tips, × 2.   X. papillosa var. exserta: 32.
Head, × 2.   35. Sepal, × 4.

Spathe

46

48

45

47

49

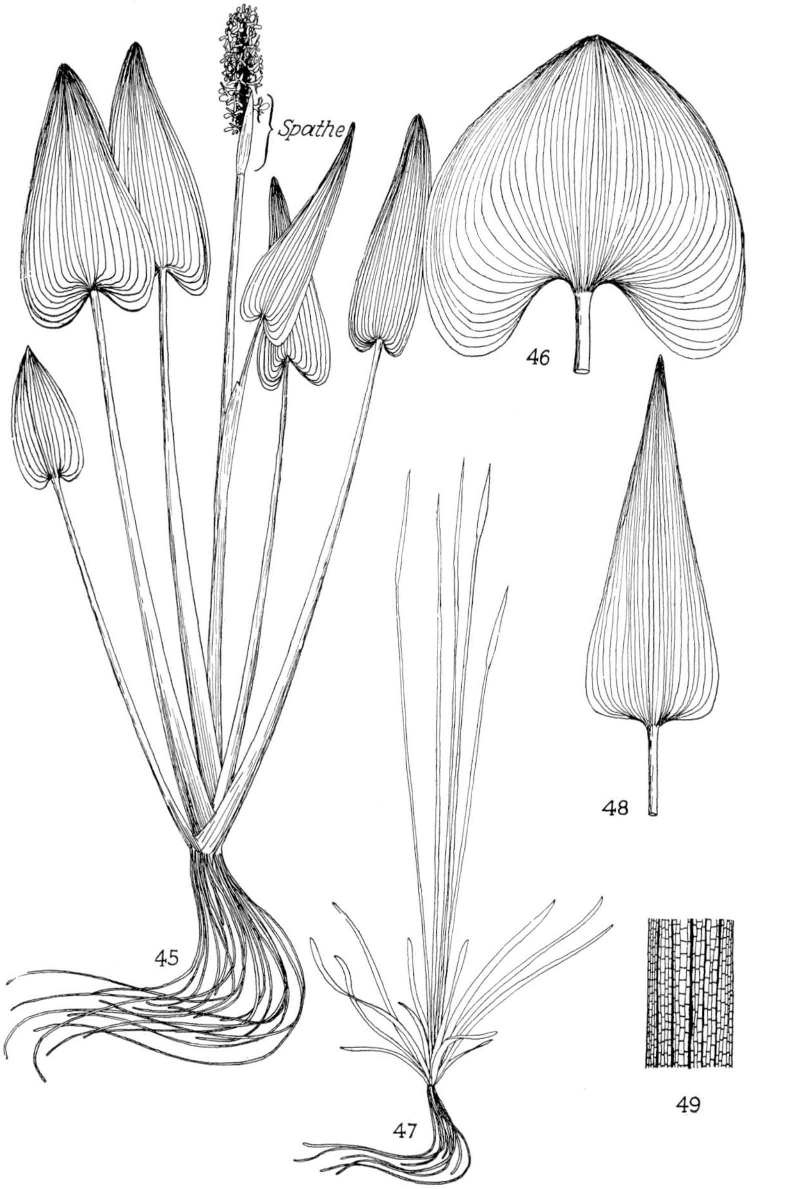

P. cordata: 45. Plant, $\times \frac{1}{4}$. **P. cordata** f. **latifolia**: 46. Leaf blade, $\times \frac{1}{4}$. **P. cordata** f. **angustifolia**: 48. Leaf blade, $\times \frac{1}{4}$. **P. cordata** f. **taenia**: 47. Plant, $\times \frac{1}{2}$. 49. Venation of submersed leaf, $\times 2$.

HETERANTHERA

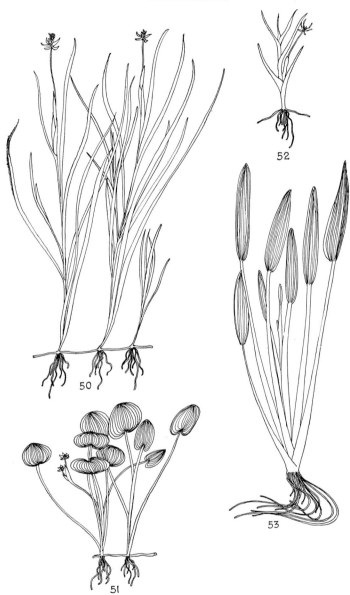

H. dubia: 50. Plant, × ½.   H. dubia f. terrestris: 52. Plant, × 1.   H. reniformis: 51 Plant, × ½.   H. limosa: 53. Plant, × ½.

4. Scales minutely toothed; sepals tipped by two points (Figs. 35, 36); leaves abruptly narrowed to rounded or blunt tips which are usually turned to one side (Figs. 41–44), the outermost dotted with minute projections toward the base. . **5**

5. Sepals irregularly toothed or appearing as if gnawed on the back above the middle (Fig. 36), their tips hidden behind the scales (Fig. 30); lowest scale thickened but rarely keeled.   **X. papillòsa** Fassett.   Sandy shores; Sawyer County, Wis.

5. Sepals nearly or quite entire (Fig. 35), the tips slightly extended beyond the scales (Fig. 32); lowest scale with a green keel.   **X. papillosa** var. **exsérta** Fassett.   Woodruff, Wis.

References: Harper, Torreya, **5**, 128–130 (1905)—*X. torta;* Fernald, Rhodora, **36**, 194 (1934)—*X. caroliniana* f. *phyllolepis;* Fassett, Rhodora, **39**, 459–460 (1937)— *X. papillosa.*

PICKERELWEED FAMILY   **PONTEDERIÀCEAE**

PICKERELWEED   **Pontedèria**

P. cordàta L.   Leaves mostly crowded toward the base, from a stout rootstock, the blade heart-shaped, the petiole with a sheath at base; flowering stem usually with one leaf and an inflated bladeless sheath (the spathe); flowers blue, in a dense spike. Muddy shores, often abundant in shallow water; N.S. to Minn., s. to the Gulf of Mexico & Okla.   The typical form has leaves of moderate width and edges of blades tapering with straight sides (Fig. 45).   **P. cordata** forma **angustifòlia** (Pursh) Solms is the narrow-leaved extreme with blades scarcely heart-shaped at base (Fig. 48).   Forma **latifòlia** (Raf.) House has blades broad and rounded (Fig. 46).   Forma **taènia** Fassett has a rosette of submersed ribbon-like leaves distinguished from those of *Sagittaria* and *Sparganium* by their venation (Fig. 49); emersed leaves with very narrow blades may also be present (Fig. 47).   Forma **albiflòra** (Raf.) House has white flowers.

MUD PLANTAIN   **Heteranthèra**

1. Leaves ribbon-like 2
1. Leaves with broad blades 3
   2. Stems and leaves long, flexible, trailing through the water.   **H. dùbia** (Jacq.) MacM.   Water Star Grass.   Fig. 50.   Submersed in still or flowing water; Que. & w. N.E. to Minn. & Ore., s. to Cuba & Mex.   Closely resembles a *Potamogeton,* from which it may be distinguished, in absence of its yellow star-like flowers, by the almost complete lack of midvein in the leaves.
   2. Stems and leaves short and somewhat rigid (Fig. 52).   **H. dubia** forma **terréstris** (Farwell) Vict.   A form appearing when stranded on the mud; this is more apt to flower than is the aquatic form.
3. Leaf blades about as broad as long, heart-shaped at base (Fig. 51).   **H. renifórmis** R. & P.   Mud and shallow water; Conn. to Neb., s. to S.A.
3. Leaf blades longer than broad, blunt at both ends (Fig. 53).   **H. limòsa** (Sw.) Willd.   Va. to Neb., s. to S.A.
   Eichòrnia cràssipes (Mart.) Solms, the Water Hyacinth, is recorded as introduced in s.e. Mo. & Va.

References: Fernald, Rhodora, **27**, 80 (1925)—*P. cordata* f. *latifolia* and f. *angustifolia;* Victorin, Les Liliiflores du Québec.   Contrib. Lab. Bot. Univ. Montréal, No. 14, 32 (1929)—*H. dubia* f. *terrestris;* Fassett, Rhodora, **39**, 274 (1937)—*P. cordata* f. *taenia.*

RUSH FAMILY   JUNCÀCEAE

Rush   Júncus

Plants looking like grasses or sedges, and frequently mistaken for such. The flowers (Fig. 1), however, are not grass-like, but resemble those of a lily. They have 3 sepals, 3 petals, 3 or 6 stamens and a capsule, as do the lilies, but the sepals and petals are narrow, stiff and scale-like. The leaf does not have a ligule at the junction of blade and sheath, as is characteristic of a grass leaf (see Fig. 1, p. 9).

The species are many and often distinguished with some difficulty, and there are occasional hybrids. Most easily recognized is the Soft Rush, *J. effusus*, which occurs in large clumps of several hundred stems (Fig. 3), and is abundant in meadows in many regions. The true aquatics are *J. militaris*, *J. repens*, *J. subtilis* and dwarfed forms of *J. marginatus* and *J. pelocarpus;* the others are littoral plants sometimes entering shallow water.

1. Inflorescence apparently borne on the side of the stem, which appears to continue beyond it (Figs. 3, 8), the continuing part being actually an involucral leaf.....**2**

1. Inflorescence obviously terminating the stem, with a more or less conspicuous involucral leaf coming from its base (Figs. 14, 15, 24, 28)....................**5**

    2. Sheaths at the base of the stem (Fig. 3) not terminated by blades..........**3**

    2. Sheaths terminated by a cylindrical blade about 1 mm. thick. **J. coriàceus** Mack. Low ground; Del. to the Gulf of Mexico. (*J. setaceus*, G, B.)

3. Stems borne in large clumps (Fig. 3). **J. effùsus** L. Soft Rush. Fig. 2. Meadows and lake shores, pools, etc.; widespread, but sometimes locally absent. Variable in size of flowers, compactness of inflorescence and diameter of stems.

3. Stems in a row, from underground rootstocks (Fig. 8).....................**4**

    4. Flowers brown, 3.5–5 mm. long; involucral leaf much shorter than the stem. **J. bálticus** Willd., var. **littoràlis** Engelm. Fig. 8. Salt marshes along the Atlantic Coast, s. to Pa.; inland in brackish places and on sandy shores about the Great Lakes, and w. to the Pacific.

    4. Flowers green, 2–3 mm. long; involucral leaf nearly or quite as long as the stem below the inflorescence. **J. filifórmis** L. Wet shores, etc.; Labrador to B.C., s. to Pa., Mich., Wis., Minn., Utah & Colo.

5. Leaves without hard cross partitions.......................................**6**

5. Leaves with hard cross partitions at regular intervals (Figs. 24, 31).........**11**

    6. Flowers many in close heads (Figs. 5, 6); leaves flat, usually 1.5–2 mm. wide..**7**

    6. Flowers borne singly or in loose heads (Figs. 9, 12); leaves 1 mm. or less wide, usually closely rolled...**8**

7. Flowers 3.5 mm. long; heads nearly spherical (Fig. 5); stems erect. **J. marginàtus** Rostk. Sandy shores and shallow water; N.S. to Ont. & Neb., s. to Fla. Small sterile plants, with a stout hard black rootstock covered with remains cf leaf bases (Fig. 7), may be found in shallow water.

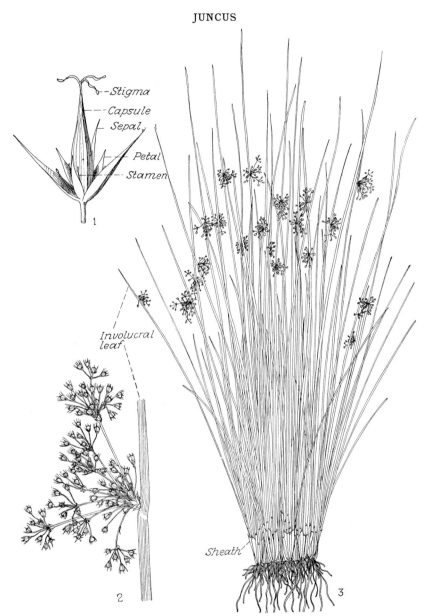

Stigma
Capsule
Sepal
Petal
Stamen

1

Involucral leaf

Sheath

2

3

1. A flower, generalized, × about 10.   **J. effusus:** 2. Inflorescence, × 1.   3. Plant, × ¼.

J. **Gerardi**: 4. Plant, × ½. **J. marginatus**: 5. Inflorescence, × 1. 7. Sterile submersed plant, × ½. **J. repens**: 6. Branch, × ½. **J. balticus** var. **littoralis**: 8. Plant, × ½.

JUNCUS

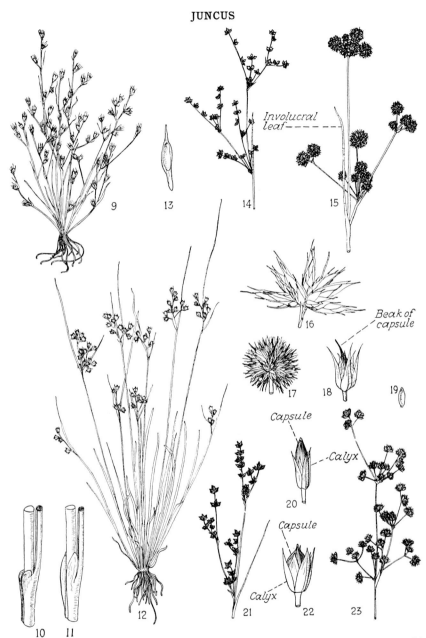

J. bufonius: 9. Plant, $\times \frac{1}{2}$.　J. macer: 11. Summit of sheath, $\times 5$.　12. Plant, $\times \frac{1}{2}$.
J. Dudleyi: 10. Summit of sheath, $\times 5$.　J. canadensis: 13. Seed, $\times 10$.　15. Inflorescence,
$\times \frac{1}{2}$.　J. brachycephalus: 14. Inflorescence, $\times \frac{1}{2}$.　J. brevicaudatus: 16. Insect gall, $\times 1$.
21. Inflorescence, $\times \frac{1}{2}$.　J. scirpoides: 17. Head, $\times 2$.　18. Flower, $\times 4$.　J. acuminatus:
19. Seed, $\times 10$.　22. Flower, $\times 4$.　23. Inflorescence, $\times \frac{1}{2}$.　J. debilis: 20. Flower, $\times 4$.

177

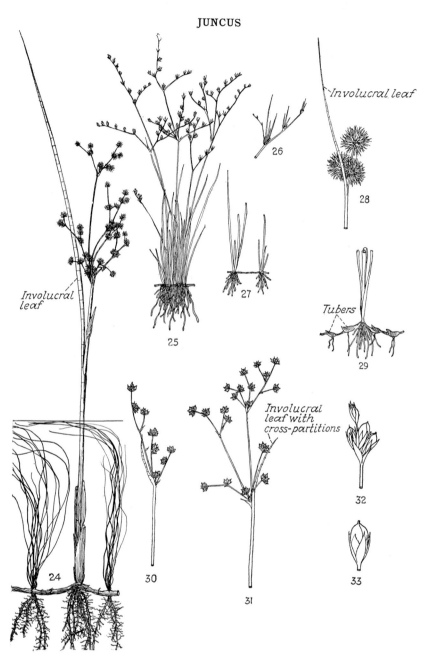

J. militaris: 24. Plant, × ¼. **J. pelocarpus:** 25. Plant, × ½. 26. Branch bearing bulblets, × 2. **J. pelocarpus f. submersus:** 27. Plant, × ½. **J. Torreyi:** 28. Inflorescence, × ½. 29. Base of plant, × ½. **J. alpinus:** 30. Inflorescence, × ½. **J. alpinus** var. **rariflorus:** 32. Head of flowers, × 3. **J. articulatus:** 31. Inflorescence, × ½. **J. articulatus** var. **obtusatus:** 33. Flower, × 4.

7. Flowers 5–10 mm. long, mostly erect or spreading, so that the heads are not spherical (Fig. 6); stems partly creeping or floating. **J. rèpens** Michx. Muddy shores; Del. to Cuba & Tex.; Lower Calif.

   8. Inflorescence a third or more the height of the plant (Fig. 9); plants annual, with soft base. **J. bufònius** L. Toad Rush. Common on damp roadsides, muddy shores, etc.; very variable in size.

   8. Inflorescence a fifth or less the height of the plant; perennials with hard bases or rootstocks..**9**

9. Flowering stem with 1 or 2 leaves (Fig. 4); stems from a rootstock. **J. Gerárdi** Loisel. Black Grass. Salt marshes along the coast, Nfd. to Fla.; inland locally in n. N.E., N.Y., and about the Great Lakes; Pacific Coast.

9. Flowering stems without leaves except at base (Fig. 12)....................**10**

   10. Summit of sheath extended into 2 papery lobes (Fig. 11). **J. màcer** S. F. Gray. Fig. 12. Very common, in wet or dry soil. (*J. tenuis*, G, B.)

   10. Summit of sheath extended into 2 short rounded firm copper-colored lobes (Fig. 10). **J. Dudlèyi** Wiegand. Dry fields, swamps and lake shores; Nfd. to Sask., s. to Va., Tenn., Kans. & Mex.

11. Stamens 3, one behind each sepal.......................................**12**

11. Stamens 6, one behind each sepal and one behind each petal...............**18**

   12. Seeds long-tailed at each end (Fig. 13)...............................**13**

   12. Seeds not long-tailed (Fig. 19).....................................**15**

13. Flowers 2.5 mm. long. **J. brachycéphalus** (Engelm.) Bucheneau. Fig. 14. Wet shores; n. Maine to Wis., s. to Conn., Pa. & Ill.

13. Flowers or capsules 4 mm. long.......................................**14**

   14. Heads nearly spherical, some on spreading rays (Fig. 15). **J. canadénsis** J. Gay. Marshy places and shallow water; Nfd. to Minn., s. to Ga. & La. Insect galls (Fig. 16) are frequent on this species and on the next.

   14. Heads narrow, on ascending rays (Fig. 21). **J. brevicaudàtus** (Engelm.) Fernald. Common on lake shores and margins of pools; Nfd. to Minn., s. to Conn., Pa. & W.Va.

15. Heads spherical, with the lower flowers reflexed (Fig. 17); capsule tipped by a slender beak (Fig. 18). **J. scirpoìdes** Lam. Wet sand; N.Y. to Fla., Wis. & Mo.

15. Heads hemispherical, the lower flowers spreading (Fig. 23); capsule abruptly narrowed to a short tip (Fig. 22).......................................**16**

   16. Heads 1–50 on each plant; flowers 3–3.5 mm. long.....................**17**

   16. Heads 200–500 on each plant; flowers 2–2.5 mm. long. **J. nodàtus** Coville. Swamps; Ill. to La. & Tex. (*J. robustus*, G.)

17. Capsule about equaling the calyx (Fig. 22). **J. acuminàtus** Michx. Fig. 23. Wet places; N.S. to Minn., s. to Ga. & Tex.

17. Capsule longer than the calyx (Fig. 20). **J. débilis** Gray. Wet sandy shores; R.I. to Ark., s. to the Gulf of Mexico.

   18. Flowers borne singly or in pairs (Fig. 25), or partly replaced by bulblets (Fig. 26)...**19**

   18. Flowers in heads (Figs. 28, 31).....................................**20**

19. Stem erect from a rootstock. **J. pelocárpus** Mey. Figs. 25, 26. Sandy lake shores; Nfd. to N.J., w. to Minn. **J. pelocarpus** forma **submérsus** Fassett, a submersed sterile form in which the cross markings of the leaves are more scattered and often incomplete (Fig. 27), occurs with the typical form or sometimes by itself.

19. Stem creeping or floating, with little bunches of leaves.   **J. subtìlis** Mey.   Creeping Rush. Maine, and northward.

    20. Heads spherical, the lower flowers reflexed (Fig. 28); rootstocks bearing tubers (Fig. 29) **21**

    20. Heads hemispherical, the lowest flowers spreading or ascending (Fig. 31) . . **22**

21. Flowers 3–4 mm. long.   **J. nodòsus** L.   Muddy and gravelly shores; Nfd. to B.C., s. to Va., Ill. & Neb.

21. Flowers 4–5 mm. long.   **J. Torrèyi** Coville.   Figs. 28, 29.   Wet ground; Mass. to Sask. & Wash., s. to Ala., Tex. & Ariz.

    22. Stems stout, the middle leaf overtopping the inflorescence (Fig. 24).   J. **militàris** Bigel.   Bayonet Rush.   Shallow water; N.S. to n. N.Y., s. to Md.; n. Mich.   The stout rootstocks often produce quantities of submersed thread-like leaves.

    22. Stems slender, with leaves not overtopping the inflorescence **23**

23. Branches of the inflorescence ascending (Fig. 30) **24**

23. Branches of the inflorescence spreading (Fig. 31) **25**

    24. None of the flowers stalked within the heads (Fig. 30).   **J. alpìnus** Vill.   Sandy shores; northern regions, s. to Maine, N.Y. & Wis.   (*J. alpinus* var. *fuscescens*, G.)

    24. One or more flowers in most heads on a slender stalk (Fig. 32).   **J. alpinus** var. **rariflórus** (Fries) Hartm.   Similar situations.   (*J. alpinus* var. *insignis*, G.)

25. Sepals and petals brown; capsule tapering to the tip.   **J. articulàtus** L.   Fig. 31.   Wet places; Nfd. to Mich. & B.C., s. to Mass. & N.Y.

25. Sepals and petals green; capsule abruptly narrowed to a small tip (Fig. 33).   J. **articulatus** var. **obtusàtus** Engelm.   N.S. to N.J.

References: Fassett, Trans. Wis. Acad., **25**, 161 (1930)—*J. pelocarpus* f. *submersus;* Fernald, Journ. Bot., **68**, 364–367 (1930)—*J. macer;* Mackenzie, Bull. Torr. Bot. Club, **56**, 28 (1929)—*J. coriaceus;* Fernald, Rhodora **35**, 233–235 (1933)—*J. alpinus* var *rariflorus.*

IRIS FAMILY   IRIDÀCEAE

Ìris

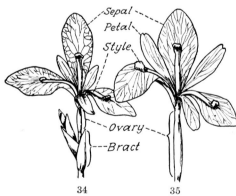

Perennial from stout underground rootstocks; leaves flat, sword-like, usually erect; flowers large and conspicuous; sepals and petals borne above the ovary; sepals larger than the petals; styles flat, lying close to the sepals.

I. versicolor: 34. Flower, × ⅔.   **I. virginica** var. **Shrevei:** 35. Flower, × ⅔.

1. Flowers blue...2
1. Flowers brown or yellow...6
  2. Flowers overtopping most or all of the leaves.........................3
  2. Leaves much overtopping the flowers. **I. foliòsa** Mack. & Bush. Ky. to Ill., Mo. & Kans. (*I. hexagona*, G.)
3. Leaves 0.5–5 cm. wide; ovary and seed pod bluntly angled...................4
3. Leaves 3–7 mm. wide; ovary and seed pod sharply 3-angled. **I. prismática** Pursh. Wet ground near the coast; N.S. to Ga.
  4. Petals shorter than the styles; sepals roughened at the base of the blade with minute bumps, which are scarcely visible to the naked eye, and with a dull greenish-yellow spot; outermost bracts with margins darkened and somewhat varnished; seeds regularly pitted and as if varnished. **I. versícolor** L. Fig. 34. Blue Flag. Meadows and shores, usually in drier places than the next; Labrador & James Bay to Va., s. to n.e. Ohio, n. Mich., n. Minn. & Man.
  4. Petals longer than the styles; sepals hairy at the base of the blade, with a bright yellow spot; outermost bracts with margins green and soft; seeds irregularly, or not at all pitted, and dull...5
5. Seed pods about as broad as long. **I. virgínica** L. Wet meadows and shores, often in shallow water; e. Va. to the Gulf of Mexico.
5. Seed pods much longer than broad. **I. virginica** var. **Shrèvei** (Small) Anderson. Fig. 35. Minn. & s. Ont., s. to W.Va., n. Ala. & n. Ark.
  6. Flowers copper-colored or reddish-brown. **I. fúlva** Ker. S. Ill. & Mo. to Tex. & Ga.
  6. Flowers yellow. **I. pseudácorus** L. Marshes; Mass. to Va. and Wis.; naturalized from Europe.

Reference: Anderson, Ann. Mo. Bot. Gard., **23**, 459–469 (1936)—*I. versicolor* and *I. virginica*.

PEPPER FAMILY   **PIPERÂCEAE**

Lizard's Tail   **Saurùrus**

S. cérnuus L. Stems 0.5–1.5 m. tall, from slender rootstocks; leaves scattered, heart-shaped; flowers white, in slong slender spikes which often droop near the tip. Wet woods, muddy shores, and shallow water; R.I. to s. Ont., Ind., Ill., Mo. & s.e. Kans., s. to Fla. & Tex.

WILLOW FAMILY **SALICÀCEAE**

WILLOW **Sàlix**

Shrubs or trees, with simple alternate leaves several times as long as wide (in our species of wet soil); flowers in catkins (Figs. 21, 45), which in many cases appear before the leaves, the staminate (male) and pistillate (female) catkins on separate plants; staminate catkins with many flowers, each of which consists of 2 to several stamens borne back of a scale (Figs. 35, 40); pistillate catkins with many flowers, each of which consists of a pistil in the axil of a scale (Figs. 23, 25); in each sex the scales are often hairy, the many overlapping silky hairs of the expanding catkins making the familiar "Pussies." The species are difficult to determine, showing much variation in width of leaf and in hairiness; furthermore, hybrids between species are common. The following key is arranged to deal with mature foliage, or with well-developed catkins of either sex.

Frequently there are found on willow twigs structures that look like little pine cones; they are insect galls and will be found at some seasons to contain living grubs.

1. Plants with mature leaves **2**

1. Plants with catkins **26**

2. Leaf margins with sharp teeth **3**

2. Leaf margins without teeth (Figs. 17, 18, 20), or with indentations (Fig. 19) . . **21**

3. Leaves strongly whitened beneath **9**

3. Leaves green beneath **4**

4. Margins with 3–5 teeth per centimeter **5**

4. Margins with 8–12 teeth per centimeter **6**

5. Leaves 1 cm. or less wide, with few hairs or none. **S. longifòlia** Muhl. Sand-bar Willow. Figs. 2, 5, 33, 36. Usually a shrub, rarely a small tree, on river banks and lake shores, often the first woody plant to appear on sand bars; N.B. to B.C., s. to La., Tex., Kans., N.D. & Man.; rather local eastward, very common in the interior.

5. Leaves more than 1 cm. wide, more or less silky beneath. **S. longifolia** forma **Wheelèri** (Rowlee) Schneid. Fig. 3. Of similar range. (*S. Wheeleri*, B.)

6. Leaves glossy above, abruptly narrowed to a long tip (Fig. 6) **7**

6. Leaves not glossy, gradually tapered to the tip **8**

7. Leaves without hairs beneath. **S. lùcida** Muhl. Shining Willow. Figs. 6, 23. Usually a shrub, sometimes a small tree, on lake shores, stream banks and in swamps; Nfd. to Man., s. to N.D., Iowa & Del.

7. Leaves with reddish hairs beneath. **S. lucida** var. **intònsa** Fernald. N.B. to Ind.

8. Leaves often curved (Fig. 7), as dark green beneath as above. **S. nìgra** Marsh. Black Willow. Figs. 7, 34, 38. Shrubs, or often large trees, on stream banks; N.B. to N.D. & the Gulf of Mexico.

8. Leaves essentially straight (Figs. 12–14), lighter green beneath **20**

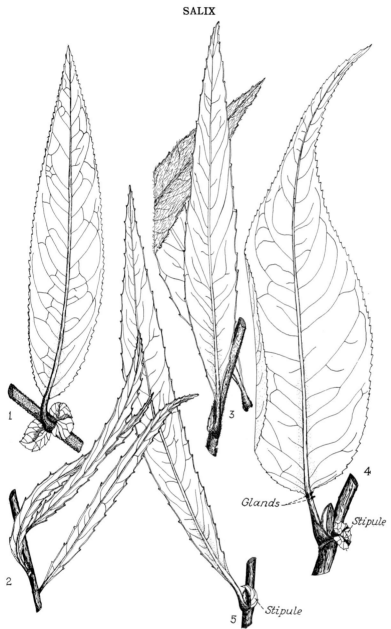

Leaves and portions of twigs, × 1.   1. **S. longipes.**   2, 5. **S. longifolia.**   3. **S. longifolia** f. **Wheeleri.**   4. **S. serissima.**

Leaves and portions of twigs, ✕ 1.   6. **S. lucida.**   7. **S. nigra.**   8. **S. amygdaloides.**

9. Petioles with conspicuous glands at tip (Fig. 4). **S. seríssima** (Bailey) Fernald. Autumn Willow. Figs. 4, 21, 35. A shrub in swamps and bogs; Nfd. & James Bay to Alberta & Mont., s. to Mass., n. Ind. & Colo.

9. Petioles without glands, or with obscure small glands......................**10**

   10. Leaves 7–15 cm. long, narrowed from below the middle.................**11**

   10. Leaves 10 cm. or less long, generally narrowed from above the middle....**12**

11. Leaves evenly narrowed from the broadest portion to the tip (Fig. 1), with petioles 2–7 mm. long; stipules 5–15 mm. broad, persistent. **S. lóngipes** Shuttleworth. Ward's Willow. Fig. 40. A small tree growing along stream beds where often covered at high water; D.C. to s. Ill., s.e. Kans., & s. to Cuba & Tex. (*S. Wardii*, G, B.)

11. Leaves tapered to a long whip-like tip (Fig. 8), with petioles 1–3 cm. long; stipules about 1 mm. broad, soon falling. **S. amygdaloìdes** Anders. Peach-leaved Willow. Figs. 29, 39. A tree growing on the borders of streams and lakes, often with the Black Willow, with which it hybridizes; Que. to Wis., Nev. & B.C., s. to N.Y., Mo. & Tex.

   12. Leaves with the entire lower surface covered with close silky hairs. **S. serícea** Marsh. Silky Willow. Figs. 9, 18, 27, 37. Streams, springs, etc.; N.B. to Iowa, s. to S.C. & Mo. Often the same plant may have both finely toothed leaves and leaves without teeth.

   12. Leaves without hairs, or with spreading hairs beneath.................**13**

13. Twigs and leaves thinly silky-hairy or without hairs.......................**14**

13. Twigs and lower leaf surfaces thickly covered with white velvety hairs. **S. missouriénsis** Bebb. Fig. 15. A small tree, along the Mississippi River and its tributaries; Ill. & Ky. to Iowa & Neb.

   14. Leaves light-green beneath and not strongly veined, often heart-shaped at base (Fig. 12) on the sterile shoots, usually tapered at base (Figs. 13, 14) on the flowering branches..**20**

   14. Leaves strongly whitened beneath and strongly veiny on both sides, tapered at base..**15**

15. Leaves not over 4 times as long as broad (Figs. 10, 11)....................**17**

15. Leaves at least 5 times as long as broad (Fig. 16).........................**16**

   16. Young twigs with silky hairs. **S. sericea** forma **glàbra** Palmer & Steyermark. Mo. & probably elsewhere within the range of *S. sericea;* see **12.**

   16. Young twigs without hairs. **S. petiolàris** Sm. Figs. 16, 30, 46. A shrub in alluvial soils; N.B. to N.D. & Sask., s. to N.J., Iowa & S.D.

17. Leaf margins without teeth, or wavy, or with 1–3 teeth per centimeter (Fig. 10) **18**

17. Leaf margins with 5–10 teeth per centimeter. (Fig. 11)....................**19**

Leaves and portions of twigs, ✕ 1.   9. **S. sericea.**   10. **S. discolor.**

Leaves and portions of twigs, × 1.   11. **S. glaucophylloides** var. **glaucophylla.**   12. **S. cordata**, sterile shoot.   13,14. **S. cordata**, flowering shoot.   15. **S. missouriensis.**   16. **S. petiolaris.**

Leaves and portions of twigs, × 1.   17. **S. pedicellaris.**   18. **S. sericea.**   19. **S. Bebbiana.**
20. **S. candida.**

18. Twigs without hairs. **S. díscolor** Muhl. Pussy Willow. Figs. 10, 32, 42, 45. Usually a shrub, sometimes a small tree, on stream banks and low wet meadows; Nfd. to N.D. & Sask., s. to Del., Ky. & Mo.

18. Twigs with fine spreading hairs. **S. discolor** var. **latifòlia** Anders. In the same regions. (*S. discolor* var. *eriocephala*, G.)

19. Leaves of rather thin texture. **S. glaucophyllòides** Fernald. A shrub on river shores; Nfd. to n. Maine.

19. Leaves of thicker texture; for further differences see **36. S. glaucophylloides** var. **glaucophýlla** (Bebb) Schneider. Blue-leaf Willow. Figs. 11, 22. Sandy shores mostly near the Great Lakes; n. Ohio to cent. Wis. (*S. glaucophylla*, G, B.)

20. Twigs without hairs, or with a few hairs when young; leaves with heart-shaped base common. **S. cordàta** Muhl. Heart-leafed Willow. Figs. 12–14, 26, 43, 48. A shrub along streams, etc.; N.B. to Man., s. to Md. & Kans. Variable and freely hybridizing with other species.

20. Twigs permanently hairy; leaves narrow, seldom heart-shaped. **S. cordata** var. **myricoìdes** (Muhl.) Carey. Perhaps a hybrid with *S. sericea.*

21. Leaves closely covered beneath with thick wool so that the veins are not visible . **24**

21. Leaves with scattered wool beneath exposing the veins, or without hairs **22**

22. Leaves very smooth above and below, the margin evenly curved with no teeth or indentations (Fig. 17). **S. pedicellàris** Pursh. Figs. 25, 47. A shrub in bogs and wet meadows; Que. to B.C., s. to N.J., n. Iowa, Idaho & Wash.

22. Leaves somewhat roughened beneath by the thickened veins, the margins sometimes with irregular indentations or low teeth (Fig. 19) **23**

23. Leaves white-hairy beneath. **S. Bebbiàna** Sargent. Figs. 19, 31, 41. A shrub; Nfd. to James Bay & Alaska, s. to N.J., n. Ind., Iowa, S.D., N.M. & Calif. (*S. rostrata*, G.)

23. Leaves without hairs beneath, or with a fine sprinkling of hairs which, particularly on the youngest foliage, are reddish. See **S. discolor, 18.**

24. Wool silky, shining **25**

24. Wool cottony, dull. **S. cándida** Flügge. Sage Willow. Figs. 20, 24, 44. A small shrub, mostly in bogs; Nfd. to B.C., s. to N.J., Pa., Ohio, Ind., S.D. & Colo.

25. Hairs on lower leaf surfaces nearly straight and lying flat on the surface; leaf margins but slightly inrolled (Fig. 18); capsule rounded at tip (Fig. 27). See **S. sericea, 12.**

25. Hairs on lower leaf surfaces tangled and erect; margins strongly inrolled; capsule pointed (Fig. 28). **S. pellìta** Anders. A shrub or small tree; Que. to s. Man., s. to Maine, Vt. & Mich.

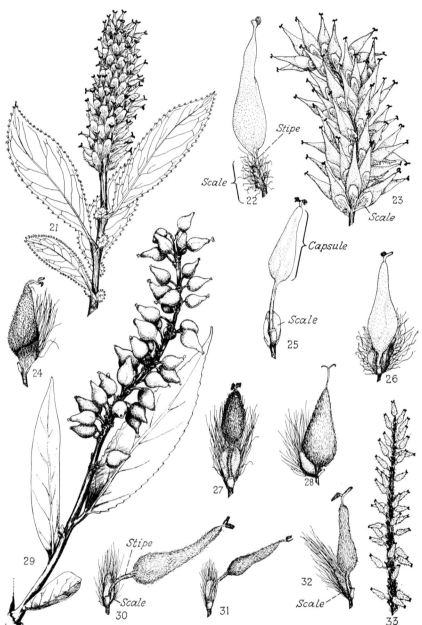

Pistillate catkins, × 2, and flowers, × 4. 21. S. serissima. 22. S. glaucophylloides var. glaucophylla. 23. S. lucida. 24. S. candida. 25. S. pedicellaris. 26. S. cordata. 27. S. sericea. 28. S. pellita. 29. S. amygdaloides. 30. S. petiolaris. 31. S. Bebbiana. 32. S. discolor. 33. S. longifolia.

26. Catkins with pistils or capsules (Figs. 21–33)..........................27
26. Catkins with stamens (Figs. 34–48)...............................42
27. Pistils or capsules hairy...38
27. Pistils or capsules without hairs.......................................28
    28. Scale at the base of each pistil (Figs. 22, 23) light-colored and dropping before the capsule matures..29
    28. Scale dark-brown or reddish at least at tip, persistent...................34
29. Stipules and bases of young leaves margined with white glands (Fig. 21)......30
29. Stipules and leaves not gland margined................................32
    30. Flowering in May and June...31
    30. Flowering in July and August. **S. serissima, 9.**
31. Leaves shining above; stipe shorter than the width of the capsule (Fig. 23). **S. lucida, 7.**
31. Leaves not shining above; stipe longer than the width of the capsule (Fig. 29). **S. amygdaloides, 11.**
    32. Leaves strongly whitened beneath. **S. longipes, 11.**
    32. Leaves green on both sides..33
33. Usually large trees. **S. nigra, 8.**
33. Usually shrubs. **S. longifolia, 5.**
    34. Scales and rachis of catkin densely hairy............................35
    34. Scales and rachis of catkin with few or no hairs (Fig. 25). **S. pedicellaris, 22.**
35. Catkins forming before the leaves.......................................36
35. Catkins forming at the same time as the leaves..........................37
    36. Stipe 1.5 mm. long at the maturity of the capsule, shorter than the scale. **S. glaucophylloides, 19.**
    36. Stipe 2–3 mm. long at the maturity of the capsule, longer than the scale (Fig. 22). **S. glaucophylloides** var. **glaucophylla, 19.**
37. Young twigs and bud scales slightly fuzzy or smooth; fruiting catkins 2.5–6 cm. long. **S. cordata, 20.**
37. Young twigs and bud scales velvety; fruiting catkins 6–10 cm. long. **S. missouriensis, 13.**
    38. Capsule densely woolly with tangled hairs; leaves with dull wool. **S. candida, 24.**
    38. Capsules silky with straight hairs; leaves with shining hairs or none......39
39. Capsule not beaked (Figs. 27, 28). **S. sericea** and **S. pellita, 25.**
39. Capsule tapered to a beak (Figs. 30–32)...............................40
    40. Scales yellow or orange. **S. Bebbiana, 23.**
    40. Scales dark-brown or nearly black..................................41
41. Stipe shorter than the scale (Fig. 32). **S. discolor, 18.**
41. Stipe longer than the scale (Fig. 30). **S. petiolaris, 16.**

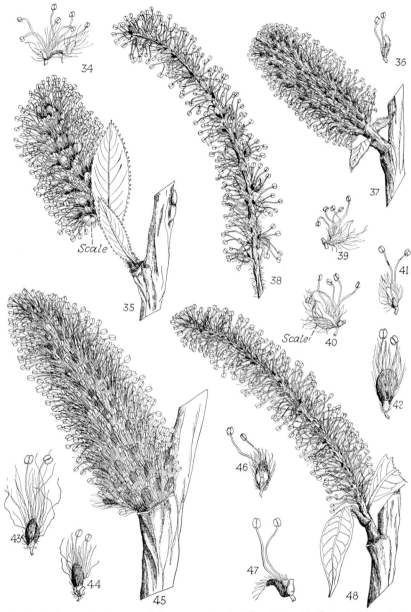

Staminate catkins, $\times$ 2, and flowers, $\times$ 4. 34. **S. nigra**. 35. **S. serissima**. 36. **S. longifolia.** 37. **S. sericea.** 38. **S. nigra.** 39. **S. amygdaloides.** 40. **S. longipes.** 41. **S. Bebbiana.** 42. **S. discolor.** 43. **S. cordata.** 44. **S. candida.** 45. **S. discolor.** 46. **S. petiolaris.** 47. **S. pedicellaris.** 48. **S. cordata.**

42. Each flower with 3–9 stamens (Figs. 34, 39, 40) **43**
42. Each flower with 2 stamens (Figs. 36, 43, 44, 46, 47) **46**
43. Catkins 3–7 cm. long, loosely flowered (Fig. 38) **44**
43. Catkins 1–5 cm. long, densely flowered (Fig. 37) **45**
44. Outer surface of scales with silky hairs nearly or quite to the tip (Fig. 34). **S. nigra, 8.**
44. Outer surface of scales without hairs on the upper half (Figs. 39, 40). **S. longipes** and **S. amygdaloides, 11.**
45. Leaves whitened beneath. **S. serissima, 9.**
45. Leaves green beneath. **S. lucida, 7.**
46. Scales yellowish **47**
46. Scales dark-orange, brown or black **48**
47. Scales covered with long, straight, silky hairs (Fig. 41). **S. Bebbiana, 23.**
47. Scales with a few cottony hairs toward the base, otherwise smooth (Fig. 36). **S. longifolia, 5.**
48. Scales with long silky hairs on the outer side (Figs. 42, 43) **49**
48. Scales nearly or quite without hairs on the outer side, and with a few on the inner side (Fig. 47). **S. pedicellaris, 22.**
49. Twigs with a close cobwebby covering. **S. candida, 24.**
49. Twigs with very short spreading hairs or none **50**
50. Catkins without stalks (Fig. 45). **51,** or **17** which deals with the same species.
50. Catkins on short stalks (Fig. 37) **52**
51. Twigs dark red-brown to black. **S. discolor, 18.**
51. Twigs yellowish or greenish to chestnut-brown. **S. glaucophylloides, 19.**
52. Twigs velvety with shining white hairs. **S. missouriensis, 13.**
52. Twigs not velvety, sometimes with full gray hairs **53**
53. Young leaves silky-hairy. **S. sericea, 12,** and **S. petiolaris, 16.** These species are not distinguishable by staminate flowers alone.
53. Young leaves slightly or not at all silky-hairy. **S. cordata, 20.**

References: Schneider, Journ. Arn. Arb., **3,** 61–125 (1921)—conspectus of North American species; Palmer and Steyermark, Ann. Mo. Bot. Gard., **25,** 769 (1938)—*S. sericea* f. *glabra.*

SWEET GALE FAMILY MYRICÀCEAE

SWEET GALE Myrìca

M. Gàle L. Figs. 49, 50. Shrub with scattered leaves which are wedge-shaped at base and toothed toward the rounded tip, sprinkled beneath with little round resinous dots, which are visible with a lens, and with white hairs; flowers in catkins, coming before the leaves, the staminate (Fig. 50) and pistillate (Fig. 49) on different plants, the pistillate developing nutlets which are covered with resinous dots; whole plant with a spicy odor when bruised. Bogs and wet shores; Lab. to Conn., s. in the mountains to Va., and along the northern Great Lakes to Minn. & Sask. In some regions replaced by **M. Gale** var. **subglàbra** (Chevalier) Fernald, the leaves of which lack the white hairs beneath.

Reference: Fernald, Rhodora, **16,** 167 (1914)—*M. Gale* var. *subglabra.*

M. Gale: 49. Branch with pistillate catkins, × 1.   50. Branch with staminate catkins, before the leaves have appeared, × 1.   **A. crispa** var. **mollis**: 51. Twig with staminate catkins, × 1.   52. Inflorescence of pistillate cones, × 1.   53. Leaf and bud, × 1.

ALNUS

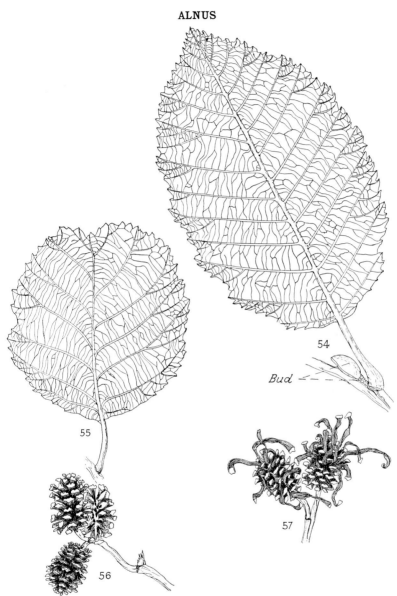

54

Bud

55

56

57

A. incana: 54. Leaf and bud, × 1.   56. Inflorescence of pistillate cones, × 1.   57. Pistillate cones infected with **Taphrina,** × 1.   **A. vulgaris:** 55. Leaf, × 1.

A. rugosa: 58. Leaf and bud, × 1. 60. Pistillate cones, × 1. **P. aquatica:** 59. Twig with leaves and fruit, × 1.

BIRCH FAMILY   BETULÀCEAE

ALDER   Álnus

Large shrubs or small trees; leaves scattered, toothed; flowers in catkins, coming in the spring (except in the fall-flowering *A. maritima*); staminate catkins long and flexible (Fig. 51); pistillate catkins becoming barrel-shaped cones with spreading scales (Fig. 52). All the species are commonly found in swamps, on the shores of lakes and rivers and in the beds of small streams.

1. Leaves dull above; flowers coming in the spring...........................2
1. Leaves glossy above; flowers coming in the fall. **A. marítima** (Marsh.) Muhl. Seaside Alder. Del. & Md. near the coast, and along the Red River in s. Okla.
    2. Leaves with crowded sharp teeth (Fig. 53); buds not stalked, the scales unequal in length; catkins opening at the same time as the leaves; pistillate cones on slender stalks which fork at a narrow angle (Fig. 52).......................3
    2. Leaves with less crowded blunt teeth (Figs. 54, 55, 58); buds stalked, with scales nearly equal in length (Fig. 54); catkins opening before the leaves; pistillate cones on short stalks which fork almost at right angles (Figs. 56, 60)........4
3. Leaves without hairs beneath. **A. críspa** (Ait.) Pursh. Green Alder. Shores and mountains, mostly in cold regions; Labrador to Alaska, s. to N.B. and on the higher mountains to N.C.; in n. Mich., Wis., Minn. & B.C. (*A. Alnobetula*, B.)
3. Leaves velvety beneath. **A. crispa** var. **móllis** Fernald. Downy Alder. Figs. 51–53. More common eastward and in less alpine regions; Nfd. to s. Maine & w. Mass.; in the same regions as the Green Alder about the Great Lakes. (*A. mollis*, G.)
    4. Buds, young catkins and leaves (especially when unfolding) very sticky; leaves nearly as broad as long, rounded or slightly indented at tip (Fig. 55). **A. vulgàris** Hill. European Alder. Sometimes escaping from cultivation.
    4. Buds, etc., not sticky, or slightly so in the Smooth Alder; leaves longer than broad, pointed at tip (Figs. 54, 58).....................................5
    The two following species are often infected with a fungus (*Taphrina*) which causes the scales of the pistillate cone to elongate (Fig. 57), becoming pink and fleshy when young and dark-brown and brittle when old.
5. Leaves doubly toothed (Fig. 54), strongly whitened beneath; cones stalked (Fig. 56). **A. incàna** (L.) Moench. Speckled Alder; Tag Alder. Nfd. to Sask., s. to N.Y., Pa., Wis., Iowa & Neb.
5. Leaves simply toothed (Fig. 58), green beneath; some cones often without stalks (Fig. 60). **A. rugòsa** (Ehrh.) Spreng. Smooth Alder. Maine to Ind. & southward; sometimes intergrading with the Speckled Alder.

Reference: Fernald, Rhodora, **15**, 44 (1913)—*A. crispa* var. *mollis*.

ELM FAMILY   URTICÀCEAE

WATER ELM   Planèra

P. **aquática** (Walt.) Gmel. Fig. 59. Small tree, often in shallow water; leaves elm-like, with low teeth and often with an unsymmetrical base; fruit enclosed in a bur-like husk. Swamps and river bottoms in the Mississippi Valley from Ky. & Mo. s. to the Gulf of Mexico.

BUCKWHEAT FAMILY   POLYGONÀCEAE

The most conspicuous vegetative character of this family is the sheath at the base of each petiole (Figs. 3, 12, 14). The flowers, in our aquatic and subaquatic forms at least, are in spikes or whorls and have a calyx of 4–6 sepals which are green, white or colored and which persist about the fruit; fruit a single nutlet, 3-sided or 2-sided. Some of the species are erect plants growing on shores; others trail through the water with floating leaves resembling those of Pondweeds, from which they may be distinguished by the fact that the leaves have a network of veins branching from the midrib, and there is a sheath at the base of the petiole. The species are difficult to determine in the absence of mature fruit.

1. Leaves finely toothed or crinkled (except in the Swamp Dock); flowers in whorls (Figs. 4, 6); sepals 6 in number, the 3 inner ones becoming enlarged in fruit, one or more with thickened midribs (Figs. 1, 5). **Rumex.**

1. Leaves not toothed; flowers in spikes (Figs. 10, 20, 23, 31); sepals 4–5, not enlarging in fruit. **Polygonum.**

Dock   Rùmex

1. Margins of sepals fringed (Fig. 5)..**3**
1. Margins of sepals not fringed (Fig. 1)...................................**2**
    2. Leaves crinkled or finely toothed on the margins (Fig. 2), often tinged with red; pedicels about twice as long as the mature sepals (Fig. 1). **R. Británnica** L. Great Water Dock. Swamps; Nfd. to Ont. & Minn., s. to N.J. & Kans.
    2. Leaves not crinkled or toothed (Fig. 7), not red-tinged; pedicels 3–4 times as long as the mature sepals (Fig. 4). **R. verticillàtus** L. Swamp Dock. Marshes, in rich soil; Que. & Vt. to Wis. & Iowa, s. to Fla. & Tex.
3. Leaves 1.5 times as long as broad (Fig. 3); fringes not so long as the width of the body of the sepal. **R. obtusifòlius** L. Bitter Dock. A weed of rich soils, barnyards, ditches, etc.; naturalized from Europe.
3. Leaves 5–10 times as long as broad (Fig. 6); fringes longer than the width of the body of the sepal (Fig. 5). **R. marítimus** L., var. **fuegìnus** (Phil.) Dusen. Mostly on brackish shores near the sea; Que. to R.I.; and along lakes and rivers; Wis. to B.C., s. to n. Ill., Kans. & Lower Calif. (*R. persicarioides*, in part, G, B.)

Smartweed   Polýgonum

The species here described all have the flowers in the axils of bracts (which resemble the sheaths at the bases of petioles), arranged in spikes at the tips of branches. Other species (the Matweeds), not normally found in the water, have them in the axils of foliage leaves. The Smartweeds are by some authors put in a separate genus, *Persicaria;* they are so treated by Britton and Brown.

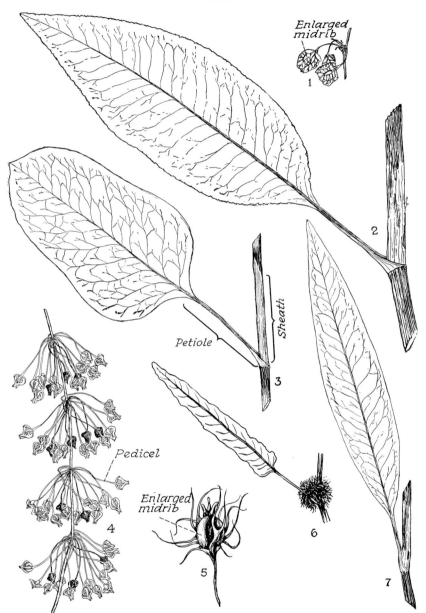

R. **Britannica**: 1. A whorl of fruits, $\times$ 1. 2. Leaf, $\times \frac{1}{2}$, with torn remains of sheath.
R. obtusifolius: 3. Leaf, $\times \frac{1}{2}$. **R. verticillatus**: 4. Part of inflorescence, $\times$ 1. 7. Leaf,
$\times \frac{1}{2}$. **R. maritimus** var. **fueginus**: 5. Fruit surrounded by sepals, $\times$ 5. 6. Leaf and whorl
of flowers, $\times \frac{1}{2}$.

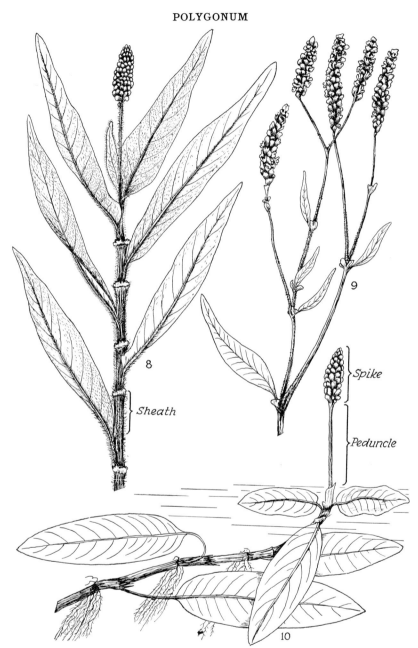

Spike

Peduncle

Sheath

8

9

10

P. natans f. Hartwrightii: 8. Upper part of plant, × ½.   P. densiflorum: 9. Upper part of plant, × ½.   P. natans f. genuinum: 10. A partly submersed branch, × ½.

1. Plants perennial, with stems prostrate and rooting at the nodes for 1 dm. or more, trailing through the water (Fig. 10) or sprawling on the mud (Fig. 20)........2
1. Plants annual, with stem nearly or quite erect from a short taproot (Figs. 32, 33), or with roots from 2 or 3 lower nodes...................................12
  2. Spike 8–14 mm. thick, very dense (Figs. 8, 13).........................3
  2. Spike loose and slender (Fig. 20)......................................9
3. Spikes 1 or 2 at the tip of the stem (Figs. 11–13).   **P. natans** and **P. coccineum,** the "Amphibious Polygonums." The same species may appear as a very different plant in different habitats; even in the same habitat there is great variation in shape and size of leaves, nature of hairy covering, etc. Frequent hybrids add confusion..4
3. Spikes several at the tips of branches (Fig. 9).   **P. densiflòrum** Meisn. Wet soil; s. N.J. to W.I., S.A., Tex. & s. Mo.   (*P. portoricensis*, B.)
  4. Sheaths with a spreading green border (Fig. 8).   **P. nàtans** forma **Hartwríghtii** (Gray) Stanford. This is a terrestrial form of *P. natans*, sometimes growing by itself in damp or dry places, and sometimes occurring as part of the same plant as the aquatic *P. natans* forma *genuinum*, when an individual grows partly in the water and partly on the shore.   (*P. amphibium* var. *Hartwrightii*, G.)
  4. Sheaths without a green border (Fig. 12)...............................5
5. Peduncle without hairs (Fig. 10); spike ovoid in shape, 1–5 cm. long.   **P. nàtans** Eaton, forma **genuìnum** Stanford. An aquatic form, with leaves ordinarily floating; Nfd. to Minn., Sask., Mont. & Wash., s. to Conn., Pa. & Calif.   (*P. amphibium*, G; *Persicaria amphibia*, B.)
5. Peduncle hairy (Fig. 11); spike cylindrical in shape, 3–18 cm. long............6
  6. Plants with floating leaves and stems....................................7
  6. Plants with leaves in the air, and stems erect or ascending or on the ground. .8
7. Leaves mostly heart-shaped at base (Fig. 11); stem 2–5 mm. thick.   **P. coccíneum** Muhl. forma **nàtans** (Wieg.) Stanford. Not very common; probably with the same range as f. *terrestre*.
7. Leaves rounded at base (Fig. 12); stems 5–15 mm. thick.   **P. coccineum** var. **rigídulum** (Sheldon) Stanford. Rare; has been found in Ont., Minn., Man. & S.D.
  8. Petioles attached near the middle of the sheath (Fig. 14).   **P. coccineum** var. **pratíncola** (Greene) Stanford. Fig. 13. Ind., Wis. & N.D. to Wash., s. to Tex. & Mex.
  8. Petioles attached near the base of the sheath (Fig. 15).   **P. coccineum** forma **terréstre** (Willd.) Stanford. Que. to Mont. & Wash., s. to Va., Ark. & Calif.; less common in the Middle West than var. *pratincola*.   (*P. Muhlenbergii*, G.)
9. Sepals gland-dotted (Fig. 30); leaves sometimes as much as 4 cm. wide.......23
9. Sepals not gland-dotted; leaves mostly narrower. The following 3 species are often difficult to differentiate. See also **P. Persicaria, 22,** which sometimes roots at the lower nodes..10
  10. Leaves with nearly parallel sides, not long pointed (Fig. 20), usually slightly or not at all hairy, sometimes covered with little stiff hairs..............11
  10. Leaves tapering to both ends from a point well below the middle (Fig. 16), covered with close stiff hairs.   **P. setàceum** Baldw. Mass. to N.Y., s. to Okla. & Gulf of Mexico.

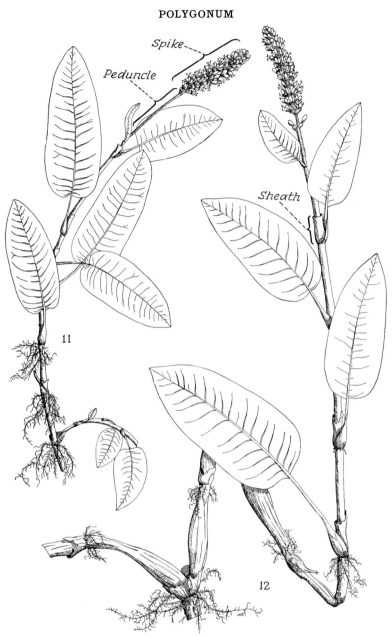

Spike

Peduncle

Sheath

11

12

P. coccineum f. **natans :** 11. Branch, × ½. **P. coccineum** var. **rigidulum :** 12. Branch, × ½.

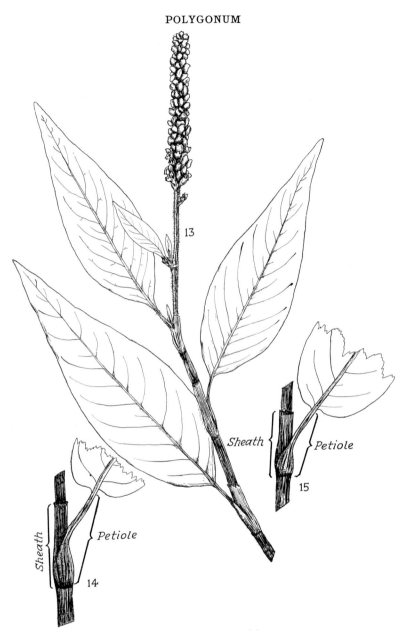

P. **coccineum** var. **pratincola**: 13. Part of a branch, $\times \frac{1}{2}$.　14. Structures at a node, $\times$ 1.
P. **coccineum** f. **terrestre**: 15. Structures at a node, $\times$ 1.

POLYGONUM

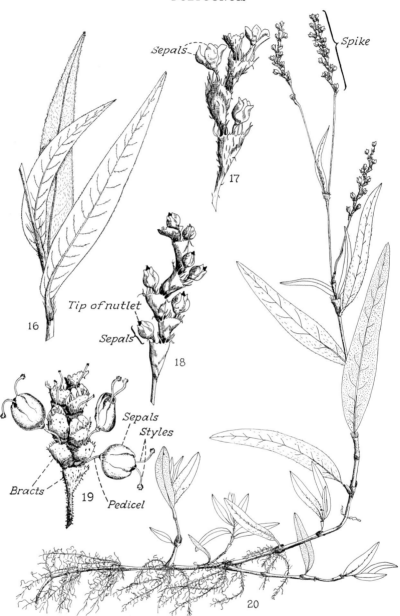

P. setaceum: 16. Part of stem, $\times \frac{1}{2}$. **P. hydropiperoides:** 17. Part of spike, $\times 3$. 20. Plant, $\times \frac{1}{2}$. **P. opelousanum:** 18. Part of spike, $\times 3$. **P. longistylum:** 19. Part of spike, $\times 3$; in this individual the styles are long and the stamens short; in others the stamens are long and the styles short.

11. Spike loose, the white or pink sepals petal-like and completely hiding the mature nutlet (Fig. 17). **P. hydropiperoìdes** Michx. Fig. 20. Mild Water Pepper. Wet soil and in shallow water; N.S. to Minn. & Calif., s. to the Gulf of Mexico.

11. Spike more dense, the green sepals not hiding the tip of the mature nutlet (Fig. 18). **P. opelousànum** Riddell. Wet sand and peat; Mass. & N.J., s. to the Gulf of Mexico, n. in the Mississippi Valley to Mo. & Ill.

> **12.** Stem with glands (Figs. 21, 27), at least just below the spike............**13**
> **12.** Stem without glands...**20**

13. Styles (Fig. 19) or stamens conspicuously longer than the sepals. **P. longistỳlum** Small. S. Ill. & Mo. to Kans., s. to La. & N.M.

13. Styles and stamens hidden by the sepals....................................**14**

> **14.** Glands not stalked, appearing as a gummy covering (Fig. 21); leaves usually white-woolly beneath. **P. scàbrum** Moench. Fig. 22. Low ground, shores, etc.; Nfd. to B.C., s. to N.J., the Great Lakes & Calif. (*P. tomentosum*, G.)
> **14.** Glands stalked (Fig. 25); leaves with close nearly straight hairs or none. .**15**

15. Sheaths fringed with bristles (Fig. 24); lower part of stem with hairs about 1 mm. long (Fig. 26); stalked glands 0.5 mm. long (Fig. 25); flowers 1.5–2.5 mm. long. **P. Carèyi** Olney. Swamps, clearings and moist roadsides; Maine to Wis., s. to Pa.

15. Sheaths not fringed with bristles (Fig. 23); lower part of stem without hairs; stalked glands about 0.1 mm. long (Fig. 27); flowers 3–4 mm. long...........**16**

> **16.** Leaves nearly or quite without hairs beneath; nutlets 2.5–3.5 mm. broad. .**17**
> **16.** Leaves hairy beneath; nutlets 2.2–2.8 mm. broad. **P. pensylvánicum** L., var. **genuìnum** Fernald. Wet places; Mass. to the Gulf of Mexico, n. in the Mississippi Valley to Mo.; s. Ont.; n. Ind.

17. Stem erect; leaves pointed..**18**

17. Stem depressed; leaves oval, blunt. **P. pensylvanicum** var. **nesóphilum** Fernald. Nantucket Is., Mass. & Block Is., R.I.

> **18.** Flowers pink to red..**19**
> **18.** Flowers white or nearly so. **P. pensylvanicum** var. **laevigàtum** forma **albíneum** Farwell. Not common.

19. Glands red. **P. pensylvanicum** var. **laevigàtum** Fernald. Figs. 23, 27. The common plant of the interior; N.B. to S.D. & Colo., & southward.

19. Glands yellowish. **P. pensylvanicum** var. **laevigatum** forma **palléscens** Stanford. Sometimes with the preceding, or replacing it locally.

> **20.** Sheaths fringed with bristles (Fig. 31)................................**21**
> **20.** Sheaths not fringed with bristles (Fig. 33)...........................**26**

21. Sepals dotted with dark glands (Fig. 29).................................**23**

21. Sepals not dotted with glands (Fig. 18).................................**22**

> **22.** Spike loose (Fig. 18). **P. opelousanum, 11**; this is often an annual in the northern part of its range.
> **22.** Spike dense (Fig. 28). **P. Persicària** L. Lady's Thumb. Mostly in waste places as a common weed; naturalized from Europe.

23. Nutlet dull (Fig. 29); internodes 2–4 cm. long...........................**24**

23. Nutlet shining; internodes 3–8 cm. long................................**25**

> **24.** Pedicels hidden by the bracts (Fig. 29); nutlets mostly 3–3.5 mm. long. **P. Hydrópiper** L. Water Pepper. Fig. 31. Mostly on damp roadsides and waste places; introduced from Europe.
> **24.** Pedicels extending from the bracts (Fig. 30); nutlets 2–2.5 mm. long. **P. Hydropiper** var. **projéctum** Stanford. Water Pepper. Wet places; Que. to Wis., s. to Okla. & Ga.

P. **scabrum**: 21. Part of stem, × 12, to show glands. 22. Part of branch, × ½. **P. pensylvanicum** var. **laevigatum**: 23. Part of branch, × ½. 27. Part of stem, × 12, to show glands. **P. Careyi**: 24. Part of branch, × ½. 25. Upper part of stem, × 12. 26. Lower part of stem, × 12.

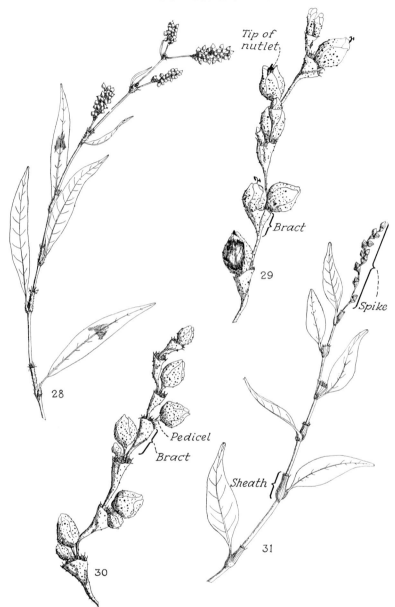

Tip of nutlet

Bract

29

Spike

Pedicel

Bract

30

Sheath

31

28

P. Persicaria: 28. Part of branch, × ½. **P. Hydropiper:** 29. Part of spike, × 3; in the lowermost flower the sepals are broken away to expose the nutlet.   31. Part of branch, × ½. **P. Hydropiper** var. **projectum:** 30. Part of spike, × 3.

Sheath

32

33

P. punctatum: 32. A small plant, $\times \frac{1}{2}$.   **P. lapathifolium:** 33. A small plant, $\times \frac{1}{2}$.

25. Plant annual, not usually prostrate at base, with fibrous roots; stamens 3–8; nutlets mostly flat on one side and rounded on the other. **P. punctàtum** Ell. Water Smartweed, Dotted Smartweed. Fig. 32. Marshes, and often in the water, widespread and common. (*P. acre*, G.)

25. Plant perennial, with tufts of roots from the nodes of the long prostrate base; stamens 8; nutlets mostly 3-angled. **P. punctatum** var. **robústius** Small. Shallow water; N.S. & Mass. to the Gulf of Mexico, and n. in the Mississippi Valley to Mo.

    26. Leaves not woolly beneath. **P. lapathifòlium** L. Fig. 33. Very common and abundant in wet soils.

    26. Leaves woolly beneath. **P. lapathifolium** var. **salicifòlium** Sibth. Nearly as common as *P. lapathifolium*, of which it is perhaps only a juvenile state.

References: St. John, Rhodora, **17**, 81 (1915)—*Rumex maritimus* var. *fueginus;* Stanford, Rhodora, **27**, 109–112, 125–130, 146–152, 156–166 (1925)—the Amphibious Polygonums; 173–184—*P. pensylvanicum* and its varieties and forms; **28**, 22–29 (1926)—*P. hydropiperoides* and *P. opelousanum;* **29**, 77–87 (1927)—*P. Hydropiper* var. *projectum;* Fernald, Rhodora, **23**, 259 (1921)—*P. lapathifolium* var. *salicifolium;* Farwell, Papers Mich. Acad. Sci., **2**, 21 (1923)—*P. pensylvanicum* var. *laevigatum* f. *albineum;* Bicknell, Bull. Torr. Bot. Club, **36**, 452–455 (1909)—notes on *Persicaria pennsylvanica, P. opelousana, P. setacea, P. punctata* and others.

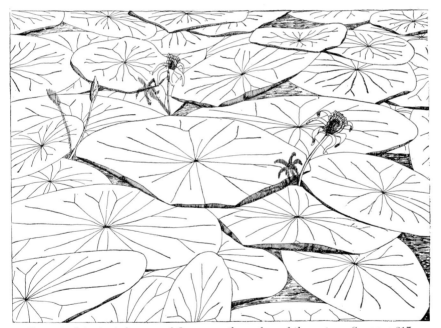

Brasenia Schreberi: leaves and flowers on the surface of the water. See page 217.

HORNWORT FAMILY   **CERATOPHYLLÀCEAE**

COONTAIL   **Ceratophýllum**

C. demérsum L.   Figs. 1–3.   Plants entirely submersed, without roots, appearing olive-green when seen through the water; leaves in whorls, stiff, repeatedly forked, the thread-like divisions with teeth along one side (Fig. 4); leaves more crowded toward the tip, giving the "coontail" appearance; fruits nut-like, borne without stalks in the leaf axils (Fig. 3).   Widely distributed, usually in hard water.   Very variable in length and degree of crowding of leaves.

WATER LILY FAMILY   **NYMPHAEÀCEAE**

1. Leaves simple..**2**
1. Leaves cut into thread-like divisions (Fig. 25).   **Cabomba.**
> **2.** Petiole attached at the summit of a deep notch (Figs. 5, 21); stem horizontal, under the mud, 5 cm. or more thick except in very young individuals........**3**
> **2.** Petioles attached at the middle of the blade (Figs. 23, 26); stem, buried in the mud or trailing through the water, 1 cm. or less thick.......................**4**
3. Flowers white or pink; veins of blade mostly radiating from the summit of the petiole and much forked (Fig. 8).   **Nymphaea.**
3. Flowers yellow; veins mostly coming from the midrib and often running nearly to the margin without forking (Fig. 20).   **Nuphar.**
> **4.** Leaves 2–4 dm. in diameter, coming from a buried rootstock and standing above the surface (Fig. 23) or floating.   **Nelumbo.**
> **4.** Leaves 1 dm. or less in diameter, scattered on a stem which trails through the water (Fig. 26).   **Brasenia.**

WATER LILY   **Nymphaèa**

Underground stem (rootstock) several centimeters thick, fleshy, with little tuberous offshots which detach and form new plants; leaves sometimes as much as 4 dm. wide, floating on the surface or in shallow places slightly elevated above it; flowers floating, the inner petals gradually passing into stamens.   (*Castalia*, G, B.)

1. Flowers 7–23 cm. wide; leaves as broad as long (Fig. 8).......................**2**
1. Flowers 2–5 cm. wide; leaves longer than broad (Fig. 5).   **N. tetragòna** Georgi. Small White Water Lily.   Rare; n. Maine. & Que. to Lake Superior and northwestward; Idaho.
> **2.** Flowers open from 7 A.M. to 1 P.M., fragrant, seldom more than 12 cm. wide, with petals blunt or pointed (Fig. 6); sepals and lower surfaces of leaves often purple; petiole not streaked with purple.   **N. odoràta** Ait.   Nfd. to Man., s. to Fla., La. & Kans.
> **2.** Flowers open from 8 A.M. to 3 P.M., not fragrant, often larger, with petals broader, rounded (Fig. 7); sepals and lower surfaces green, usually without any purple coloring; petioles with 4 or 5 purple streaks.   **N. tuberòsa** Paine.   Fig. 8. Less common; N.Y. & N.J., w. to Neb. & Ark.   Some botanists profess difficulty in telling this from *N. odorata*, while others consider the two species as quite distinct.

CERATOPHYLLUM

Fruit

C. demersum: 1,2,3. Plants, × 1.   4. Part of a leaf, × 4.

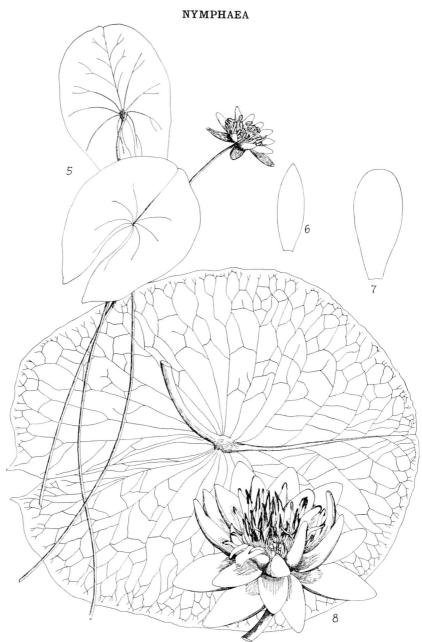

N. tetragona: 5. Leaves and flower, × ¼. **N. odorata**: 6. Petal, × ½. **N. tuberosa**: 7. Petal, × ½. 8. Leaf and flower, × ¼.

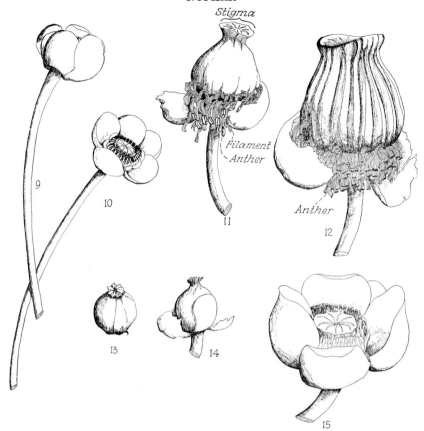

Stigma

Filament
Anther

Anther

9

10

11

12

13

14

15

N. microphyllum: 9,10. Flowers, × ½. 13. Fruit, removed from calyx, × ½. 14. Fruit, surrounded by calyx, × ½. **N. rubrodiscum:** 11. Fruit and remains of calyx and stamens, × ½. **N. variegatum:** 12. Fruit and remains of calyx and stamens, × ½. 15. Flower, × ½.

Yellow Water Lily   Nùphar

Underground stem very thick and spongy, with semicircular scars of petioles and circular scars of flower stalks; flowers with 5–6 fleshy cupped sepals; petals small and scale-like; stamens many; fruit fleshy, somewhat globular, many-seeded. (*Nymphaea*, G, Б.)

Also known as Spatterdock and Cow Lily.

1. Anthers shorter than the filaments (Figs. 10, 11); blades of leaves 3.5–20 cm. long, 3.5–14.5 cm. wide...**2**

1. Anthers equaling or exceeding the filaments (Fig. 12) blades (10–) 17–22 cm. long, (8.5–) 11–25 cm. wide...**3**

  2. Flower 2 cm. or less wide; rays of stigma less than 10 (Fig. 10); blades 3.5–10 cm. long, 3.5–7.5 cm. wide, with a notch two-thirds or more the length of the midrib (Fig. 16 ; young fruit without a ring of decaying stamens (Figs. 13, 14). **N. microphýllum** (Pers.) Fernald. Figs. 9, 10. Que. to e. Pa. & N.J., w. to n. Wis.

  2. Flower 3 cm. or more wide; rays of stigma more than 10 (Fig. 11); blades 7.5–20 cm. long, 5.5–14.5 cm. wide, the notch about half the length of the midrib (Fig. 17); young fruit with a ring of decaying stamens (Fig. 11). **N. rubrodíscum** Morong. Que. to e. Pa. & N.J., w. to n.e. Minn.

213

N. microphyllum: 16. Leaves, × ½. **N. rubrodiscum**: 17. Leaf, × ½. **N. variegatum**: 18. Leaf, lower surface, × ½. 19. Leaf, upper surface, × ½.

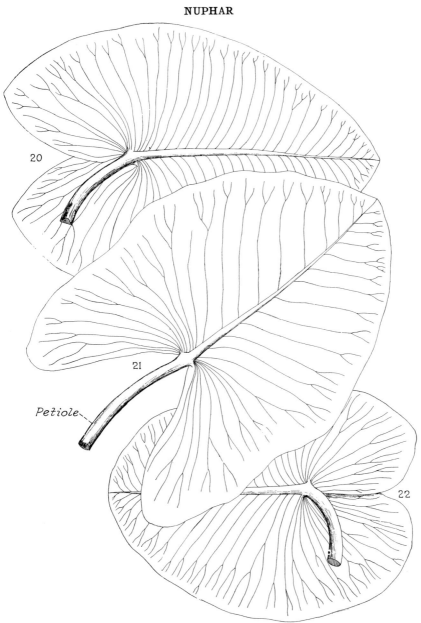

Petiole

N. fraternum: 20. Leaf, lower surface, $\times$ ½.   N. advena: 21. Leaf, lower surface, $\times$ ½.
N. ozarkanum: 22. Leaf, lower surface, $\times$ ½.

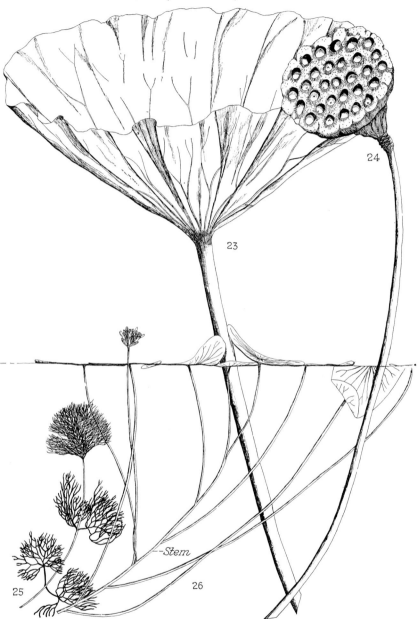

C. caroliniana: 25. Foliage, $\times \frac{1}{4}$. B. Schreberi: 26. Upper part of plant, $\times \frac{1}{4}$. N. lutea: 23. Leaf, $\times \frac{1}{4}$. 24. Fruit, $\times \frac{1}{4}$.

3. Notch broad, the somewhat pointed lobes spreading at an angle of 45–80 deg. (Fig. 21); petiole oval in cross section and holding the leaves in a nearly erect position; sepals and fruit rarely marked with red.............................**4**

3. Notch usually narrow, often closed by the overlapping rounded lobes (Figs. 19, 20); petiole (except in *N. ozarkanum*) flattened on the upper surface, with a narrow wing down each side (Figs. 18–20); fruit and base of sepals within marked with red...**5**

    4. Blade 16–33 cm. long, 4–25 cm. wide, with a notch 4.5–10 cm. deep; flowers 3–4 cm. wide. **N. ádvena** Ait. Fig. 21. S. Maine; s.e. Conn. & N.Y. to N.C., w. to e. Wis., s.e. Neb. & Tex.

    4. Blade 10–11.5 cm. long, 8.5–9 cm. wide, with a notch 4 cm. long; flowers 3.5 cm. wide. **N. advena** var. **brevifòlium** Standl. S. Ill.

5. Flower 4.5 cm. or more wide. **N. variegàtum** Engelm. Figs. 12, 15, 18, 19. N.S. & Que. to Minn. & Mont., s. to N.J., Ohio, Ind. & Neb. (*Nymphaea advena* var. *variegata*, G.) In some regions this is difficult to distinguish from *N. advena*.

5. Flower 2.2–2.6 cm. wide..**6**

    6. Tip of leaf pointed (Fig. 20); petiole flattened on upper surface. **N. fratérnum** (Mill. & Standl.) Standl. Local, cent. N.J.

    6. Tip of leaf rounded (Fig. 22); petiole oval in cross section. **N. ozarkànum** (Mill. & Standl.) Standl. S. Mo. & Ark.

AMERICAN LOTUS   Nelúmbo

N. lùtea (Willd.) Pers. Leaves large, bluish-green, (Fig. 23), shedding water if pushed below the surface; flowers up to 2 dm. wide, with many yellow petals; fruits round and nut-like, in a light spongy top-shaped structure (Fig. 24). Shallow water and muddy shores; e. Mass. to the Gulf of Mexico, n. in the Mississippi Valley to Minn. & Wis.; Lake Erie & Lake Ontario.

In many places this is the subject of such misstatements as "The only place outside of Egypt where the Lotus grows" (it has no relation to the Lotus of Egypt) and "The only place in the world where the American Lotus grows."

WATER SHIELD   Brasènia

B. Schreberi Gmel. Fig. 26 & p. 209. Branches rising through the water from an underground horizontal stem, thickly coated with a gelatinous slime; leaves scattered, with petioles of varying lengths so that the blades float; sepals about 1.5 cm. long. Somewhat local; N.S. to Man., s. to Fla., Tex. & Neb.

FANWORT   Cabómba

C. caroliniàna Gray. Fig. 25. Stem slender, with a gelatinous slime; leaves opposite or whorled, the thread-like divisions branching in fan-like fashion, superficially resembling those of Coontail; flowers about 1 cm. long. W. Mass. (where perhaps introduced); N.J., s. Mich., s. Ill. & e. Mo. to the Gulf of Mexico.

References: Wiegand and Eames, Cornell Univ. Agric. Exp. Sta. Mem., **92**, 206 (1926)—*Nymphaea;* Miller and Standley, Contrib. U.S. Nat. Herb., **16**, 63–108 (1912)—*Nuphar;* Standley, Rhodora, **31**, 37 (1929)—*Nuphar advena* var. *brevifolium;* Standley, Field Mus. Pub. Bot. **8**, 310 (1931)—*Nuphar fraternum* & *N. ozarkanum*. The text for *Nymphaea* has been written with the help of Dr. H. S. Conard.

BUTTERCUP FAMILY   RANUNCULÂCEAE

Sepals 5 to many, sometimes green, or often white or colored like petals; stamens many; fruits dry pods or nutlets, many together.

1. Fruit a nutlet (Fig. 5); both sepals and petals present (Fig. 22)................**2**
1. Fruit a several-seeded pod (Figs. 31, 32); only sepals present, these white or colored (usually yellow) like petals. **Caltha.**
  2. Sepals swollen at base to form a spur; nutlets in a long spike (Fig. 36). **Myosurus.**
  2. Sepals not spurred; nutlets in a roundish (Figs. 3, 23) or slightly elongate (Fig. 29) head. **Ranunculus.**

BUTTERCUP; CROWFOOT   Ranúnculus

1. Leaves cut into many thread-like or narrow ribbon-like divisions (Fig. 1)......**2**
1. Leaves cut into a few flat lobes (Fig. 26) or simple (Figs. 18, 23)............**12**
  2. Divisions of leaves thread-like; flowers white; nutlets wrinkled (Fig. 5)....**3**
  2. Divisions of leaves ribbon-like and flat (Fig. 9), but often very narrow (Fig. 8); nutlets smooth (Fig. 16)...**7**
3. Plants with all leaves submersed and cut into thread-like divisions............**4**
3. Plants with flat floating leaves (Fig. 7) in addition to the submersed divided ones. **R. hederàceus** L.   Nfd.; e. Pa. to S.C.   Perhaps naturalized from Europe. The submersed leaves are often absent.
  4. Leaves with a petiole above the sheath (Fig. 2); leaves limp, collapsing when taken from the water...**5**
  4. Leaves dividing into 3 parts directly above the sheath, the petiole missing (Fig. 1); leaves rather stiff, keeping their shape when taken from the water..**6**
5. Receptacle hairy beneath the nutlets (Fig. 3); nutlets 1.5 mm. or less long, usually hairy. **R. trichophy̆llus** Chaix. Water Crowfoot. Figs. 2, 3. Common in shallow water; Labrador & N.S. to N.J., w. to Alaska & Lower Calif. (*R. aquatilis* var. *capillaceus*, G; *Batrachium trichophyllum*, B.)
5. Receptacle without hairs (Fig. 4); nutlet 1.5 mm. or more long, without hairs. **R. trichophyllus** var. **calvéscens** Drew. N.B. to Mich., s. to Conn. & Pa.
  6. Nutlets 8–30, usually about 16, in a head, each with a beak about 1 mm. long (Fig. 5). **R. longiróstris** Godron. Fig. 1. Stiff Water Crowfoot. In calcareous water; w. Que. to Conn., w. to Minn., s. to Del., Tenn., Kans., Tex. & N.M. (*R. circinatus*, G; *Batrachium circinatum*, B.)
  6. Nutlets 30–80, usually about 40, in a head, short-beaked (Fig. 6). **R. subrígidus** Drew. Nfd. to w. N.E.; Mich. to n. Mex. & w. to the Pacific.
7. Most leaves shallowly lobed (Fig. 7). See **R. hederaceus, 3,** which often lacks completely the submersed dissected leaves.
7. Leaves mostly dissected, or lobed at least half way to the base. The Yellow Water Crowfoot group, superficially similar to the preceding group of species, but with more variable leaves the segments of which, no matter how narrow, are always flat...**8**
  8. Petals 6–17 mm. long; fruits (including the beak) 2.5–3.5 mm. long, with a corky wing margin (Fig. 11)...**9**
  8. Petals 3.5–5 mm. long; fruits (including the beak) 1.5–2 mm. long, not wing-margined (Fig. 16)...**10**

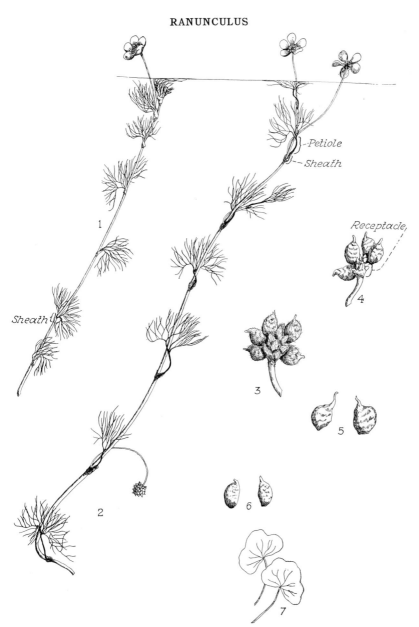

R. longirostris: 1. Part of plant, $\times \frac{1}{2}$.   5. Nutlets, $\times 5$.   **R. trichophyllus:** 2. Part of plant, $\times \frac{1}{2}$.   3. Head of nutlets, from which some have fallen, $\times 5$.   **R. trichophyllus** var. **calvescens:** 4. Head of nutlets, from which some have fallen, $\times 5$.   **R. subrigidus:** 6. Nutlets, $\times 5$.   **R. hederaceus:** 7. Floating leaves, $\times 1$.

R. flabellaris: 8,10. Upper part of plants, × ½. 9. Part of leaf, × 3. 11. Nutlet, × 10. **R. flabellaris** f. **riparius:** 12. Part of leaf, × 3. 13,14. Parts of plants, × ½. **R. Purshii:** 15. Upper part of plant, × ½. 16. Nutlet, × 10. **R. Purshii** f. **terrestris:** 17. Upper part of plant, × ½.

9. Leaves dissected into many very narrow segments (Figs. 8, 9, 10), not hairy. **R. flabellàris** Raf. Maine to Wash., s. to N.C., Ark., Kans. & Calif. A submersed form. (*R. delphinifolius*, G, B.)
9. Leaves with broader divisions (Figs. 12–14), often closely hairy. **R. flabellaris** forma **ripàrius** Fernald. A terrestrial form, appearing when left stranded on the mud and often grading into the aquatic form. (*R. delphinifolius* var. *terrestris*, G.)
   10. Generally submersed form; leaves dissected into very narrow segments (Fig. 15). **R. Púrshii** Richards. Labrador to Alaska, s. to N.S., n. Maine, Mich., Iowa, N.D., N.M. & Ore.
   10. Generally emersed forms; leaves with broader segments (Fig. 17), sometimes hairy...**11**
11. Branches prostrate or creeping, with 1–4 flowers. **R. Purshii** forma **terréstris** (Ledeb.) Glück. Fig. 17. About as common as the aquatic form.
11. Branches ascending, 7–50-flowered. **R. Purshii** var. **prolíficus** Fernald. Rare; e. Que., Mich., Mont., Colo. and perhaps elsewhere. Somewhat approaching **R. sceleratus, 21.**
   12. Leaves not lobed (Figs. 18, 19, 22, 23)................................**13**
   12. Leaves deeply lobed (Fig. 26), or cut into several leaflets (Fig. 27).......**18**
13. Stems prostrate, rooting at nearly all the nodes, even the flowering stems coming from nodes that touch the soil (Fig. 18)...................................**14**
13. Stems erect or prostrate and rooting only near the base with the flowers on erect or ascending stems..**15**
   14. Leaves thread-like. **R. réptans** L. Creeping Spearwort. Fig. 18. Wet shores; from Arctic regions s. to N.J., Pa., Mich., Colo. & Calif. (*R. Flammula* var. *filiformis*, G.)
   14. Leaves flat, several millimeters broad. **R. reptans** var. **ovàlis** (Bigel.) T. & G. Fig. 19. Similar situations to *R. reptans;* intermediate individuals occur. (*R. Flammula* var. *reptans*, G.)
15. Stems reclining below and rooting at several nodes (Fig. 23)...............**16**
15. Stems erect (Fig. 22), often developing creeping branches from the base; nutlet not beaked...**17**
   16. Nutlet with a large beak (Fig. 24). **R. ámbigens** Wats. Water Plantain Spearwort. Fig. 23. Muddy places; s. Maine, s. Ont., Ind. & Mo., s. to Del., Md. & Tenn. (*R. laxicaulis*, G; *R. obtusiusculus*, B.)
   16. Nutlet with little or no beak. **R. Flámmula** L. Nfd. & s. N.S.
17. Petals 3–6 mm. long; stamens 12–20. **R. laxicaúlis** (T. & G.) Darby. Spearwort. Fig. 22. Wet ground; Del. to the Gulf of Mexico, n. in the Mississippi Valley to Mo. & Ill. (*R. oblongifolius*, G, B.)
17. Petals 2 mm. long; stamens 3–10. **R. pusíllus** Poir. S. N.Y. to the Gulf of Mexico, n. in the Mississippi Valley to Mo. & Tenn.
   18. Stems prostrate, rooting at most of the nodes (Fig. 21)...............**19**
   18. Stems erect (Figs. 26, 27, 29), or in submersed forms, the leaves in a rosette without an elongate stem (Fig. 25).....................................**20**
19. Leaves divided into 3 leaflets each of which has a distinct stalk of texture different from that of the blade (Fig. 20). **R. rèpens** L. Creeping Buttercup. Fig. 21. Moist soil, ditches, etc.; naturalized from Europe.
19. Leaves variously divided, sometimes even on the same plant (Figs. 15, 17), the base of each leaflet often somewhat narrowed, but not of different texture from the blade (Figs. 12, 14). Subterrestrial forms, usually stranded on the mud, of **R. flabellaris, R. Purshii** and **R. hederaceus.** See **7.**

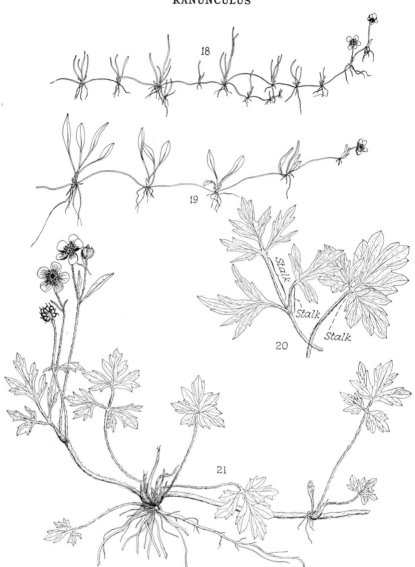

R. reptans: 18. Plant, × ½. R. reptans var. ovalis: 19. Plant, × ½. R. repens: 20.
Leaves, × 1.   21. Plant, × ½.

Sepal
Petal

Head of
nutlets

22

23

Beak

24

R. laxicaulis: 22. Plant, $\times \frac{1}{2}$.   **R. ambigens:** 23. Plant, $\times \frac{1}{2}$.   24. Nutlets, $\times$ 10.

223

R. sceleratus: 26. Plant, $\times \frac{1}{2}$.   **R. sceleratus f. natans**: 25. Plant, $\times \frac{1}{2}$.

RANUNCULUS

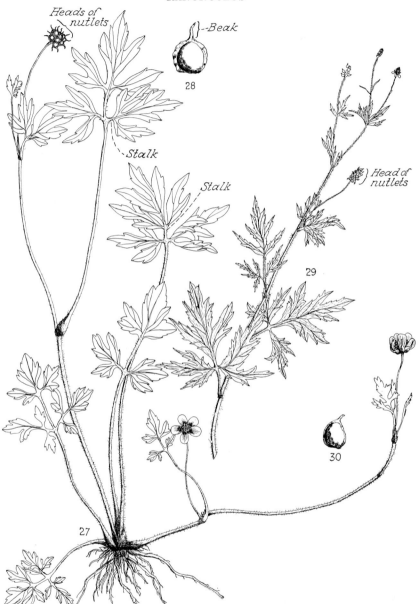

Heads of nutlets

Beak

28

Stalk

Stalk

Head of nutlets

29

30

27

R. septentrionalis var. caricetorum : 27. Plant, × ½.   28. Nutlet, × 10.   R. pensylvanicus :
29. Upper part of plant, × ½.   30. Nutlet, × 10.

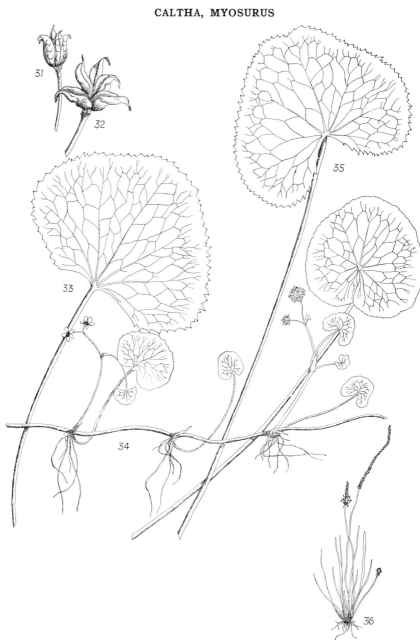

C. palustris: 31. Young fruit, $\times \frac{1}{2}$. 32. Opened fruit, $\times \frac{1}{2}$. 35. Leaf, $\times \frac{1}{2}$. **C. palustris** var. **flabellifolia:** 33. Leaf, $\times \frac{1}{2}$. **C. natans:** 34. Part of plant, $\times \frac{1}{2}$. **M. minimus:** 36. A small plant, $\times \frac{1}{2}$.

20. Leaves cut into leaflets each of which has a stalk (Figs. 27, 29).........**22**
20. Leaves only deeply lobed (Figs. 25, 26)..............................**21**
21. Leaves supported on a stem. **R. sceleràtus** L. Cursed Crowfoot. Fig. 26. A common weed on muddy shores.
21. Leaves all rising from the base of the plant, with blades floating on the surface. **R. sceleratus** forma **nàtans** Glück. Fig. 25. An aquatic state of *R. sceleratus;* many individuals are intermediate.
    22. Flowers 1.5–3 cm. broad; nutlets in a round head (Fig. 27), each with a long beak (Fig. 28); branches produced at flowering time spreading over the ground..**23**
    22. Flowers about 7 mm. broad; nutlets in a cylindrical head (Fig. 29), each short-beaked (Fig. 30); all stems erect. **R. pensylvánicus** L.f. Bristly Crowfoot. Common on wet shores, etc.; Nfd. to B.C., s. to Ga., Kans. & Colo.
23. Hairs of stem and petioles close or absent. **R. septentrionàlis** Poir. Swamp Buttercup. Common in wet places, mostly in the woods; N.B. to Man., s. to Ga. & Neb.
23. Plants with copious spreading or slightly downwardly pointing hairs. **R. septentrionalis** var. **caricetórum** (Greene) Fernald. Fig. 27. Ohio to Minn., s. to Ark.

MARSH MARIGOLD   Cáltha

Leaves basal and on the stem, rounded, finely toothed or nearly without teeth; flowers with 5–9 colored or white petal-like sepals; fruit a several-seeded pod with papery walls.

1. Stems erect or nearly so...**2**
1. Stems prostrate or floating, rooting at the nodes..........................**3**
    2. Leaves with a deep notch (Fig. 35); flowers 2–4 cm. broad. **C. palústris** L. Cowslip. Figs. 31, 32, 35. Very abundant in swamps; Nfd. to Sask., s. to S.C., Tenn. & Neb.
    2. Leaves with a shallow notch or cut nearly square across the base (Fig. 33); flowers 1–2 cm. broad. **C. palustris** var. **flabellifòlia** (Pursh) T. & G. Cool mountain springs; s.e. N.Y. to Md.
3. Stems prostrate; flowers yellow; fruits 1 cm. or more long. **C. palustris** var. **radìcans** (Forst.) Hartm. Arctic America, & s.e. N.Y.
3. Stems floating; flowers white or pink; fruits 5 mm. or less long. **C. nàtans** Pall. Fig. 34. West end of Lake Superior and northwestward.

MOUSETAIL   Myosùrus

M. mínimus L. Fig. 36. Leaves narrow and ribbon-like or tongue-like, all from the base of the plant; flowers several, each on a naked stem, with 5 spurred sepals and 5 small white petals. Wet and muddy places; s. Ont. & Ill. to B.C., s. to the Gulf of Mexico.

References: Glück, Biol. und Morph. Untersuch. uber Wasser- und Stumpfge-wächse, **3**, 511 (1911)—*R. sceleratus* f. *natans;* Fernald, Rhodora, **19**, 135–137 (1917)—*R. reptans;* Drew, Rhodora, **38**, 1–47 (1936)—the Water Crowfoot group; Fernald, Rhodora, **38**, 171–178 (1936)—the Yellow Water Crowfoot group, *R. septentrionalis* var. *caricetorum* and *R. ambigens;* Fernald, Rhodora, **41**, 541–542 (1939)—*R. laxicaulis.*

CRESS FAMILY   CRUCÍFERAE

The flowers and pods of the members of this family are remarkably similar in general structure, so that with but a little familiarity a plant in flower or fruit can be recognized as one of the Cruciferae. The fruits, moreover, although possessing a general similarity, are quite characteristic for each genus. The leaves, on the other hand, present, even in a single species, a confusing series of variations.

The flowers (Fig. 35) have 4 sepals, 4 petals and 6 stamens, of which 2 are shorter than the 4 others. The petals have their upper portions sharply bent and spreading, giving the flower a cross-like appearance when viewed from above. The fruit is a pod, of 2 more or less cupped plates joined by their edges and enclosing the seeds; usually there is a flat membranous partition between them (Figs. 17, 18).

The following key to genera will be more or less helpful, but may not be satisfactory in the case of some individuals. The drawings of leaves show much of the range of variation within each species and may serve to place some of the plants, especially if they are studied in the field and several of the leaf forms observed.

1. Leaves quill-like, in a rosette (Fig. 11). **Subulària.**
1. Leaves flat, simple or variously divided, scattered on the stem.................2
    2. Flowers yellow or greenish. **Rorippa.**
    2. Flowers white or purple...3
3. Pods only a little longer than broad (Fig. 25). **Neobeckia.**
3. Pods several times as long as broad (Figs. 22, 39).........................4
    4. Pods flattened. **Cardamine.**
    4. Pods round in cross section. **Nasturtium.**

AWLWORT   Subulària

S. aquática L. Fig. 11. Small tufted plants; leaves slender, 1–3 cm. long, tapering to a fine point; flowers minute, on a naked stem; pod roundish, 1–4 mm. long. At edges of lakes and streams, usually in shallow water; Nfd. to B.C., s. to Maine, N.H., Ont., Wyo. & Calif.; not common.

R. palustris var. **hispida**: 1. Upper part of plant, $\times \frac{1}{2}$. 3. Leaf, $\times \frac{1}{3}$. **R. palustris**: 2. Upper branch, $\times \frac{1}{2}$. **R. palustris** var. **glabrata**: 4. Leaf and part of inflorescence, $\times \frac{1}{2}$.

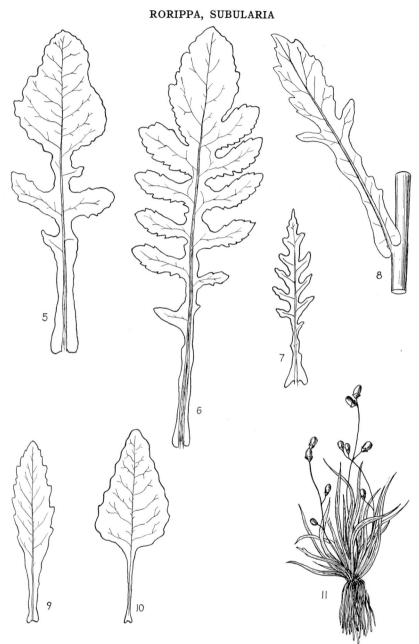

R. palustris var. hispida & var. glabrata: 5, 6, 7, 8. Leaves, $\times \frac{1}{2}$. R. palustris var. hispida f. inundata & var. glabrata f. aquatica: 9, 10. Leaves, $\times \frac{1}{2}$. S. aquatica: 11. Plant, $\times \frac{1}{2}$.

Roríppa

Flowers often very small and inconspicuous; leaves mostly pinnately divided, or lobed and coarsely toothed. The commonest species is the Marsh Cress, **5** and **6**; the others are not very often found. The drawings of leaves must not be taken too literally, since they are extremely variable in each species. (*Radicula*, G, B.)

1. Petals shorter than the sepals (Figs. 4, 12, 14); mature pod with a style less than 1 mm. long; plants annual or biennial, without a rootstock...................**2**
1. Petals longer than the sepals (Figs. 16, 19); mature pod with a style 1–3 mm. long; plants perennial, from a horizontal rootstock..............................**10**
   2. Pedicels 6–8 mm. long, mostly longer than the pods (Figs. 1, 4); leaves or their lobes mostly sharp-toothed (Figs. 1–10).................................**3**
   2. Pedicels 4 mm. or less long, shorter than the mature pods (Figs. 12, 14); leaves or their lobes mostly with blunt teeth....................................**7**
3. Plants weak-stemmed and delicate; all leaves pinnately divided to about the same degree. **R. palústris** (L.) Bess. Yellow Cress. Fig. 2. A native of Europe, occasionally found as an adventive in America, and perhaps native about the Gulf of St. Lawrence.
3. Plant stout; upper leaves less deeply lobed than the lower, or none of the leaves deeply lobed...**4**
   4. Stems hairy (Fig. 3); pod nearly as thick as long (Fig. 1)..................**5**
   4. Stem with few or no hairs; pod 2–3 times as long as thick (Fig. 4)..........**6**
5. Lower leaves pinnately divided. **R. palustris** var. **híspida** (Desv.) Rydb. Marsh Cress. Figs. 1, 3, 5–8. Very common on wet shores.
5. Lower leaves nearly simple, the lateral lobes much reduced, and the terminal lobe much the largest. **R. palustris** var. **hispida** forma **inundàta** Vict. Fig. 9; the leaf illustrated in Fig. 10 may also occur. A form found occasionally in the water.
   6. Lower leaves pinnately divided. **R. palustris** var. **glabràta** (Lunell) Vict. Marsh Cress. Figs. 4–8. Very common on wet shores.
   6. Lower leaves nearly simple. **R. palustris** var. **glabrata** forma **aquática** Vict. Fig. 10; the leaf illustrated in Fig. 9 may also occur. A form appearing in the water.
7. Pedicels about as long as the thickness of the pods (Fig. 14); seeds, when viewed under the microscope, with little bumps (Fig. 15)..........................**8**
7. Pedicels shorter than the thickness of the pods (Fig. 12), sometimes almost none; seeds pitted (Fig. 13). **R. sessiliflòra** (Nutt.) Hitchc. Muddy shores; mostly in the Mississippi Valley, n. to Wis.
   8. Pods 3–4 times as long as thick......................................**9**
   8. Pods nearly as thick as long. **R. obtùsa** var. **sphaerocárpa** (Gray) Cory. Mostly in the Mississippi Valley, n. to Wis.
9. Leaves deeply lobed. **R. obtùsa** (Nutt.) Britton. Fig. 14. Mich. & Ill. to Kans., Tex. & Calif.
9. Leaves not lobed. **R. obtusa** var. **intégra** (Rydb.) Vict. Utah; said to occur along the St. Lawrence River at Montreal.

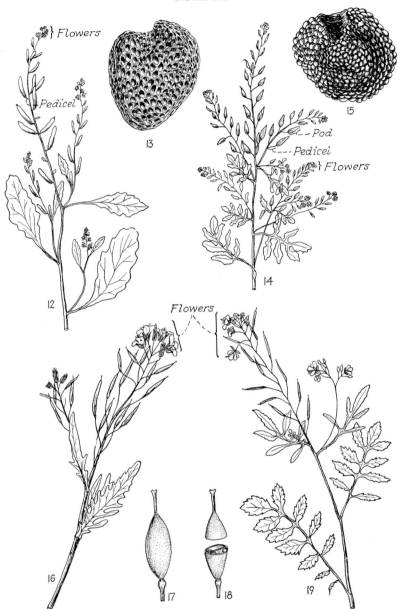

Flowers

Pedicel

13

Pod

Pedicel

Flowers

15

14

12

Flowers

16

17

18

19

R. sessiliflora: 12. Upper branch, × ½. 13. Seed, × 100. R. obtusa: 14. Upper branch, × ½. 15. Seed, × 100. R. sinuata: 16. Upper branch, × ½. R. amphibia: 17. Fruit, × 3. 18. Same, cut across. R. sylvestris: 19. Upper branch, × ½.

Pod

Rachis

C. **bulbosa:** 20. Plant, × ½. **C. bulbosa** var. **purpurea:** 21. Basal leaf, × ½. **C.
pensylvanica:** 22. Plant, × ½. **C. pratensis** var. **palustris:** 23. Plant, × ½.

233

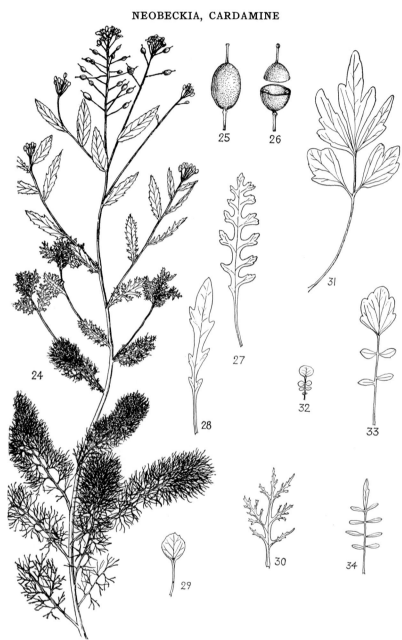

N. aquatica: 24. Upper part of stem, $\times \frac{1}{2}$. 25. Fruit, $\times 3$. 26. Same, cut across. 27–30. Variations in leaves, $\times \frac{1}{2}$. **C. pensylvanica:** 31–34. Variations in leaves, $\times \frac{1}{2}$.

10. Upper leaves with clasping bases (Fig. 16). **R. sinuàta** (Nutt.) Hitchc. Wet or dry shores, railroads, etc.; from Ill. & Mo. westward.

10. Upper leaves not clasping...**11**

11. Leaves with slender petioles, deeply divided, none of them clasping. **R. sylvéstris** (L.) Bess. Creeping Yellow Cress. Fig. 19. Wet shores, widely distributed; naturalized from Europe.

11. Leaves without petioles, or the lower with winged petioles, the lower often somewhat clasping....................................**12**

    12. None of the leaves lobed. **R. amphíbia** (L.) Bess. Great Water Cress. A native of Europe, naturalized about Montreal, Can., in Lake Androscoggin, Auburn, Maine, in Ridgefield, Conn., and perhaps elsewhere. Sometimes grows nearly a meter high. This resembles superficially some forms of the Lake Cress, below, which differs in having white flowers, whereas *R. amphibia* has yellow flowers. Fruits may be distinguished by the central partition in those of *R. amphibia* (Fig. 18), which is lacking in the Lake Cress (Fig. 26).

    12. Lower leaves pinnately divided. **R. amphibia** forma **variifòlia** (DC.) Vict. A variation found with the typical form.

BITTER CRESS   Cardámine

1. Leaves pinnately parted (Figs. 22, 23)....................................**2**

1. Leaves simple (Figs. 20, 21), or sometimes with 1 or 2 very small lateral leaflets..**3**

    2. Leaflets rarely stalked (Figs. 22, 31–34), variable in size, the base running more or less into a wing on the rachis; flowers white, inconspicuous, with petals about 3 mm. long. **C. pensylvánica** Muhl. Streams and swamps, common; Labrador to Minn. & Mont., s. to Fla. & Kans.

    2. Leaflets mostly stalked, those of the upper leaves narrower than those of the lower; flowers white or pinkish, with petals 1 cm. or more long. **C. praténsis** L. var. **palústris** Wimm. & Grab. Cuckoo Flower. Fig. 23. Swamps; from subarctic regions s. to Conn., N.J., Ind., Minn. & n. B.C.

3. Plants mostly erect, with a bulb-like base (Fig. 20), except in a form in the Ozarks.**4**

3. Plants mostly prostrate and rooting below, sometimes with long runners, and not bulb-like at base. **C. rotundifòlia** Michx. Mountain Water Cress. Cool shaded springs; N.Y. & N.J. southward.

    4. Flowers white; stem without hairs; basal leaves longer than wide. **C. bulbòsa** (Schreb.) BSP. Spring Cress. Fig. 20. Wet meadows, springs, brooks, etc.; Vt. & e. Mass. to Minn., s. to Fla. & Tex. **C. bulbosa** forma **fontinàlis** Palmer & Steyermark has the lower leaves rounder and lacks the bulbs at base. It occurs about springs in the Ozarks and simulates the more eastern *C. rotundifolia*.

    4. Flowers purplish; stem somewhat hairy; basal leaves as wide as long (Fig. 21). **C. bulbòsa** var. **purpùrea** (Torr.) BSP. Purple Cress. Springs, wet woods, etc.; Conn. to s. Ont., & Wis., s. to Va. & Ky. (*C. Douglassii*, G, B.)

LAKE CRESS   Neobéckia

N. aquática (Eaton) Britton. Figs. 24–30. Stems usually trailing through the water, the upper end emerging; leaves very variable, even on the same plant, but the emersed generally less deeply lobed or divided than the submersed; flowers white; fruit plump, without a central partition (Fig. 26). Not very common; Que. & Vt. to Minn., s. to Fla., La. & Ark. (*Radicula aquatica*, G.) See *Rorippa amphibia*, above, which resembles this.

N. officinale : 35. Flower, with one petal removed, × 5.   39. Upper branch, × 1.   N. officinale f. nanum : 36. Branch, × 1.   N. officinale f. terrestre : 37. Stem and leaves, × 1. N. officinale f. submersum : 38. Leaf and stem, × 1.   N. officinale f. siifolium : 40. Leaf and stem, × 1.

WATER CRESS   Nastúrtium

N. officinàle R. Br. Figs. 35–40.   Often growing in solid tangled masses in shallow water; stems partly prostrate and rooting at the nodes, floating or sinking, erect toward the tip which supports the flowers and fruits; leaves usually pinnately compound, with rounded leaflets, but variable in shape and size and sometimes even simple; flowers white; fruits spreading, often curved.   Common in wet places, particularly in springs; said to be naturalized from Europe, but often appearing as if native.   (*Radicula Nasturtium-aquaticum*, G; *Sisymbrium Nasturtium-aquaticum*, B.)

Many forms and varieties of the Water Cress have been named, of which the following seem to be the most strongly marked, but they are probably due to habitat alone.

1. Leaves compound **2**
1. Leaves mostly reduced to a single leaflet (Fig. 36).   **N. officinale** forma **nànum** Glück.   Usually submersed, sometimes growing along the submersed portion of a stem which produces another form in other portions.
   2. Stems 1–3 m. long, 5–8 mm. thick, with a large central cavity; lateral leaflets 3–10 cm. long, the margins remotely and bluntly toothed (Fig. 40).   **N. officinale** forma **siifòlium** Steud.   The most robust development of the species, trailing in shallow water with the leaf-bearing stems emersed.
   2. Stems mostly less than 1 m. long, 5 mm. or less thick, with a small central cavity; lateral leaflets 3 cm. or less long, the margins curved or angled **3**
3. Leaflets firm enough to keep their shape in the air, opaque, essentially flat **4**
3. Leaflets limp, translucent, curling (Fig. 38).   **N. officinale** forma **submérsum** Glück.   A deep-water form, usually sterile.
   4. Most lateral leaflets less than 1 cm. long (Fig. 37).   **N. officinale** forma **terréstre** Glück.   Often to be found on the higher and drier ground.
   4. Most lateral leaflets more than 1 cm. long (Fig. 39).   **N. officinàle** R. Br.   The common form, usually submersed below and emersed and flowering above.

References: Glück, Biol. und Morph. Untersuch. Wasser- und Stumpfgew., **3**, 178–184 (1911); in Pascher, Die Süsswasser-Flora Mitteleuropas, **15**, 268–271 (1936); Hegi, Ill. Fl. Mitt.-Eur., **4**, pt. 1: 320 (1926?)—forms of *Nasturtium officinale;* Wiegand and Eames, Cornell Univ. Agric. Exp. Sta. Mem., **92**, 233 (1926)—*Cardamine bulbosa* var. *purpurea;* Marie-Victorin, Le genre Rorippa dans le Québec, Contrib. Lab. Bot. Univ. Montréal, No. 17—*Rorippa* species, varieties, and forms; Palmer and Steyermark, Ann. Mo. Bot. Gard., **25**, 771 (1938)—*Cardamine bulbosa* f. *fontinalis;* Cory, Rhodora, **38**, 406 (1936)—*Rorippa obtusa* var. *sphaerocarpa;* Fernald, Rhodora, **22**, 11–14 (1920)—*Cardamine pratensis* var. *palustris.*

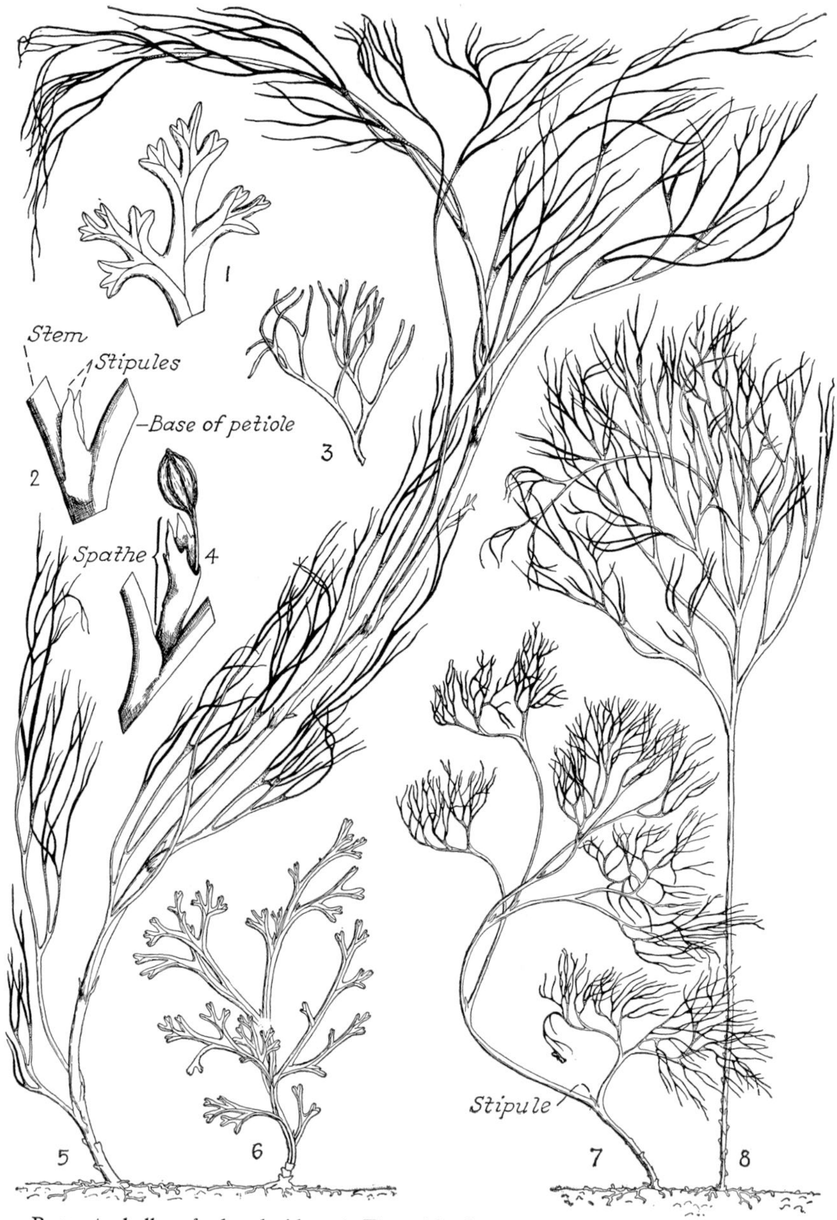

Stem

Stipules

—Base of petiole

Spathe

Stipule

P. ceratophyllum f. **chondroides**: 1. Tips of leaflets, × 3.   2. Base of leaf and stipules, × 2.   6. Plant, × 1.   **P. ceratophyllum** f. **abrotanoides**: 3. Tips of leaflets, × 3.   4. Fruit and spathe, × 3.   5. Plant, × 1.   **P. ceratophyllum**: 7, 8. Plants, × 1.

RIVER WEED FAMILY   PODOSTEMÀCEAE

A large family in the Southern Hemisphere, of alga-like plants, usually olive-green, attached to rocks by disk-like processes (Figs. 5–8) in running water.   But one genus in the United States.

River Weed **Podostêmum**

Leaves long-petioled, repeatedly forking, often with stipules at base (Fig. 2); flowers inconspicuous, in the axils of the upper leaves or clustered toward the tip of the stem; fruit a pod composed of 2 valves, one of which soon drops, leaving a persistent 5-ribbed one (Fig. 4). These plants are seldom seen, since they occur in swift, usually "white," water as a thick moss-like covering on the rocks. They are very variable in appearance and may be separated, rather arbitrarily, into forms. *Podostemum* is found in streams, from the Ottawa River and central Maine s. through N.Y., Conn., N.J., Ky. & w. N.C. to s. Ala. & s. Okla.

1. Terminal divisions of leaves $\frac{1}{3}$–$\frac{2}{3}$ mm. wide, 1–3 times as long as broad (Fig. 1). **P. ceratophýllum** forma **chondroìdes** Fassett. Fig. 6.
1. Terminal divisions of leaves $\frac{1}{10}$–$\frac{1}{3}$ mm. wide, about as thick as a coarse horsehair (actually flat), many times as long as broad (Fig. 3)...................2
  2. Plant rather rigid; stems rarely exceeding 1.5 dm. in length; leaves spreading at an angle, often absent from the lower part of the stem. **P. ceratophýllum** Michx. Figs. 7, 8.
  2. Plant lax; stems reaching 8 dm. in length; leaves loosely ascending, usually borne along the whole length of the stem. **P. ceratophyllum** forma **abrotanoìdes** (Nutt.) Fassett. Fig. 5.

Reference: Fassett, Rhodora, **41**, 525–529 (1939)—forms of *P. ceratophyllum*.

ORPINE FAMILY   CRASSULÂCEAE

Most members of this family (Stonecrop, Live-forever, Hen and Chickens, etc.) have thick fleshy leaves and grow in dry places. The flowers of the Crassulaceae are characterized by their perfect symmetry, the number of stamens being equal to or twice that of the sepals, the petals and the carpels.

Ditch Stonecrop **Pénthorum**

P. sedoìdes L. Fig. 12. Plants 2–6 dm. high; leaves scattered, finely toothed; flowers on gland-covered branches which are coiled and straighten as the flowers unfold; sepals 5, petals none, stamens 10; fruit a 5-beaked pod (Fig. 9). Muddy places, ditches, swamps, etc.: N.B. to Minn., s. to Fla., Kans. & Tex.

Tillaèa

T. aquática L. Figs. 13, 14. Inconspicuous dwarf plants, 1–8 cm. high; leaves opposite, their bases joined about the stem, 2–6 mm. long; flowers in the upper leaf axils, short-stalked. Mostly in fresh tidewater, sometimes in brackish sand; Nfd. to Md., and from Mont. & Calif. to Mex. & e. Tex.; also widespread in the Old World. **T. Vaillántii** Willd., differing only in having the flowers on slender stalks, grows on P.E.I. & Nantucket Is., Mass.

Potentilla palustris: 10. Flower, $\times 2\frac{1}{2}$. 11. Plant, $\times \frac{1}{2}$. **Penthorum sedoides:** 9. Fruit, $\times 2\frac{1}{2}$. 12. Upper part of plant, $\times \frac{1}{2}$. **T. aquatica:** 13. Upper part of plant, $\times 3$. 14. Plant, $\times 1$.

ROSE FAMILY   **ROSÁCEAE**

This large family has many species in meadows and swamps, but only one that regularly enters the water.

CINQUEFOIL   **Potentílla**

P. palústris (L.) Scop. Marsh Cinquefoil. Figs. 10, 11. Stem trailing in the mud or water, rooting at the nodes; leaves pinnately compound, with sheathing stipules at the base, the lower long-petioled; petals purple. Swales, brooks and bogs; Greenland & Labrador to Alaska, s. to N.J., Pa., Ohio, Ind., Ill., Iowa, Wyo. & Calif. Individuals are variable in amount and length of hairy covering, distribution of glands and shape of leaflets; on these characters varieties and forms have been based by Fernald and Long, Rhodora, **16**, 5–11 (1914), and by Gunnarsson, Bot. Notiser for 1914, 217–224.

WATER STARWORT FAMILY   **CALLITRICHÀCEAE**

WATER STARWORT   **Callítriche**

Delicate plants with slender limp stems, usually supported by the water or sometimes prostrate on the mud; leaves opposite, the submersed ones ribbon-like, the floating ones (usually present in all but the last species) crowded toward the tip and forming rosettes on the surface; flowers inconspicuous; fruits in the axils of leaves, the size of a pin head, separating into 2 portions. Occasionally plants grow stranded on the mud, with all the leaves of the broader type. For determination of the species mature fruits are necessary; these are rarely absent from well-developed plants.

1. Fruits on short stalks (Fig. 17), without bracts. **C. defléxa** A. Br., var. **Austìni** (Engelm.) Hegelm. Fig. 20. Damp soil, pathways, muddy shores, etc.; w. Mass. to Va., s. Ind., & Ill., s. to La. & Mex. (*C. Austini*, B.)

1. Fruits without stalks, fitting snugly in the axil of the leaf, with 2 bracts in some species (Figs. 16, 18)..2

  2. Fruits 1 mm. or less broad..3

  2. Fruits about 2 mm. broad...4

3. Fruits higher than broad, with keeled lobes (Fig. 16). **C. palústris** L. Fig. 15. Very common in quiet or slowly flowing water.

3. Fruits about as broad as high, with lobes rounded on the edge (Fig. 18). **C. heterophýlla** Pursh. Nfd. to Man., s. to Fla., La. & Colo.

  4. Commonly with the upper leaves roundish and in a floating rosette; the 2 portions of the fruit grown closely together. **C. stagnàlis** Scop. Rare in N. A.; reported from e. Mass. to Pa.; tidal shores of the St. Lawrence River; springs in Wis. Some care is necessary to distinguish this from the next. If round floating leaves are present, the plant is definitely *C. stagnalis;* if only ribbon-like leaves are present it is probably *C. hermaphroditica,* but the fruit should be dissected under a lens for certainty.

  4. All the leaves narrow and ribbon-like; the 2 portions of the fruit easily separable without tearing any tissue. **C. hermaphrodítica** L. Fig. 19. Not common; w. Mass. & N.B. to Wis., Man. & Ore., s. to Colo. (*C. autumnalis*, G, B.)

References: Schinz and Thellung, Vierteljahrs. Naturforsch. Gesell. Zurich, **53**, 548 (1909)—*C. hermaphroditica;* Svenson, Rhodora, **32**, 37–38 (1932), and Fernald, *ibid.*, 39–40—*C. stagnalis.*

C. palustris: 15. Plants, × 1.   16. Fruits, × 5.   C. deflexa var. Austini: 17. Fruits, × 5.
20. Plant, × 1.   C. heterophylla: 18. Fruits, × 5.   C. hermaphroditica: 19. Fruits, × 5.

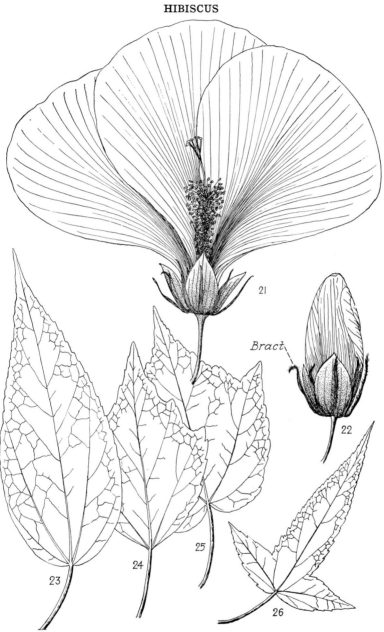

Bract

H. palustris: 21. Flower, with 2 petals removed, × ½. 23–25. Leaves, × ½. H. lasiocarpos: 22. Expanding bud, × ½. H. militaris: 26. Leaf, × ½.

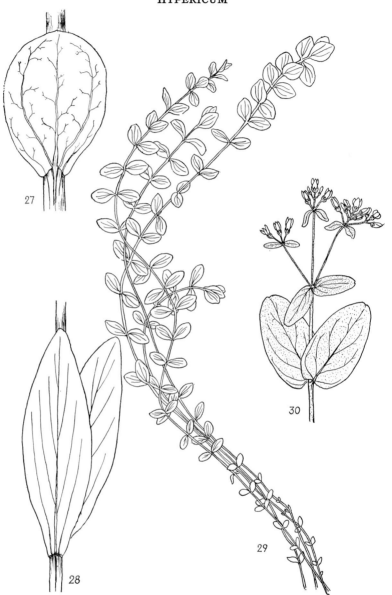

H. boreale f. **callitrichoides**: 27. Leaf, × 2. 29. Plants, × 1. **H. ellipticum** f. **aquaticum**:
28. Leaves, × 2. **H. punctatum**: 30. Branch, × 1.

MALLOW FAMILY   MALVÀCEAE

Rose Mallow   Hibíscus

Stout herbs often 2 m. tall, growing in swales and on the margins of ponds and sloughs; flowers very showy, the petals often 1 dm. or more long; calyx united into a short tube with 5 lobes and numerous narrow bracts just beneath it; stamens united below into a tube and appearing as a fringe along its upper part, as in all the *Malvaceae*.

1. Leaves velvety beneath..2
1. Leaves not velvety. **H. militàris** Cav. Fig. 26. Pa. to Fla., w. to Minn., Neb. & La.
   2. Leaves without hairs on the upper surface. **H. palústris** L. Figs. 21, 23–25. E. Mass. to Ont., s. to the Gulf of Mexico. (*H. Moscheutos*, G, B.)
   2. Leaves with hairs on the upper surface...................................3
3. Bracts fringed with hairs (Fig. 22). **H. lasiocárpos** Cav. Ga. to Tex., n. to Ky., Ind., Ill. & Mo.
3. Bracts not fringed. **H. incànus** Wendland. Md. to Fla. & La.

Reference: Fernald, Rhodora, **41**, 112 (1939)—*H. palustris*.

ST. JOHN'S-WORT FAMILY   HYPERICÀCEAE

St. John's-wort   Hyperìcum

Leaves opposite with scattered translucent dots (except in submersed forms) which are best seen by holding the leaf so that the light shines through it and examining with a lens; flowers yellow in most kinds, purplish in two, with 5 sepals, 5 petals and many stamens; fruit a pod.

1. Plants definitely aquatic, with limp stems (Fig. 29); leaves without translucent dots; flowers usually absent; plants looking like *Callitriche*. Usually the terrestrial form may be found on the shore nearby, or a partly submersed plant may have the emersed part characteristic of the land plant.........................2
1. Plants on land or at the water's edge, with erect stems; leaves with translucent dots; flowers usually present..3
   2. Leaves 3-nerved from the base (Fig. 27). **H. boreàle** forma **callitrichoìdes** Fassett. Fig. 29.
   2. Leaves feather-veined (Fig. 28). **H. ellípticum** forma **aquáticum** Fassett.
3. Stem, and often the sepals and petals, rarely even the pod, marked with black dots (Fig. 30). **H. punctàtum** Lam. Damp places; Que. to Minn., s. to Fla. & Tex.
3. Stem and flowers without black dots......................................4
   4. Leaves with 3–7 strong veins from the base (Figs. 31, 33, 34) or narrow and with no veins in addition to the midrib (Fig. 32)...........................5
   4. Leaves with veins coming at intervals along the midrib (Figs. 35–43).........11
5. Leaves mostly 3 or more times as long as broad, their bases not clasping the stem (Figs. 31, 32). Along the Atlantic Coast the following is a rather complex group, treated in **7**. Inland, there are but two species, described in **6**.
5. Leaves about half as broad as long, their bases clasping the stem (Figs. 33, 34)..**10**
   6. Leaves 5–7-veined, somewhat rounded at base, 2–11 mm. wide; upper flowers somewhat crowded into a flat-topped mass; sepals 5–7 mm. long; pods 5–6.5 mm. long. **H. màjus** (Gray) Britton. Fig. 31. Common on lake shores; e. Que. to Man., s. to n. N.J., Ill., Iowa & S.D.; e. Wash.

Bracts

Leaves

31

32

33

34

H. majus: 31. Upper part of plant, × 1.   H. canadense: 32. Upper part of plant, × 1.
H. boreale: 33. Upper part of plant, × 1.   H. mutilum: 34. Upper part of plant, × 1.

H. **Ascyron**: 35. Upper branch, × 1. **H. denticulatum**: 36. Leaves and stem, × 1. H. **denticulatum** var. **ovalifolium**: 37. Leaves and stem, × 1. **H. ellipticum**: 38. Leaves and stem, × 1. 39. Upper branch, × 1.

H. **adpressum** var. **spongiosum**: 40. Plant, $\times \frac{1}{4}$. **H. tubulosum** var. **Walteri**: 41. Upper branch, $\times \frac{1}{2}$. **H. adpressum**: 42. Leaves and stem, $\times 1$. **H. virginicum**: 43. Summit of stem, $\times \frac{1}{2}$.

6. Leaves 1–3-veined, tapered to the base, 1–4 mm. wide; flowers not crowded; sepals 2.5–5 mm. long; pods 3.5–5.5 mm. long. **H. canadénse** L. Fig. 32. Wet shores; Nfd. to Ga., w. to n. Ind., s. Ont. & locally to Wis.

7. Sepals 5–7 mm. long; pods 5–6.5 mm. long; leaves 5–7-nerved. **H. majus,** 6.

7. Sepals 2.5–5 mm. long; pods 2–5.5 mm. long; sepals 1–5-nerved.............. 8

8. Leaves 1–4 mm. wide, 1–3-veined; pods 3.5–5.5 mm. long..................9
8. Leaves 2–6 mm. wide, 3–5-veined; pods 2–4 mm. long. **H. dissimulàtum** Bicknell. N.S. to N.C.
9. Petals broadly rounded at tip, orange-yellow, with obscure nerves. **H. canadense,** 6
9. Petals pointed, tinged with red toward the tips and along the conspicuous nerves. **H. canadense** var. **magninsulàre** Weatherby. Known only from Grand Manan, N.B.
    10. Upper flowers with reduced, but rounded, leaves at base. **H. boreàle** (Britton) Bicknell. Fig. 33. Pond margins and bogs; Nfd. to Ont., s. to N.J., Pa. & Ind.
    10. Upper flowers with narrow pointed scale-like bracts at base. **H. mùtilum** L. Fig. 34. Low grounds; N.S. to Man., Kans., Tex. & Fla.
11. Plants 5–15 dm. high; leaves 4–9 cm. long; flowers 2.5–5 cm. broad; pod 2–3 cm. long. **H. Áscyron** L. Great St. John's-wort. Fig. 35. Vt. to Man., s. to Pa., Ill., Mo. & Kans.
11. Plants smaller in all parts..12
    12. Petals yellow (rarely very small and red); leaves 1 cm. or less wide.......13
    12. Petals purplish; leaves 1.5 cm. or more wide...........................17
13. Stem square in cross section just below each pair of leaves, with a sharp ridge running down from the bases of the leaf margins and of the midrib (Figs. 36, 37) **14**
13. Stem nearly round or slightly flattened below each pair of leaves, a rounded ridge running down from the base of the midrib only (Fig. 38)..............**15**
    14. Leaves several times as long as wide (Fig. 36). **H. denticulàtum** Walt. Low ground; N.J. to Ill., s. to Fla. & Tenn.   (*H. virgatum*, G, B.)
    14. Leaves about twice as long as wide (Fig. 37). **H. denticulatum** var. **ovalifòlium** (Britton) Blake. N.J. & N.C.
15. Leaf margins flat (Fig. 38). **H. ellípticum** Hook. Fig. 39. Wet shores; N.B. to Man., s. to Pa., Mich., Wis. & Minn.
15. Leaf margins slightly inrolled (Fig. 42)..................................16
    16. Stem not thickened at base. **H. adpréssum** Bart. Wet shores; e. Mass. to Ga. & La.
    16. Stem thickened and spongy at base. **H. adpressum** var. **spongiòsum** Robinson. Fig. 40. E. Mass.
17. Leaves narrowed to a short petiole (Fig. 41). **H. tubulòsum** var. **Waltèri** (Gmel.) Lott. Cypress swamps and bayous; N.J. to Ind. & Mo., s. to Fla. & La.   (*H. petiolatum*, G, B.)
17. Leaves without petioles, rounded or slightly heart-shaped at base (Fig. 43)...**18**
    18. Filaments united to above the middle. **H. tubulòsum** Walt. S. Va., s. Ohio, s. Ind. & Mo., s. to Fla. & La.
    18. Filaments united only at base.....................................**19**
19. Styles on mature fruits 2–3 mm. long; sepals pointed, 5–7 mm. long. **H. virgínicum** L. Fig. 43. Wet places; N.S. & Maine, w. to Ohio, s. to Fla.
19. Styles 0.5–1, rarely –2, mm. long; sepals blunt, 2.5–5 mm. long. **H. virginicum** var. **Fasèri** (Spach) Fernald. Nfd. & Labrador to Man., s. to N.S., Conn., Pa., n. Ind., n. Ill., Iowa & Neb. Where their ranges overlap these two intergrade.

References: Bicknell, Bull. Torr. Bot. Club, **40**, 610 (1913)—*H. dissimulatum;* Blake, Rhodora, **17**, 134 (1915)—*H. denticulatum;* Weatherby, Rhodora, **30**, 188 (1928)—*H. canadense* var. *magninsulare;* Fernald, Rhodora, **38**, 433–436 (1936)— varieties of *H. virginianum* and *H. tubulosum* (as *H. petiolatum*); Lott, Journ. Arn. Arb., **19**, 279 (1938)—*H. tubulosum;* Fassett, Rhodora, **41**, 376 (1939)—forms of *H. boreale* and *H. ellipticum.*

WATERWORT FAMILY   ELATINÀCEAE
WATERWORT   Elatìne

Dwarf plants with opposite leaves usually 1 cm. or less long. When growing under water the stems are erect, a few centimeters long and rather limp; when on land they are prostrate, rooting at the nodes, with leaves nearly flat on the ground. The fruit is a pod as large as a pinhead, with such thin walls that the mature seeds may be seen inside.

The differentiation of species is based almost entirely on mature seeds, which must be examined with a compound microscope. Superficially all species look alike; in fact, members of the same species growing on mud or in the water will differ more in appearance than will different species growing in the same habitat. In other words, the general form of the plant is greatly influenced by submergence or emergence, only the seeds remaining diagnostic.

Of our two species, *E. minima* is the more common and usually grows on sandy lake shores.

1. Pod usually made up of 2 parts (Fig. 46); pits on seed coat with rounded ends, scarcely reduced in size toward the ends of the seed and regularly arranged in distinct lines (Fig. 44). **E. mínima** (Nutt.) Fisch. & Meyer. Fig. 45. On sandy lake shores and tidal river mouths from Nfd. to n. Va.; westward on sandy shores through N.Y., s. Ont., Mich., n.w. Wis. & e. Minn. (*E. americana*, in part, G, B.)
1. Pod 3-parted (Fig. 50); pits on the seed coat with angled ends, reduced in size toward the ends of the seed, in irregular rows (Figs. 52, 53)..................**2**
   2. Pits numbering 16–25 in each row (Fig. 52)..............................**3**
   2. Pits numbering 9–15 in each row (Fig. 53). **E. triándra** var. **brachyspérma** (Gray) Fassett. In drying pools, Pickaway County, Ohio; near Springfield, Ill.; very locally westward to Ore. & Calif. (*E. brachysperma*, G, B.)
3. Leaves elongate, often minutely indented at tip (Fig. 51). **E. triándra** Schkuhr. Rare, in muddy ponds; Adams and Polk Counties, Wis.; Sask., S.D. & Colo. westward; introduced at Skowhegan, Maine. This species has 3 distinct-appearing forms..**4**
3. Leaves broader and rounded at tip (Fig. 54). **E. triandra** var. **americàna** (Pursh) Fassett. Mostly on tidal shores, river mouths from the St. Lawrence River to Del.; in the Ottawa River at Hull, Que.; pond shores in Conn.; Jackson County, Mo.
   4. Aquatic forms with erect limp branches from creeping stems; internodes (1.5–) 3.5–14 mm. long and 0.5–1.5 mm. thick; leaves 2.8–13 mm. long and (0.5–) 0.8–2 mm. broad, bright-green, translucent..............................**5**
   4. Terrestrial forms with much-branched creeping stems and prostrate leaves; internodes 0.5–5 mm. long and 0.3–0.8 mm. thick; leaves 2–5 (–6) mm. long and 0.5–1 (–1.8) mm. broad, dark-green, often shining and reddish, opaque. **E. triandra** forma **terréstris** Seubert. Fig. 49.
5. Internodes (1.5–) 3–8.5 mm. long; leaves 2.8–6.5 mm. long. **E. triandra** forma **intermèdia** Seubert. Fig. 48. A form in very shallow water.
5. Internodes (1.5–) 3.5–14 mm. long; leaves 3–13 mm. long. **E. triandra** forma **submérsa** Seubert. Fig. 47. Usually in 0.5 m. or more of water.

References: Fernald, Rhodora, **19**, 10–15 (1917)—*E. minima;* Fassett, Rhodora, **41**, 367–376 (1939)—revision of *Elatine* in N. A.

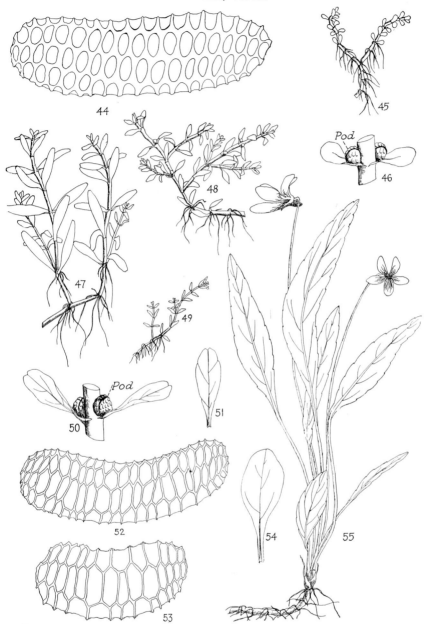

E. **minima**: 44. Seed, × 100.   45. Plant, × 1.   46. Leaves and 2 pods, × 10.   **E. triandra** f. **submersa**: 47. Plant, × 1.   **E. triandra** f. **intermedia**: 48. Plant, × 1.   **E. triandra** f. **terrestris**: 49. Plant, × 1.   **E. triandra**: 50. Leaves and 2 pods, × 10.   51. Leaf, × 10.   52. Seed, × 100.   **E. triandra** var. **brachysperma**: 53. Seed, × 100.   **E. triandra** var. **americana**: 54. Leaf, × 10.   **V. lanceolata**: 55. Plant, × 1.

VIOLET FAMILY   **VIOLÀCEAE**

VIOLET   **Vìola**

V. lanceolàta L.   Lance-leaved Violet.   Fig. 55.   Leaves long and narrow, in a rosette; flowers white; plants spreading by stolons.   Sandy and muddy shores; N.S. to Minn. and southward.

LOOSESTRIFE FAMILY   LYTHRÀCEAE

This family should not be confused with the yellow Loosestrife (*Lysimachia*) in the Primrose Family.

The only conspicuous members of the Lythraceae are the Spiked Loosestrife, with its showy spike of brilliant purple flowers, and the Swamp Loosestrife, which attracts attention by the spongy thickening of the submersed parts of the stem.

Technically, the character of the family rests in its thin-walled many-seeded pod, which is closely surrounded by the cup-like calyx tube (Figs. 4, 13). Superficially these structures closely resemble those of the False Loosestrife, p. 259, but in that group the calyx tube is actually fused with the pod, only the lobes being free (Fig. 19). Specimens of *Ammannia*, *Rotala* and *Didiplis* may be confused with *Ludwigia* and can sometimes be separated with certainty only by dissecting mature fruits under a lens. These 4 genera are separated by vegetative characters on p. 22.

1. Calyx with 4 teeth or lobes...3
1. Calyx with 5–7 teeth...2
    2. Stems woody, often spongy at base, with leaves opposite or in 3's (the upper rarely alternate), tapered to a petiole (Fig. 2); calyx tube about as broad as long (Fig. 3). **Decodon.**
    2. Stems slender, not woody, with leaves all opposite or alternate, not petioled (Figs. 5, 7); calyx tube much longer than broad (Fig. 6). **Lythrum.**
3. Leaves with somewhat clasping broadened bases (Figs. 8, 9). **Ammannia.**
3. Leaf bases not broadened and clasping....................................4
    4. Calyx with little outgrowths alternating with the lobes (Fig. 11); leaves narrowed to a petiole (Fig. 12). **Rotala.**
    4. Calyx without outgrowths; leaves not narrowed to a petiole (Figs. 14, 15). **Didiplis.**

SWAMP LOOSESTRIFE   Décodon

D. verticillàtus (L.) Ell. Figs. 1–4. Plants mostly at the edge and trailing or arching into the water, the stems with a light corky thickening where submersed, 6–25 dm. long, angled; leaves opposite or in 3's, the uppermost with clusters of purple flowers. Also called Water Willow. Typical *D. verticillatus* has the stems and leaves finely downy and occurs from Maine to Minn., s. to Fla. & La. **D. verticillatus** var. **laevigàtus** T. & G. is not downy, and grows from N.S. to Wis., s. to Va. & Tenn.

SPIKED LOOSESTRIFE   Lýthrum

1. Leaves opposite, somewhat heart-shaped at base, the upper much shorter than the flowers; flowers numerous, in long spikes; calyx more or less downy; petals purple, showy. **L. Salicària** L. Figs. 5, 6. Wet meadows, shores, etc., abundant eastward; N.S. to Ont., rarely to e. Wis., s. to N.Y., D.C., Ind. & Mo. Naturalized from Europe.
1. Leaves alternate on the branches, on the stem either alternate, opposite or rarely in 3's; flowers mostly shorter than the leaves, not hairy; petals deep purple. **L. alàtum** Pursh. Fig. 7. Swamps and wet shores; Ont. to Minn., s. to Ga., La. & Colo.; also e. Mass. & Conn.

D. verticillatus: 1. Plant, $\times \frac{1}{6}$. 2. Upper part of branch, $\times \frac{1}{2}$. 3. Calyx, external view, $\times 5$. 4. Flower dissected, $\times 5$; the calyx has been broken away to show the pod within, which is breaking into 3 parts.

Petal

Calyx-lobe

Calyx-tube

L. Salicaria: 5. Upper branch, × ½. 6. Flower, dissected, × 3. **L. alatum:** 7. Upper branch, × ½. **A. auriculata:** 8. Upper branch, × ½. **A. coccinea:** 9. Upper branch, × ½.

Ammánnia

1. Flowers on stalks about as long as the calyx.   **A. auriculàta** Willd.   Fig. 8.   Pond margins, ditches, etc.; w. Mo. & Neb. to Tex., s. to Brazil.
1. Flowers practically without stalks....................................2
    2. Style long and slender. **A. coccínea** Roth. Fig. 9. N.J. to Ohio, Iowa, & S.D., s. to Fla., Tex. & Brazil.
    2. Style very short. **A. Koèhnei** Britton. N.J. to Fla. & Miss.

Rotàla

1. Leaves 1.5–4 (rarely –5) mm. broad; fruits 2–4 mm. long, 2–3.3 mm. broad. **R. ramòsior** (L.) Koehne. Figs. 10, 11, 13. Pond margins; Mass. to Fla. & Tex. near the coast; s. Mich. to Minn., Mo.; Wash. & Ore.
1. Leaves 5–10 mm. broad; fruits 3.5–5 mm. long, 3.8–4.4 mm. broad. **R. ramosior** var. **intèrior** Fern. & Grisc. Fig. 12. N.Y. to Iowa, s. to Fla., La. & Okla.

Water Purslane   Dídiplis

D. diandra (Nutt.) Wood. Minn. & Wis. to Tex., e. to N.C. & Fla. Plants growing in the water or on wet shores, with quite different aspect in the two habitats.
1. The submersed form, with leaves ribbon-like, thin, limp, not narrowed at base (Fig. 14). **D. diándra** forma **aquática** (Koehne) Fassett.
1. The emersed form, with leaves broader, somewhat tapered to each end (Fig. 15) and of firmer texture. **D. diandra** forma **terréstris** (Koehne) Fassett.

References: Fernald, Rhodora, **19**, 154–155 (1917)—*Decodon verticillatus* var. *laevigatus;* Fernald and Griscom, Rhodora, **37**, 169 (1935)—*Rotala ramosior* var. *interior;* Fassett, Rhodora, **41**, 376–377 (1939)—forms of *Didiplis diandra.*

EVENING PRIMROSE FAMILY   ONAGRÀCEAE

A large family with many terrestrial genera and a few aquatic ones. The Evening Primrose has many species, all of which grow on dry land, and there are many kinds of Willow Herb (*Epilobium*) which occur in moist places but rarely in the water.

The principal characters of this family lie in the flowers, whose sepals, petals and stamens are borne at the summit of the ovary (Fig. 19), to which the calyx tube is fused. The lobes of the calyx, usually 4 in number, are often persistent on the fruit, which is in most cases a many-seeded pod.

R. ramosior: 10. Upper branch, × 1.   11. Flower, external view, × 5.   13. Flower'
diagrammatic section, × 5; the pod is surrounded by but not fused with the calyx tube
and contains the young seeds. **R. ramosior** var. **interior**: 12. Upper branch, × 1.
D. diandra f. **aquatica**: 14. Branch, × 1.   **D. diandra** f. **terrestris**: 15. Plant, × 1.

L. palustris var. **americana**: 16. Stem, base of 2 leaves, and 2 fruits, × 5.   17. Plant, × ½.

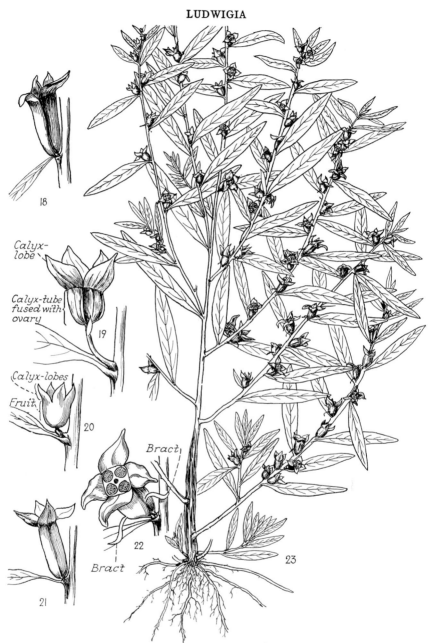

Calyx-lobe

Calyx-tube fused with ovary

Calyx-lobes

Fruit

Bract

Bract

L. linearis: 18. Pod, × 2.   L. alternifolia: 19. Pod, × 2.   L. sphaerocarpa: 20. Pod, × 2.
L. polycarpa: 22. Pod, × 2.   23. Plant, × ½.   L. glandulosa: 21. Pod, × 2.

1. Leaf blade longer than broad, not toothed.................................2
1. Leaf blade toothed, about as broad as long (Fig. 24). **Trapa.**
    2. Stamens as many as the petals; pod less than 1 cm. long. **Ludwigia.**
    2. Stamens twice as many as the petals; pod 1.5–4 cm. long. **Jussiaea.**

FALSE LOOSESTRIFE Ludwígia

1. Leaves opposite...2
1. Leaves alternate..7
    2. Leaves narrowed to a petiole (Fig. 17); flowers not stalked; petals small or absent...3
    2. Leaves without petioles; flowers stalked; petals conspicuous...............4
3. Plants nearly prostrate on the mud or in shallow water, with roots from all but the upper nodes; leaves with short petioles. **L. palústris** (L.) Ell., var. **americàna** (DC.) Fern. & Grisc. Figs. 16, 17. Common and abundant in wet places; N.S. to Man. & Ore., s. to Fla., W.I., Mex. & Calif. (*Isnardia palustris*, B.)
3. Plants in deep water, with limp stems sometimes as long as 8 dm. and held erect by the water; leaves broad, thin, long-petioled. **L. palustris** var. **americana** forma **elongàta** Fassett.
    4. Flowers on stalks longer than the leaves. **L. arcuàta** Walt. In swamps; Va. to Fla.
    4. Flowers on stalks shorter than the leaves.................................5
5. Flowers 12–13 mm. broad, on stalks 5–12 mm. long; sepals 4–6 mm. long. **L. brévipes** (Long) Eames. Moist sand; Long Beach Is., N.J.; e. Va.
5. Flowers 5.5–6.5 mm. broad, on stalks 1–2.5 mm. long; sepals 2.5–3.0 mm. long. 6
    6. Plants prostrate on the mud, rooting at the nodes. **L. lacústris** Eames. Ponds; Conn.
    6. Plants erect in deep water, with limp stems. **L. lacustris** forma **aquátilis** Eames.
7. Flowers without stalks or on stalks shorter than the ovary (Figs. 18, 20–22)....8
7. Flowers with stalks longer than the ovary (Fig. 19). **L. alternifòlia** L. Seedbox. Swamps; e. Mass. to Fla. & Tex., and from s.w. Ont. to Kans. & southward.
    8. Fruit nearly as thick as long (Figs. 20, 22).................................9
    8. Fruit several times as long as thick (Figs. 18, 21)........................10
9. Fruit nearly globular, shorter than the calyx lobes, lightly hairy, without a conspicuous bract (Fig. 20). **L. sphaerocárpa** Ell. Swamps and sloughs; e. Mass. southward along the coast to Fla. & La.; n. Ind. In this species and the next the lower part of the stem is often thick and spongy.
9. Fruit somewhat 4-sided, longer than the calyx lobes, not hairy, with 2 bracts near its base (Fig. 22). **L. polycárpa** Short & Peter. Fig. 23. S. Maine to s. Ont., Minn. & Neb., s. to Tenn. & Kans.
    10. Fruit with parallel sides, abruptly narrowed at base (Fig. 21). **L. glandulòsa** Walt. Wet woods, ponds and swamps; s. Ill. to Va., Fla. & Tex.
    10. Fruit tapered from apex to base (Fig. 18). **L. lineàris** Walt. N.J. to Fla. & Tex.

T. natans: 24. Plant, × ¼. 25. Fruit, × ½. **J. diffusa:** 26. Plant, × ¼. **J. decurrens:** 27. Stem, leaf and fruit, × ½.

WATER PRIMROSE, PRIMROSE WILLOW   **Jussiaèa**

Flowers with 4–5 conspicuous yellow petals, borne at the summit of a long slender ovary.

1. Stem erect, with wings from the leaf bases (Fig. 27); calyx lobes and petals 4. **J. decúrrens** (Walt.) DC. Ponds; Va. to Ind. & Mo., s. to Fla. & Tex.
1. Stem prostrate, rooting at the nodes, not winged; calyx lobes and petals 5. **J. diffùsa** Forsk. Fig. 26. Ky. & Ill. to Kans., s. to Fla. & Tex.; tropical America.

WATER CHESTNUT   **Tràpa**

T. nàtans L. Fig. 24. Plants floating, usually rooting in the mud, with leaves clustered toward the tip of the stem; leaves with rather 4-sided blades and long petioles which have an inflated spongy region; fruit (Fig. 25) nut-like, 4-horned, edible. Locally abundant in the Atlantic states; introduced from Eurasia.

References: Eames, Rhodora, **35**, 228 (1933)—*L. brevipes* and *L. lacustris;* Fernald and Griscom, Rhodora, **37**, 176 (1935)—*L. palustris* var. *americana;* Fassett, Rhodora, **41**, 377 (1939)—*L. palustris* var. *americana* f. *elongata.*

WATER MILFOIL FAMILY   **HALORAGIDÀCEAE**

Stems simple or slightly branched, with leaves variously arranged, often pinnately compound; flowers without stalks, in the axils of leaves or bracts. The foliage is variable even on the same plant, the leaves in the water being usually more deeply divided than are those in the air. The genera are based on floral characters but may be separated artificially as follows:

1. Leaves cut into thread-like divisions (Figs. 30, 39, 41–51)....................**2**
1. Leaves not divided..**3**
    2. Leaves whorled or scattered and crowded around the stem (Figs. 41–51). **Myriophyllum.**
    2. Leaves scattered, not crowded (Figs. 28–30). **Proserpinaca.**
3. Leaves scattered..**4**
3. Leaves whorled (Fig. 35). **Hippuris.**
    4. Leaves several centimeters long (Fig. 28). **Proserpinaca.**
    4. Leaves only about 1 mm. long (Fig. 40). **Myriophyllum.**

MERMAID WEED   **Prosperinàca**

Calyx lobes 3; petals none; stamens 3; fruit 3-sided (Figs. 32–34); stems mostly creeping at base. These plants, especially *P. palustris*, are noted for their habit of producing deeply divided leaves on the submersed portions of the stem and unlobed leaves on the emersed; changes in water level may produce alternate sets of the two types. In *P. pectinata* all the leaves tend to be deeply lobed.

P. palustris var. **amblyogona**: 28. Plant, $\times \frac{1}{2}$. 32. Fruit, $\times 3$. **P. intermedia**: 29. Part of stem and leaves, $\times \frac{1}{2}$. **P. pectinata**: 30. Part of stem and leaves, $\times \frac{1}{2}$. 34. Fruit, $\times 3$. **P. palustris** var. **crebra**: 33. Fruit, $\times 3$. **H. vulgaris** f. **fluviatilis**: 31. Part of branch, $\times \frac{1}{2}$. **H. vulgaris**: 35. Part of branch, $\times \frac{1}{2}$.

1. Flowers in the axils of unlobed leaves (Fig. 28); calyx lobes on the mature fruit as broad as long (Fig. 33)..2
1. Flowers in the axils of deeply lobed leaves; calyx lobes on the mature fruit longer than broad (Fig. 34)..4
   2. Mature fruit 4–6 mm. broad. **P. palústris** L. N.J. & Va. to W.I., w. to La. & Mo. Grades into the next variety on the Atlantic Coast.
   2. Mature fruit 2.3–4 mm. broad....................................3
3. Fruit with sharply ridged angles (Fig. 33). **P. palustris** var. **crèbra** Fern. & Grisc. Mud and shallow water; N.S. to Wis., s. to Ala. & Okla. Grades into the next variety.
3. Fruit with rounded angles (Fig. 32). **P. palustris** var. **amblyógona** Fernald. Fig. 28. Often replacing the last; Ont. & Ind., s. to Mo.
   4. Rachis of leaves about as broad as the segments (Fig. 30). **P. pectinàta** Lam. Fig. 34. Sandy swamps; N.S. to Fla. & La.
   4. Rachis of leaves broader than the segments (Fig. 29). **P. intermèdia** Mack. Rare; e. Mass. to Ga. Intermediate between *P. palustris* and *P. pectinata*, sometimes growing with both and perhaps a hybrid between them.

MARE'S-TAIL   **Hippùris**

H. vulgàris L. Fig. 35. Stems unbranched, usually with the lower part submersed and with limp ribbon-like leaves 1–10 cm. long, the upper part emersed and with flowers in the axils of shorter more rigid leaves. Some plants are wholly submersed, and others wholly emersed, with but one type of leaf. Usually in cold water; Labrador to Alaska, s. to n. N.E., N.Y., Ind., Ill., Wis., Minn., Neb., Ariz. & N.M. The typical form has a rather rigid stem and submersed leaves 3–5 cm. long; of occasional occurrence is **H. vulgaris** forma **fluviàtilis** (Hoffm.) Cosson & Germain (Fig. 31), with a weak stem and longer submersed leaves.

WATER MILFOIL   **Myriophýllum**

Calyx lobes 4; petals 4 or none; stamens 4 or 8; fruit of 4 joined nut-like bodies. With a few exceptions, the species look much alike, and some are nearly impossible to distinguish without flowers or fruits. In the foliage there are sometimes differences that may be seen by minute comparison of individuals, but these are almost impossible to describe intelligibly.

1. Leaves cut into thread-like (Fig. 47) or narrow ribbon-like (Fig. 39) segments..2
1. Leaves small, roundish, inconspicuous. **M. tenéllum** Bigel. Fig. 40. Usually submersed, in sandy ponds and streams; Nfd. to Wis., s. to N.J. & Pa.
   2. Foliage leaves all whorled (Figs. 37, 39, 41, 44)..........................3
   2. Foliage leaves partly whorled and partly scattered (Figs. 47–49, 51)........8
3. Flowers in spikes borne above the water....................................4
3. Flowers in the axils of submersed leaves. **M. brasiliénse** Cambess. Parrot-feather. A native of S.A., cultivated in aquariums and rarely escaping. (*M. proserpinacoides*, B.)
   4. Floral bracts mostly scattered (Fig. 37); leaves 5–12 mm. long. **M. alterniflòrum** DC., var. **americànum** Pugsley. Small Water Milfoil. Nfd. to n. Wis., s. to Conn. & Vt.
   4. Floral bracts whorled (Figs. 41, 43, 44)....................................5

MYRIOPHYLLUM

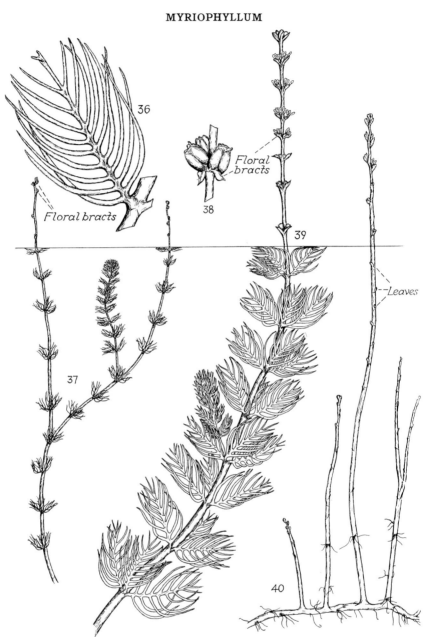

M. exalbescens: 36. Leaf, × 3.   38. Node with fruits, × 3.   39. Upper part of plant, × 1.
M. alterniflorum var. americanum: 37. Upper branches, × 1.   M. tenellum: 40. Plant, × 1.

264

MYRIOPHYLLUM

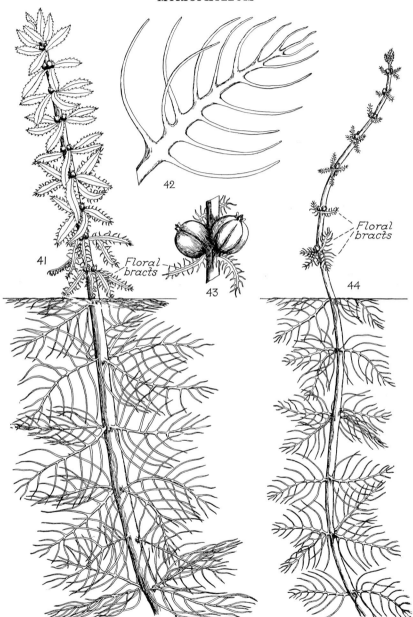

M. heterophyllum: 41. Upper part of plant, × 1.   **M. verticillatum:** 42. Leaf, × 3.   43. Node and fruits, × 3.   44. Upper part of plant, × 1.

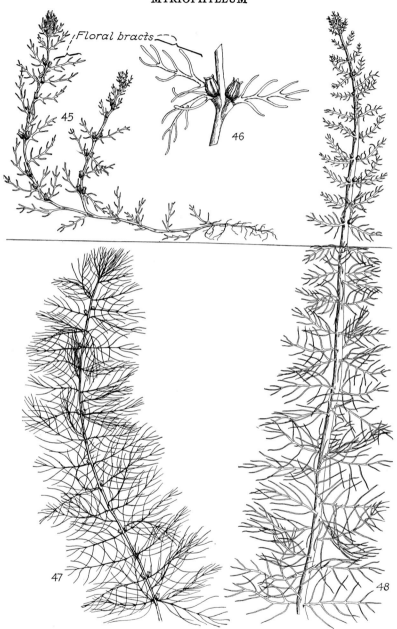

M. humile: 45. Emersed plant, × 1.   46. Node and fruits, × 3.   **M. Farwellii:** 47. Upper part of plant, × 1.   **M. humile f. natans:** 48. Upper part of plant, × 1.

M. scabratum: 49. Terrestrial form, × 1. 50. Two nodes and fruits, × 3. 51. Partly submersed form, × 1.

5. Bracts shorter than the flowers and fruits, not lobed (Fig. 38); rachis of leaves thread-like and of nearly equal diameter throughout (Fig. 36), the segments not broadened at base.   **M. exalbéscens** Fernald.   Fig. 39.   Common; Nfd. to Wash., s. to Conn., N.Y., Ill., Kans., Ariz. & Calif.   (*M. spicatum*, G, B.)

5. Bracts mostly longer than the flowers, toothed (Fig. 41) or lobed (Fig. 44); rachis of leaves flat and much broader toward the base of the leaf (Fig. 42), the segments broadened at base..6

   6. Bracts lance-ovate, toothed (Fig. 41).   **M. heterophýllum** Michx.   Va. to Fla., w. to Ont. & Minn., s. to Mo. & Tex.

   6. Bracts deeply lobed (Figs. 43, 44)......................................7

7. Stamens 8.   **M. verticillàtum** L.   Green Milfoil.   Figs. 42–44.   Greenland (?); Nfd. to B.C., s. to n. & w. N.E., N.Y., Ill. & Utah.   The bracts vary from 2–8 mm. in length and are often shorter than the flowers or fruits.

7. Stamens 4.   **M. verticillatum** var. **Chenèyi** Fassett.   Local; Hudson River, N.Y., & n. Wis.

   8. Fruit segments plump, not ridged or roughened on the back (Fig. 46)......9

   8. Fruit segments ridged or with little bumps over the back (Fig. 50)........11

9. Floral leaves borne on emersed stems, with flat segments..................10

9. Floral leaves borne on submersed stems, with thread-like segments.   **M. hùmile** forma **capillàceum** (Torr.) Fernald.   A submersed form of the next.

   10. Stems creeping, rooting in the mud.   **M. hùmile** (Raf.) Morong.   Figs. 45, 46.   N.S. to Vt., w. to Ill.

   10. Stems erect, supported by the water.   **M. humile** forma **nàtans** (DC.) Fernald.   Fig. 48.   A partly submersed form of the last.

11. Floral leaves on emersed stems, their segments flat (Fig. 50); segments of submersed leaves, when present, about 0.2 mm. wide at base, with nearly parallel sides.   **M. scabràtum** Michx.   Figs. 49–51.   E. Mass. to S.C., w. Ky., & w. Tenn., w. to Iowa & Tex.   (*M. pinnatum*, B.)

11. Flowers in the axils of ordinary submersed leaves; segments of leaves about 0.1 mm. wide at base and tapering from base to tip.   **M. Farwéllii** Morong.   Fig. 47. N.S. & Que. to cent. Wis., s. to s. Maine, N.H., Vt. & N.Y.

References: Fernald, Rhodora, **11**, 120 (1909), and Fernald and Griscom, Rhodora, **37**, 177 (1935)—varieties of *Prosperpinaca palustris;* Mackenzie, Torreya, **10**, 250 (1910)—*P. intermedia;* Fernald, Rhodora, **21**, 120–122 (1919)—*Myriophyllum exalbescens;* Hegi, Ill. Fl. von. Mitt.-Eur., **5**, pt. 2, 907 (1927?), and Glück, Süsswasser Fl. Mitteleur., **15**, 338–339 (1936)—*Hippuris vulgaris* f. *fluviatilis;* Pugsley, Journ. Bot., **76**, 51–53 (1938)—*M. alterniflorum* var. *americanum;* Fassett, Rhodora, **41**, 524 (1939)—*M. verticillatum* var. *Cheneyi.*

PARSLEY FAMILY   UMBELLÍFERAE

The fundamental characters of this family lie in the fruit, which is composed of 2 nut-like parts ordinarily crowned by the lobes of the calyx (Fig. 57), and in the arrangement of the flowers, which is usually an umbel.   An umbel is a group of flowers on radiating stalks of essentially equal length.   In most of the genera described here the umbel is compound, each ray bearing at its tip a small umbel (Figs. 64, 67), but in the Pennywort the stalks may be so short that each umbel becomes practically a head (Fig. 52).   The genera are based largely on the mature fruit, but our aquatic genera may be artificially separated as follows:

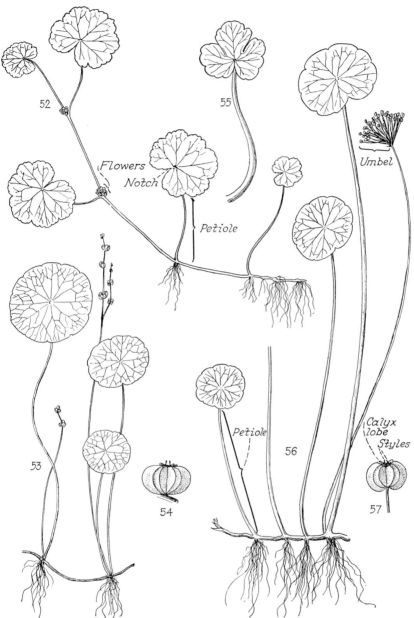

H. americana: 52. Plant, × ½. **H. ranunculoides:** 55. Leaf, × ½. **H. verticillata:** 53. Plant, × ½. 54. Fruit, × 5. **H. umbellata:** 56. Plant, × ½. 57. Fruit, × 5.

1. Leaves round (Figs. 52, 53, 56). **Hydrocotyle.**

1. Leaves cut into leaflets which are longer than wide..........................**2**

   2. Leaves palmately divided (Figs. 65, 70)...................................**4**

   2. Leaves pinnately divided (Figs. 58–61) or rarely not divided..............**3**

3. Leaflets margined by jagged teeth (Fig. 68). **Berula.**

3. Leaflets margined by regular teeth (Figs. 58, 60, 61), or deeply cut and compound (Fig. 59) or simple (Fig. 62). **Sium.**

   4. Sheaths 5–10 cm. long, inflated (Fig. 63). **Angelica.**

   4. Sheaths seldom more than 3 cm. long (Figs. 65, 70), scarcely inflated. **Cicuta.**

WATER PENNYWORT   Hydrocótyle

Leaves nearly round, with long petioles attached near the middle, the margins with broadly rounded teeth; flowers in umbels or heads which are sometimes almost without stalks; fruits rounded, each half with 3 ridges (Figs. 54, 57).

1. Petioles attached at the base of a deep notch (Fig. 52); groups of flowers on very short stalks...**2**

1. Petioles attached at the center of the blade with no notch (Figs. 53, 56); groups of flowers on long stalks..**3**

   2. Leaves shallowly lobed; groups of flowers almost without stalks. **H. americàna** L. Fig. 52. Very common in moist places; N.S. to Minn., s. to N.J., Pa. & N.C.

   2. Leaves deeply lobed (Fig. 55); groups of flowers on stalks 2.5–7.5 cm. long. **H. ranunculoìdes** L.f. Usually floating; Pa. to Fla. near the coast, w. to Ark. & s. to S.A.; Pacific Coast.

3. Fruit notched at base and apex (Fig. 57). **H. umbellàta** L. Fig. 56. Meadows, ditches, etc.; N.S. to Fla. & Tex., Mich., Ind. & perhaps Minn.

3. Fruit not notched (Fig. 54)..**4**

   4. Petioles not hairy...**5**

   4. Petioles with minute scattered hairs. **H. verticillata** var. **Fetherstoniàna** (Jennings) Mathias. In a stream, Wyoming County, N.Y.

5. Inflorescence often forking; stalks of fruits 2 mm. or less long. **H. verticillàta** Thunb. Fig. 53. Moist soil; Mass. to Fla., Mo. & Tex.; W.I.; Cent. & S. A.

5. Inflorescence rarely forking; stalks of fruits 1–10 mm. long. **H. verticillata** var. **triradiàta** (A. Richard) Fernald. Wet sand at low elevations, along and near the coast; Mass. & Calif. to S. A. (*H. Canbyi*, G, B; *H. australis*, B.)

References: Mathias, Brittonia, **2**, 240 (1936)—*H. verticillata* var. *Fetherstoniana;* Fernald, Rhodora, **41**, 437 (1939)—var. *triradiata.*

WATER PARSNIP   Sìum

Generally erect, with stout hollow fluted stems; leaves, at least on the stems, usually pinnate, with sharp-toothed leaflets, variable in width; fruits roundish, with prominent ribs. The basal rosettes are often partly or wholly submersed and may bear several types of leaves.

SIUM

S. suave: 58. Type of leaf usually found on stems, × ½. 59, 60. Types of leaf usually found submersed, × ½. **S. suave** f. **Carsonii:** 61. Leaf, × ½. **S. suave** f. **fasciculatum:** 62. Part of plant, showing basal leaves and one fascicle, × ½.

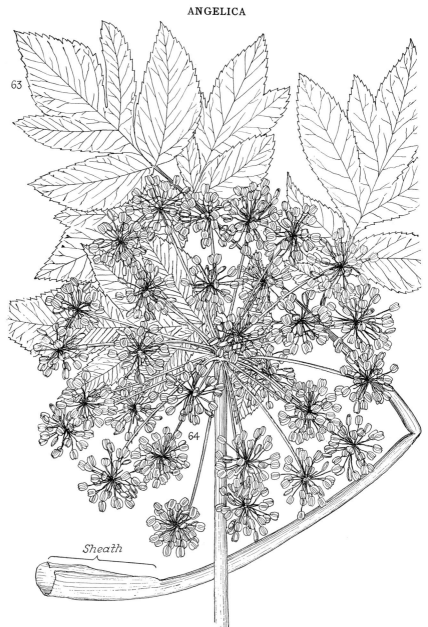

63

64

Sheath

A. **atropurpurea**: 63. Lower leaf, $\times \frac{1}{2}$.   64. Umbel, $\times \frac{1}{2}$.

C. maculata: 65. Leaf, $\times \frac{1}{2}$. 67. Umbel, $\times \frac{1}{2}$. 69. Root, and base of stem split lengthwise, $\times \frac{1}{2}$. **C. bulbifera:** 70. Upper part of plant, $\times \frac{1}{2}$. **B. erecta:** 66. Upper part of plant, $\times \frac{1}{2}$. 68. Leaf, $\times \frac{1}{2}$.

1. Stems stout, 0.8–2 m. high; leaves with 3–8 pairs of leaflets. **S. suàve** Walt. Figs. 58–60. Muddy places and shallow water; N.S. to B.C., s. to Fla., La. & Calif. (*S. cicutaefolium*, G, B.)

1. Stems weak, often reclining; leaves with 1–3 pairs of leaflets or sometimes simple. . **2**

    2. Leaves mostly compound and borne singly. **S. suave** forma **Carsònii** (Durand) Fassett. Fig. 61. A more or less submersed form. (*S. Carsonii*, G, B.)

    2. Primary leaves simple or compound, with axillary bunches of very small simple leaves. **S. suave** forma **fasciculàtum** Fassett. Fig. 62. Usually on tidal river shores, rarely in ponds and lakes with fluctuating water level.

References: Blake, Rhodora, **17**, 131 (1915)—*S. suave;* Fassett, Rhodora, **23**, 111–113 (1921)—forms of *S. suave.*

Angèlica

A. atropurpùrea L. Plants stout, often 2 m. tall; stem streaked with purple and green and with a powdery bloom; leaves 3-parted, the divisions stalked and pinnately divided (Fig. 63), without hairs; umbels spherical, long-stalked, many aggregated into a large spherical umbel (Fig. 64). Wet open ground; Nfd. to Del., w. to s.e. Wis. & Ill. **A. atropurpurea** var. **occidentàlis** Fassett has the leaflets finely hairy beneath. Upper Peninsula of Mich. to s. Wis.

Reference: Fassett, Rhodora, **33**, 74 (1931)—*A. atropurpurea* var. *occidentalis.*

Water Hemlock   Cicùta

Stems hollow, erect or somewhat reclining; leaves 3-parted, each part bearing 3 or more leaflets; flowers in a compound umbel; fruit roundish with flat corky ribs.

1. Leaflets 0.5–2 cm. wide; umbels many; plant without bulbets. **C. maculàta** L. Spotted Cowbane; Musquash Root. Figs. 65, 67. Wet places; N.B. to Man., s. to Fla. & N.M. The roots (Fig. 69) are swollen and fleshy, with an attractive odor, and are very poisonous.

1. Leaflets 1–5 mm. wide, ribbon-like; umbel usually only one, rarely producing fruits; leaves usually with bulblets in their axils. **C. bulbífera** L. Fig. 70. Wet shores and shallow water; Nfd. to N.B., s. to Md., Ind., Neb. & Ore.

Bérula

B. erécta (Huds.) Coville. Figs. 66, 68. Plants rather delicate, 2–9 dm. high, the leaves pinnate, with sharply toothed leaflets; umbels compound, on several branches; fruit round, with inconspicuous ribs. Usually in springs, often mixed with Water Cress; s. Ont. to B.C., s.to Ill., Neb., N.M. & Calif.

DOGWOOD FAMILY   CORNÀCEAE

Woody plants with simple leaves which are not toothed or rarely have a few coarse teeth; calyx tube joined to the ovary, which is capped by the minute calyx lobes that persist on the fruit (Fig. 72), the 4 or 5 white or greenish petals and 4–12 stamens.

Aborted fruit

Calyx-lobes

N. aquatica: 71. End of twig, $\times$ ½. **N. sylvatica** var. **caroliniana**: 72. End of twig, $\times$ ½. **N. sylvatica**: 73. End of twig, $\times$ ½. **N. sylvatica** var. **biflora**: 74. End of twig, $\times$ ½.

275

S<small>OUR</small> G<small>UM</small>; T<small>UPELO</small>   **Nýssa**

Trees, in wet or dry soil, often in standing water, in which case the trunks are usually enlarged at base; leaves scattered; staminate flowers numerous, with 5–12 stamens; pistillate flowers solitary or few in a cluster, with 5–10 imperfect stamens.

1. Leaves 1–3 dm. long, usually with a few teeth, often somewhat heart-shaped at base, downy beneath when young; fruits solitary at the end of each stalk, blue, 2–3 cm. long. **N. aquática** L. Cotton Gum. Fig. 71. Swamps and ponds; N.J. to Fla. & Tex., and in the Mississippi Valley to Mo. & s. Ill.

1. Leaves 2–15 cm. long, without teeth, the hairs on the lower surface scattered or none; pistillate flowers in clusters of 3–8 at the end of each stalk, so that the fruits are either in clusters or accompanied by one or more aborted fruits (Fig. 72); fruits bluish-black, 7–15 mm. long...**2**

    2. Leaves 2.5–8 (–10) cm. long, abruptly short-pointed at tip (Figs. 73, 74).....**3**

    2. Leaves 8–15 cm. long, tapering to a long point. **N. sylvática** var. **caroliniàna** (Poir.) Fernald. Fig. 72. S. Pa. to n. Ohio & s. Ont., s. to N.C., n. Miss. & e. Tex.

3. Leaves 2–6 cm. wide when well developed. **N. sylvatica** Marsh. Fig. 73. S. Maine to s. Ont. & s.e. Wis., s. to Fla., Ark. & e. Tex.

3. Leaves 1.5–3 cm. wide when well developed. **N. sylvatica** var. **biflòra** (Walt.) Sarg. Fig. 74. S.e. Md. to Fla., w. to s. La. & e. Tex.

D<small>OGWOOD</small>   **Córnus**

Mostly woodland shrubs with opposite leaves; flowers with 4 white petals, in a round-topped close cluster; fruit blue or white, fleshy, one-seeded.

C. stonífera Michx. Red-osier Dogwood. Fig. 78. Twigs dark-red, especially conspicuous in early spring; pith large; fruits white. Wet ground, often bordering bogs; Nfd. to Alaska, s. to Va., Ky., Neb., Ariz. & Calif. **C. stolonifera** var. **Bailèyi** (Coult. & Evans) Drescher. Leaves velvety beneath. Pa. & N.Y. westward about the Great Lakes, and to the Rocky Mts., sometimes appearing distinct and again grading into the species.

References: Drescher, Trans. Wis. Acad., **28**, 190 (1933)—*Cornus stolonifera* var. *Baileyi;* Fernald, Rhodora, **37**, 433–437 (1935)—varieties of *Nyssa sylvatica.*

HEATH FAMILY   **ERICÀCEAE**

A large family occurring mostly on dry soil and in bogs. In most species the corolla is bell-shaped or tubular, 4-5-lobed, conspicuous, and the fruit is a pod or a berry.

Chamaedáphne calyculàta (L.) Moench. Leatherleaf. Fig. 77. Leaves with brown shield-shaped scales on both surfaces, these appearing as dots on the upper surface and as a brown covering on the lower; flowers in the axils of reduced leaves (Fig. 75), with a white tubular corolla, appearing in early spring; fruit a dry pod (Fig. 76). Bogs, and often at the water's edge; Labrador to B.C., s. to Minn., Wis., Ill. & Ga.

Chamaedaphne calyculata: 75. Flower in axil of leaf, × 2. 76. Fruit in axil of leaf, × 2.
77. Flowering branch, × 1. **Cornus stolonifera**: 78. Fruiting branch, × ½.

H. inflata: 1. Floating plant, $\times \frac{1}{4}$.   **L. thyrsiflora:** 2. Upper part of plant, $\times \frac{1}{2}$.   **S. floribundus:** 3. Plant, $\times \frac{1}{2}$.

PRIMROSE FAMILY   **PRIMULÀCEAE**

Flowers with sepals united into a tubular or bell-shaped calyx; petals united, sometimes with 5–7 long lobes; calyx, corolla and stamens borne at the base of the ovary, which becomes a pod with the seeds borne on a central knob within.

1. Immersed leaves deeply divided (Fig. 1), the emersed stems swollen.   **Hottonia.**
1. All leaves simple; stems not inflated..**2**
   2. Leaves in a basal rosette and scattered on the stem (Fig. 3).   **Samolus.**
   2. Leaves opposite (Figs. 4–7).   **Lysimachia.**

Featherfoil   Hottònia

H. inflàta Ell.   Fig. 1.   Stem submersed, limp, bearing deeply divided leaves above and scattered roots below; flowering stems emerging, leafless, inflated.   Quiet water; s. Maine to Fla., and north in the Mississippi Valley to Mo. & Ind.

Water Pimpernel   Sámolus

S. floribúndus HBK.   Fig. 3.   Leaves in a basal rosette and scattered on the stem; flowering branches mostly from the axils of the upper leaves; flowers white, on long slender stalks.   Wet places; N.B. to B.C., s. to W.I. & S.A.

Loosestrife   Lysimàchia

Corolla yellow, united at base and with 5 (sometimes 4, 6, or 7) spreading lobes; leaves opposite, or appearing whorled on the branches.

1. Stem square in cross section; flowers 1 cm. or more broad (Figs. 4–7), not in heads..**2**
1. Stem round in cross section; flowers small and crowded into dense long-stalked heads.   **L. thyrsiflòra** L.   Tufted Loosestrife.   Fig. 2.   Swamps; Que. to Alaska, s. to e. Mass., s. N.Y., Pa., Mo., Neb., Mont. & Calif.
   2. Leaves when pressed showing black dots, often with purple-brown bulblets in their axils (Fig. 10); petals marked with black streaks.   **L. terréstris** (L.) BSP. Swampcandle.   Fig. 5.   Low wet ground; Nfd. to Man., s. to Ga. & Ark.
   2. Leaves and petals not marked with black; petioles fringed (Figs. 6–9, 13, 14); bulblets absent...**3**
   The following species are often treated as a separate genus, *Steironema* (G, B.)
3. Leaves narrow and ribbon-like, the margins inrolled and the side veins obscure (Fig. 8).   **L. quadriflòra** Sims.   Fig. 4.   Wet shores; N.Y. to Man., s. to Va. & Mo.
3. Leaves 1 cm. or more broad, the margins flat and the side veins prominent....**4**
   4. Stem sprawling or prostrate, often rooting at the nodes; calyx 3–5 mm. long.   **L. radìcans** Hook.   W. Tenn. & Mo. to Miss. & e. Tex.; Va.
   4. Stem erect or arching, not rooting at the nodes; calyx 5–10 mm. long........**5**
5. Leaves rounded or heart-shaped at the base of the blade (Fig. 9).   **L. ciliàta** L. Fringed Loosestrife.   Fig. 6.   Wet ground; N.S. to B.C., s. to Ga., Ala., Kans., N.M. & Ariz.

L. quadriflora: 4. Plant, × ½. **L. terrestris:** 5. Summit of plant, × ½. **L. ciliata:** 6. Summit of plant, × ½. **L. lanceolata:** 7. Plant, × ½.

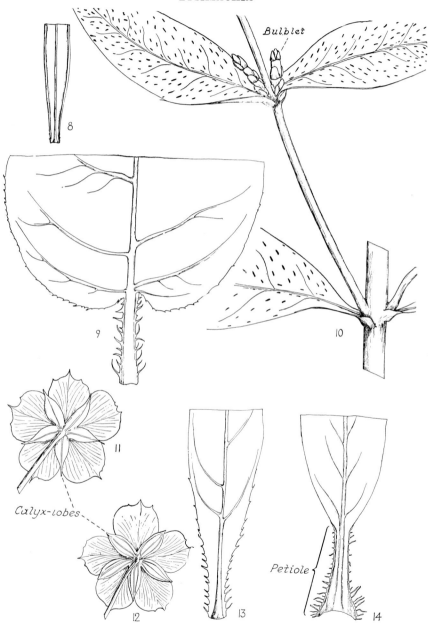

LYSIMACHIA

L. quadriflora: 8. Base of leaf, × 2. L. ciliata: 9. Base of leaf, × 2. L. terrestris: 10. Stem, leaves, and bulblets, × 2. L. lanceolata: 11. Flower, seen from below, × 2. 13. Base of leaf, × 2. L. hybrida: 12. Flower, seen from below, × 2. 14. Base of leaf, × 2.

281

5. Leaves, except the lower ones, tapered at base (Fig. 7)......................6
   6. Leaves with slender runners at base (Fig. 7); middle leaves not petioled, fringed at base (Fig. 13); calyx lobes with all but the midvein obscure (Fig. 11). **L. lanceolàta** Walt. Swamps, shores, etc.; Pa., Ohio, s. Mich. & Wis., s. to Fla. & La.
   6. Plants without runners; middle leaves petioled, the petioles but not the blades fringed (Fig. 14); calyx lobes with 3 veins (Fig. 12). **L. hỳbrida** Michx. Que. to w. Ont. & N.D., s. locally to Fla. & Tex.

Reference: Fernald, Rhodora, **39**, 438–442 (1937)—*L. lanceolata* and *L. hybrida.*

OLIVE FAMILY   OLEÀCEAE

Ash   Fráxinus

Large trees; flowers small and inconspicuous; fruits single-seeded, with a broad flat wing. Most species are upland or forest trees, but three of them occur in swamps and sometimes in shallow water.

1. Leaflets not stalked (Fig. 15); calyx absent (Fig. 17). **F. nìgra** Marsh. Black Ash. Swamps and wet woods; Nfd. to Man., s. to Va., Ill. & Ark.
1. Leaflets stalked (Figs. 16, 20); calyx present (Figs. 18, 19)....................2
   2. Petioles velvety; wing of fruit indented at tip (Fig. 18). **F. tomentòsa** Michx. f. Pumpkin Ash. Fig. 16. Swamps; Va. & w. N.Y. to Mo., s. to Fla. & Ark. (*F. profunda*, G, B.)
   2. Petioles not velvety, sometimes hairy; wing of fruit rounded or pointed at tip (Fig. 19). **F. caroliniàna** Mill. Water Ash. Fig. 20. Swamps; s. Va. to Fla., La. & Mo.

Reference: Fernald, Rhodora, **40**, 450 (1938)—*F. tomentosa.*

Swamp Privet   Forestièra

F. acuminàta (Michx.) Poir. Shrub or small tree; leaves opposite, simple, the margins with short rounded teeth except toward the base; flowers in clusters; fruit dark-purple, fleshy, one-seeded, about 1.5 cm. long. Sloughs and swamps; s.w. Ind. & s. Ill. to n. Fla. & Tex. (*Adelia acuminata*, G.)

GENTIAN FAMILY   GENTIANÀCEAE

A large family of terrestrial plants, with one genus strictly aquatic.

Floating Heart   Nymphoìdes

Stems rising from buried rootstocks, slender, limp, often bearing near the surface a cluster of roots, a single leaf and several flowers.

1. Stem with a cluster of short thick roots borne with the white flower (Fig. 21)...**2**
1. Stem without clusters of roots; flowers bright yellow. **N. peltàtum** (Gmel.) Britten & Rendle. Cultivated, escaping in D.C. & e. N.Y., and perhaps elsewhere.
   2. Floating leaves with smooth thread-like petioles, their blades 1.5–5 cm. broad and scarcely pitted beneath; corolla 0.5–1 cm. broad; capsule 3–4 mm. long; seeds smooth. **N. cordàtum** (Ell.) Fernald. Fig. 21. N.S. to Fla., w. to Ont., Minn. & La. (*N. lacunosum*, G, B.)

FRAXINUS

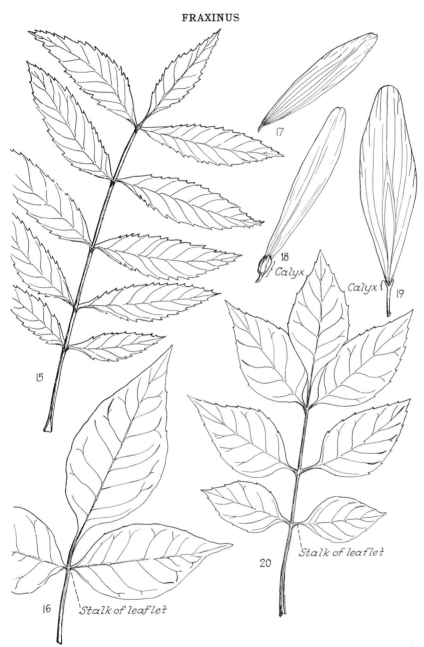

F. nigra: 15. Leaf, × ½. 17. Fruit, × 1. **F. tomentosa:** 16. Upper part of leaf, × ½.
18. Fruit, × 1. **F. caroliniana:** 19. Fruit, × 1. 20. Leaf, × ½.

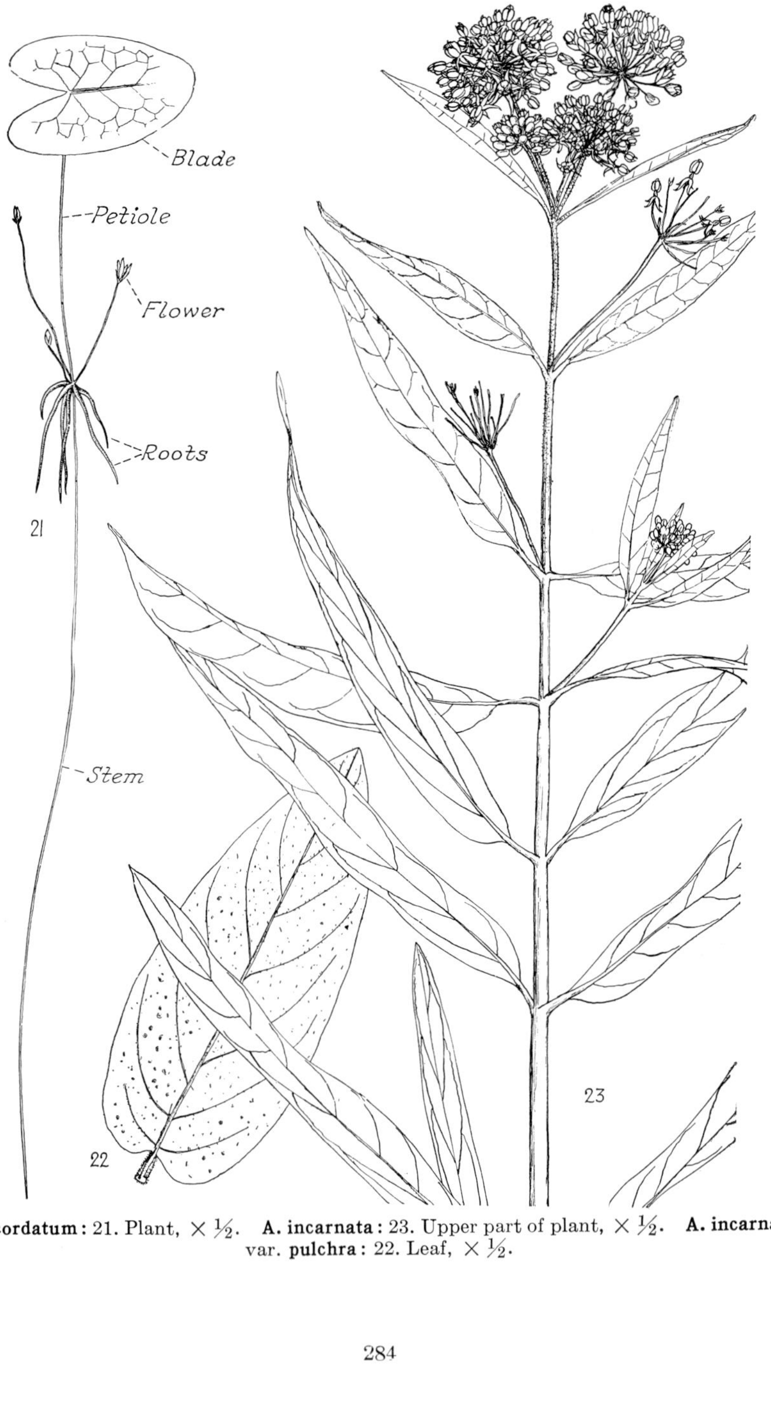

Blade

Petiole

Flower

Roots

21

Stem

22

23

N. cordatum: 21. Plant, × ½.   A. incarnata: 23. Upper part of plant, × ½.   A. incarnata var. pulchra: 22. Leaf, × ½.

2. Floating leaves with cord-like often purple-glandular petioles; blades 0.4–1.5 dm. broad, coarsely pitted beneath; corolla 1–2 cm. broad; capsule 6–9 mm. long; seeds glandular-roughened. **N. aquáticum** (Walt.) Ktze. S. N.J. & Del. to Fla. & Tex.

Reference: Fernald, Rhodora, **40**, 338 (1938)—*N. cordatum.*

MILKWEED FAMILY   ASCLEPIADÀCEAE

Plants usually with white milky juice; corolla rose-purple, or rarely white, 5-lobed, its lobes turned sharply downward; a 5-lobed crown between the corolla and the stamens. Only one species is commonly found in the water.

Milkweed   Asclèpias

1. Leaves long and narrow, nearly or quite without hairs, with petioles 1 cm. or more long. **A. incarnàta** L. Swamp Milkweed. Fig. 23. Swamps, lake shores, stream banks, etc.; N.B. to Sask., s. to Tenn., La. & Colo.
1. Leaves broader, shorter petioled (Fig. 22) 2
    2. Plants with 11–21 pairs of leaves which are finely hairy beneath and 0.9–1.8 dm. long. **A. incarnata** var. **púlchra** (Ehrh.) Pers. Fig. 22. N.S. to Minn., s. to Ga. & N.C.
    2. Plants with 7–11 pairs of leaves which are not hairy and are 4.6–6.5 cm. long. **A. incarnata** var. **neoscótica** Fernald. S.w. N.S.

Reference: Fernald, Rhodora, **23**, 288 (1921)—*A. incarnata* var. *neoscotica.*

BORAGE FAMILY   BORAGINÀCEAE

Plants often covered with harsh hairs; corolla tubular with 5 spreading lobes; ovary 4-lobed, developing into 4 nutlets with the style rising in the middle (Fig. 2); flowers on a tightly coiled branch which uncoils as they open.

Forget-me-not   Myosòtis

Flowers blue with a yellow spot in the center; nutlets smooth and shining; leaves without petioles.

1. Stem angled; flowers 6–9 mm. across; calyx lobes shorter than the tube (Fig. 2); style longer than the nutlets and about equaling the calyx tube. **M. scorpioìdes** L. Fig. 1. Wet ground; Nfd. to D.C. & Ga., w. to Wis. & La.; Pacific Coast; naturalized from Europe.
1. Stem roundish in cross section; flowers 3–6 mm. across; calyx lobes about equaling the tube (Fig. 3); style shorter than the nutlets. **M. láxa** Lehm. Fig. 4. Brooksides, etc.; Nfd. to Ga., w. to Ont. & Tenn.; Pacific Coast.

VERVAIN FAMILY   **VERBENÀCEAE**

A large family in the tropics, with a few species northward. The corolla is tubular and somewhat 2-lipped, and the fruit consists of 2 or 4 nutlets. The most common genus in this region is the Vervain (*Verbena*), which grows in dry or damp ground, but scarcely in the water.

Frog-fruit   **Líppia**

Stems reclining, often rooting at the nodes; flowers in the axils of scales in a close head on a long slender stalk.

1. Leaves narrowed from near the middle to a rather sharp point. **L. lanceolàta** Michx. Figs. 5, 6. Muddy or sandy shores; N.J. to s. Ont. & Minn., s. to the Gulf of Mexico. The width of leaves is variable from plant to plant.
1. Leaves narrowed from well above the middle to a blunt or rounded tip. **L. nodiflòra** (L.) Michx. Fig. 7. Mo. & N.C. to the Gulf of Mexico.

MINT FAMILY   **LABIÀTAE**

Herbs with opposite leaves and, in most cases, square stems; flowers with a tubular, 5-lobed, more or less 2-lipped corolla; ovary deeply 4-lobed, forming 4 nutlets. The members of this family are often difficult to identify; most easily recognized, by its odor, is the Mint, *Mentha*.

1. Corolla 1 cm. or more long (except in *Scutellaria lateriflora*), strongly 2-lipped...**2**
1. Corolla less than 1 cm. long, nearly regular.................................**5**
  2. Lips of corolla about as long as the tube (Fig. 9); calyx lobes sharp (Fig. 12). **Stachys.**
  2. Lips of corolla much shorter than the tube (Figs. 19, 23); calyx lobes blunt..**3**
3. Flowers blue, in the axils of leaves near the tip of the stem (Figs. 19, 20).......**4**
3. Flowers rose-purple or bluish, in spikes at the tip of the stem (Figs. 22, 25, 27). **Physostegia.**
  4. Bases of leaves embracing the stem and touching each other. See **Mimulus,** p. 305.
  4. Leaves petioled (petioles sometimes very short), not tuuching each other (Figs. 19, 20). **Scutellaria.**
5. Plants not fragrant; stamens 2. **Lycopus.**
5. Plants fragrant; stamens 4. **Mentha.**

Hedge Nettle   **Stàchys**

Flowers in whorls in the axils of much reduced leaves; calyx with bell-shaped tube and 4 or 5 long triangular sharp pointed teeth; corolla purple, with slender tube and 2 lips, the upper entire and the lower 3-parted. The species often grade into each other.

M. scorpioides: 1. Upper branch, × 1.   2. Calyx and fruit, × 5; the calyx tube is torn to show the nutlets and style.   **M. laxa:** 3. Calyx and fruit, × 5; the calyx tube is torn to show the nutlets and style.   4. Upper branch, × 1.   **L. lanceolata:** 5. End of branch, × 1; a narrow-leaved form.   6. Pair of leaves, × 1; a broad-leaved form.   **L. nodiflora:** 7. Stem and leaves, × 1.

287

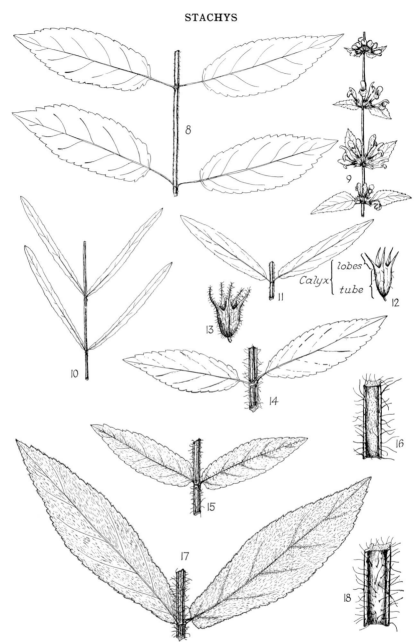

S. tenuifolia: 8. Stem and leaves, × ½. 9. Part of inflorescence, × ½. S. hyssopifolia: 10. Stem and leaves, × ½. S. aspera: 11. Stem and leaves, × ½. 12. Calyx, × 2. S. hispida: 13. Calyx, × 2. 14. Stem and leaves, × ½. S. palustris: 15. Stem and leaves, × ½. 16. Stem, × 2. S. homotricha: 17. Stem and leaves, × ½. 18. Stem, × 2.

1. Upper and lower leaves without petioles, the middle ones sometimes on petioles up to 8 mm. long; calyx with or without hairs..............................**2**
1. All leaves (except those with flowers in their axils) with petioles; calyx without hairs. **S. tenuifòlia** Willd. Figs. 8, 9. Wet ground; w. N.Y. & s.e. Pa. to Mich. & Minn., s. to Tenn., Ala., La. & Tex., w. to Okla. & Kans.
>   2. Leaves without hairs or with a few scattered hairs, the distance between hairs being greater than the length of the hairs...............................**3**
>   2. Leaves copiously hairy...**5**
3. Leaves rarely more than 6 mm. broad. **S. hyssopifòlia** Michx. Fig. 10. Wet sand, fields, etc.; e. Mass. to N.J. & Del., n. Va., & n. N.C. (Includes *S. atlantica*, B.)
3. Leaves 8 mm. or more broad..**4**
>   4. Leaves 3–13 mm. broad, the middle ones on petioles 1–3 mm. long; calyx nearly or quite without hairs (Fig. 12). **S. áspera** Michx. Fig. 11. Moist soil, local; Pa. to Fla., w. to Iowa & Mo. (*S. ambigua*, G, B.)
>   4. Leaves 1.5–4 cm. wide, the middle ones with petioles 1–8 mm. long; calyx bristly-hairy (Fig. 13). **S. híspida** Pursh. Fig. 14. Moist soil; N.E. & Ont. to Md., Ga., w. to N.D. & Ark. (*S. tenuifolia* var. *aspera*, G.)
>   5. Leaves 2 cm. or more broad, narrowed below the middle (Fig. 17); sides of stem usually bristly like the angles (Fig. 18). **S. homótricha** (Fernald) Rydb. Wet shores, etc.; N.B. to N.Y. & Minn., s. to Conn. & Mo. (*S. palustris* var. *homotricha*, G.)
>   5. Leaves usually less than 2.5 cm. broad, the sides parallel below the middle (Fig. 15); sides of stem usually without hairs or with finer hairs than are on the angles (Fig. 16). **S. palústris** L. Wet ground and lake shores; Nfd. to N.C., w. to the Mississippi River.

Reference: Epling, in Fedde, Rep. Spec. Nov. Beihefte, **80,** 1–75 (1934)—revision of North American species.

SKULLCAP   **Scutellària**

Corolla blue (rarely white or pink), tubular, arching; calyx 2-lipped, the tube with a projection on the upper side.

1. Leaves with very short petioles. **S. epilobiifòlia** Hamilton. Fig. 19. Gravel shores, wet sand, swales, etc.; Nfd. to B.C., s. to N.C., Ohio & Neb. (*S. galericulata*, G, B.)
1. Leaves with slender petioles...**2**
>   2. Flowers 5–8 mm. long, on one side of slender branches. **S. lateriflòra** L. Fig. 20. Wet places; Nfd. to B.C., s. to Fla. & N.M.
>   2. Flowers 1–1.5 cm. long, solitary in the axils of small leaves. **S. Churchilliàna** Fernald. River shores; N.B. & n. Maine.

Reference: Fernald, Rhodora, **23,** 85–86 (1921)—*S. epilobiifolia.*

FALSE DRAGONHEAD   **Physostègia**

Plants erect, sometimes 1 m. or more tall; leaves sharply toothed; flowers in one or more long spikes, rose-purple or bluish, rarely white. Identification of the species is rather difficult, and to see the characters of the calyx a magnification of at least 10 times is necessary. (*Dracocephalum*, B.)

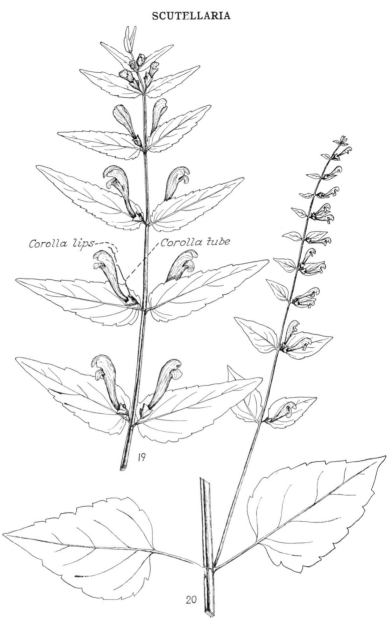

Corolla lips
Corolla tube

S. epilobiifolia: 19. Upper branch, × 1. **S. lateriflora:** 20. Leaves and branch, × 1.

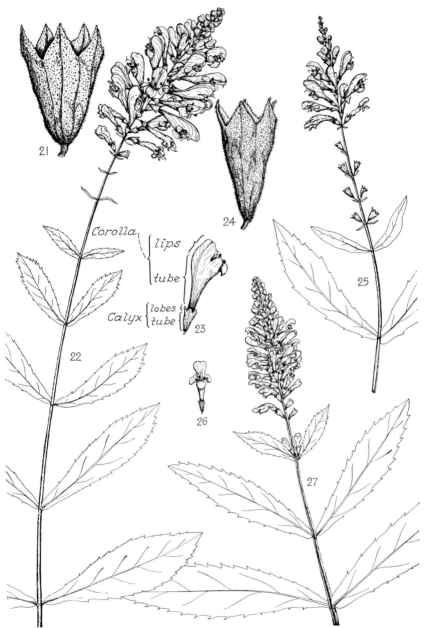

P. virginiana: 21. Calyx, × 5. 22. Upper part of plant, × ½. 23. Flower, × 1.
P. speciosa: 24. Calyx, × 5. 25. Upper part of plant, × ½. **P. parviflora:** 26. Flower,
× 1. 27. Upper part of plant, × ½.

1. Upper leaves much smaller than the middle ones (Fig. 22); calyx with minute round glands mixed with the hairs (Fig. 21); flowers 2–3 cm. long (Fig. 23). **P. virginiàna** (L.) Benth.   Moist soil along rivers, etc.; N.E. to Ill. & Iowa, s. to Fla. & Tex.
1. Upper leaves nearly as large as the middle ones (Figs. 25, 27); calyx with or without glands; flowers 1–2 cm. long...2
   2. Calyx without glands; flowers generally 1.5 cm. or more long..............3
   2. Calyx with glands; flowers generally 1.5 cm. or less long. **P. parviflora** Nutt. Figs. 26, 27.   Wis. & westward.
3. Calyx densely velvety (Fig. 24); corolla white-hairy, at least when the buds are expanding. **P. speciòsa** Sweet.   Figs. 24, 25.   Maine to s. Minn. & Iowa, s. to N.J. & Ind.
3. Calyx and corolla nearly or quite without hairs. **P. speciosa** var. **glabriflòra** Fassett.   Racine, Wis.

References: Fassett, Rhodora, **41**, 377 (1939)—*P. speciosa* var. *glabriflora*.   In preparing this treatment the manuscript of the Flora of Indiana, by Deam, has been followed.

WATER HOREHOUND   **Lýcopus**

Plants unbranched or with slender branches, mostly with stolons at base; flowers small, in dense clusters in the axils of the leaves (Figs. 43–47).

1. Leaves usually lobed at least halfway to the midrib (Fig. 38), rarely unlobed; calyx lobes tipped with a rigid spine (Fig. 43); nutlets with tip smooth-margined (Figs. 28, 29)...2
1. Leaves not lobed (Figs. 39–41); calyx lobes not spine-tipped (Figs. 44–46); nutlet with summit irregular or bumpy (Figs. 30–37)..............................3
   2. Stem with a few scattered hairs or none. **L. americànus** Muhl.   Figs. 38, 43. Wet ground, lake shores, etc.; Nfd. to B.C., s. to Fla., Tex., Utah & Calif.
   2. Stem with spreading white hairs. **L. americanus** var. **Lóngii** Benner.   N.Y. to s. Wis., s. to s. Va.
3. Calyx lobes rather blunt, mostly shorter than the mature nutlets (Figs. 44, 45)..4 The following two species are called Bugle Weed; they are usually distinct, but an occasional individual is difficult to place.
3. Calyx lobes sharp, much longer than the mature nutlets (Fig. 46)..............6
   4. Base of plant with a tuber (Fig. 42), and stolons ending in tubers; clusters of fruits not very dense, 4–9 mm. thick, with the calyx lobes not hidden by the nutlets (Fig. 45)...5
   4. Base of plant and stolons without tubers; clusters of fruits dense, 8–15 mm. thick, with the calyx lobes mostly hidden by the mature nutlets (Fig. 44). **L. virgínicus** L.   Figs. 30, 31, 39.   Maine to Neb., s. to Fla., Ala. & Mo.
5. Main stem erect, not rooting at tip. **L. uniflòrus** Michx.   Figs. 32, 35, 40, 42, 45. Labrador to B.C., s. to Va., Mich., Minn., Neb., Wyo. & Ore.
5. Main stem and branches much elongated, recurved and rooting at tip. **L. uniflorus** forma **flagellàris** Fernald.   Rare; N.S., Maine and perhaps elsewhere.
   6. Leaves not petioled (Figs. 41, 49); stem with a tuber at base..............7
   6. Leaves petioled (Fig. 48); stem without a tuber..........................8
7. Flower clusters with conspicuous bracts at base (Fig. 46); nutlets with shallow depressions and low elevations along the summit (Figs. 33, 36). **L. ásper** Greene. Mich. to B.C., s. to Kans. & Ariz.   (*L. lucidus* var. *americanus*, G.)

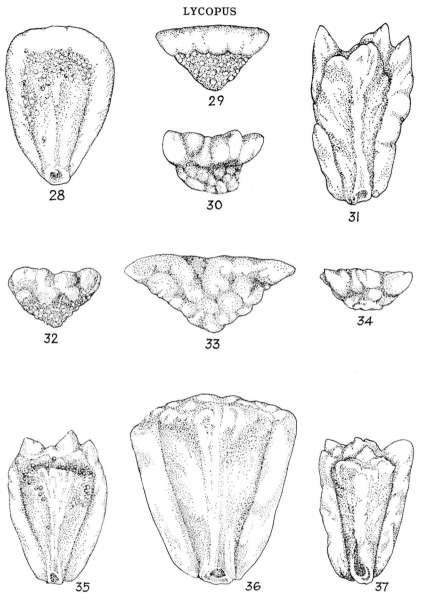

Nutlets, × 28.   **L. americanus:** 28. Face.   29. Top.   **L. virginicus:** 30. Top.   31. Face.
L. uniflorus: 32. Top.   35. Face.   **L. asper:** 33. Top.   36. Face.   **L. rubellus:** 34. Top.
37. Face.   Figs. 28–37 from Hermann, Rhodora, **38,** plate 439 (1936).

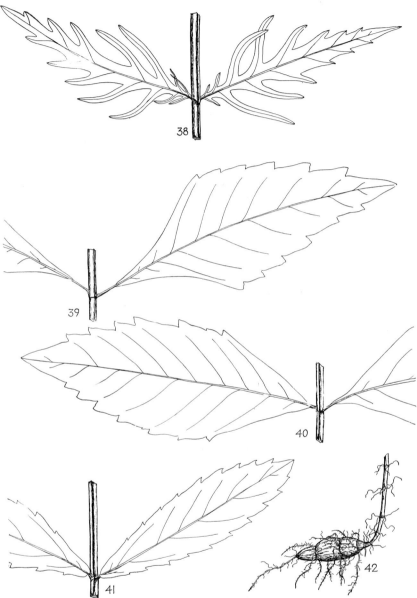

L. americanus: 38. Stem and leaves, × 1. **L. virginicus**: 39. Stem and leaves, × 1. **L. uniflorus**: 40. Stem and leaves, × 1. 42. Base of stem with tuber, × 1. **L. asper**: 41. Stem and leaves, × 1.

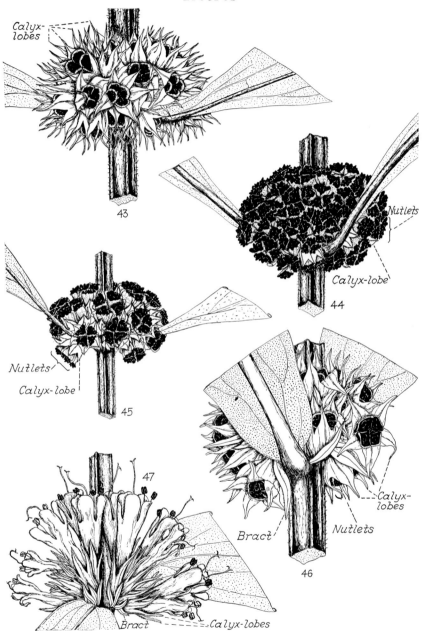

Heads of flowers or fruits, × 5. **L. americanus:** 43.   **L. virginicus:** 44.   **L. uniflorus:** 45.
L. asper: 46.   **L. sessilifolius:** 47.

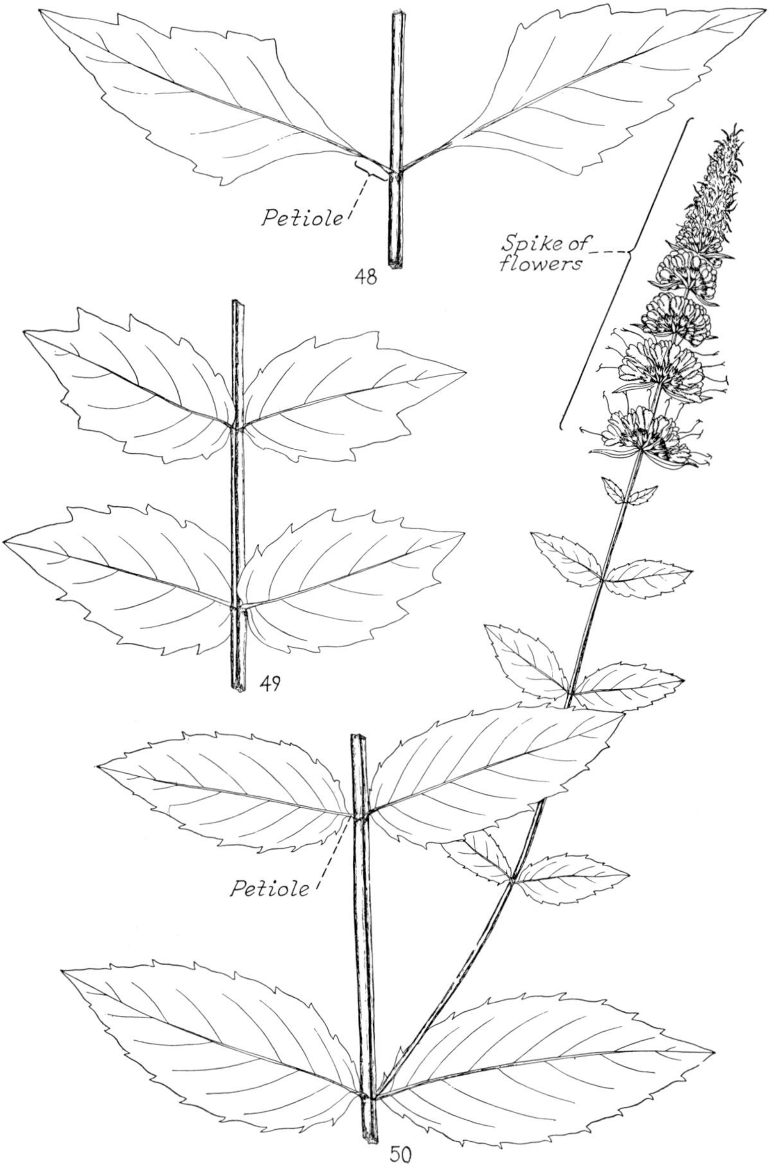

Petiole

48

Spike of flowers

49

Petiole

50

L. rubellus: 48. Stem and leaves, × 1. **L. sessilifolius**: 49. Stem and leaves, × 1. **M. spicata**: 50. Stem, leaves and flowering branch, × 1.

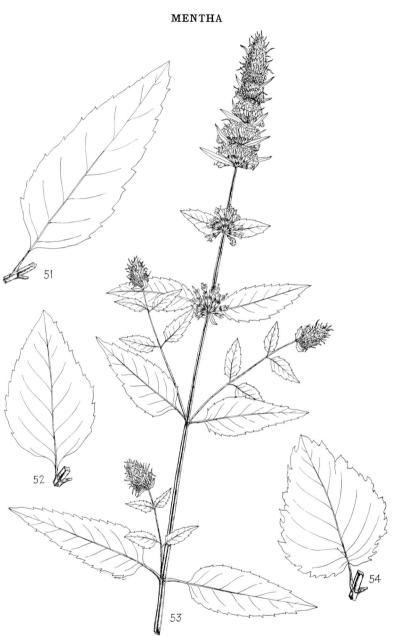

M. arvensis var. canadensis: 51. Stem and leaf, × 1.   M. aquatica: 52. Stem and leaf, × 1.
M. piperata: 53. Summit of plant, × 1.   M. crispa: 54. Stem and leaf, × 1.

M. arvensis: 56. Summit of plant, × 1.   **S. Dulcamara:** 55. Flowering branch, × 1.

7. Flower clusters with inconspicuous bracts (Fig. 47); nutlets with deep depressions and large elevations along the summit. **L. sessilifòlius** Gray. Fig. 49. Mass. to Fla. & Miss.
8. Leaves sharply toothed; stems practically without hairs. **L. rubéllus** Moench. Figs. 34, 37, 48. N.H. to Fla. & Mo. Merges into the next variety.
8. Leaves minutely and remotely toothed; stem minutely hairy. **L. rubellus** var. **arkansànus** (Fresenius) Benner. Mo. to Tex.

References: Benner, Bartonia, No. 16, 46–50 (1934)—*L. americanus* var. *Longii* and *L. rubellus* var. *arkansanus*. Fernald, Rhodora, **23**, 289–290 (1921)—*L. uniflorus* f. *flagellaris*. Figs. 28–37 are from Hermann, Rhodora, **38**, plate 439 (1936).

MINT   Méntha

1. Petioles 2 mm. or less long; flowers in leafless spikes. **M. spicàta** L. Spearmint. Fig. 50. Wet places; naturalized from Europe.
1. Petioles much longer (Figs. 51–54); flowers in clusters, at least the lower in the axils of leaves (Figs. 53, 56)...2
  2. Upper flower clusters without leaves or in the axils of much reduced leaves; flower clusters borne to the tip of the stem (Fig. 53).....................3
  2. All flower clusters in the axils of leaves which are several times as long as the clusters; tip of stem with leaves which have no flowers in their axils (Fig. 56)..5
3. Leaves more than twice as long as wide (Fig. 53); stem and leaves nearly or quite without hairs. **M. piperàta** L. Peppermint. Wet places; naturalized from Europe.
3. Leaves mostly less than twice as long as wide (Figs. 52, 54); stem and leaves usually hairy...4
  4. Leaves irregularly toothed and crisped (Fig. 54). **M. críspa** L. Curled Mint. Wet places; Mass. to N.J.; naturalized from Europe.
  4. Leaves simply toothed (Fig. 52). **M. aquática** L. Water Mint. Wet places, rare; N.S. to Ga.; naturalized from Europe.
5. Upper leaves, with flowers in their axils, much smaller than the others. **M. Cardìaca** Gerarde. N.S. to Pa.; naturalized from Europe.
5. Upper leaves nearly as large as the rest (Fig. 56). **M. arvénsis** L. The native American mints, which are extremely variable in leaf shape and amount of hairy covering; the following varieties are rather arbitrarily distinguished...........6
  6. Stems and leaves with scattered or dense white hairs.....................7
  6. Leaves nearly or quite without hairs; stems with minute hairs along the angles. **M. arvensis** var. **glabràta** (Benth.) Fernald. Very common.
7. Leaves tending to be rounded at base (Fig. 56). **M. arvénsis** L. Abundant.
7. Leaves wedge-shaped at base (Fig. 51)...................................8
  8. Leaves and stems densely woolly. **M. arvensis** var. **lanàta** Piper. Widespread.
  8. Leaves and stems with scattered hairs. **M. arvensis** var. **canadénsis** (L.) Briquet. Fig. 51. Widespread and common.

NIGHTSHADE FAMILY   SOLANÀCEAE

A large family with many cultivated plants; flower generally more nearly regular than in the next family; fruit a pod or a berry.

NIGHTSHADE   Solànum

S. Dulcamàra L. Bittersweet; Matrimony Vine. Fig. 55. A perennial vine; leaves heart-shaped and often somewhat lobed; flowers purple or blue; berries red. Moist or dry soil, stream banks, swales, drained bogs, etc.

FIGWORT FAMILY   SCROPHULARIÀCEAE

Corolla united into a tube, more or less 2-lipped; fruit a capsule with 2 compartments. It is the fruit that distinguishes the Figwort Family from the Mint Family with its 4 nutlets. The genera are separated on characters in the flowers and fruits and are often difficult to distinguish in the sterile condition.

1. Leaves in little tufts on the ground (Fig. 1). **Limosella,** p. 300.
1. Leaves opposite, borne on the stem..**2**
   2. Leaves deeply lobed (Fig. 2). **Leucospora,** p. 300.
   2. Leaves not lobed...**3**
3. Flowers borne in long racemes in the axils of leaves (Fig. 3). **Veronica,** p. 300.
3. Flowers not in axillary racemes..**4**
   4. Flowers in a close spike at the tip of the stem (Fig. 12). **Chelone,** p. 303.
   4. Flowers borne singly in the axils of leaves..............................**5**
5. Leaves toothed or wavy margined..**6**
5. Leaves not toothed or with a few obscure teeth...........................**7**
   6. Flowers pale-yellowish; stamens 4, only 2 of them with anthers. **Gratiola,** p. 307.
   6. Flowers bright-yellow or bright-blue; all 4 stamens with anthers. **Mimulus,** p. 305.
7. Flowers yellow. **Gratiola,** p. 307.
7. Flowers pale-blue, purplish, or whitish...................................**8**
   8. Stamens 4, all with anthers. **Bacopa,** p. 307.
   8. Stamens 4, only 2 with anthers. **Lindernia,** p. 307.

MUDWORT   Limosélla

Leaves in tufts which are connected by thread-like runners; flowers solitary at the tips of recurving stalks, very small.

1. Leaves flat, 10–20 in a rosette; corolla about 2 mm. wide, dull white or purplish, with pointed lobes. **L. aquática** L. Muddy shores; Labrador & Nfd.; Minn. to Hudson Bay & s. B.C., s. to Calif. & N.M.
1. Leaves thread-like, 5–10 in a rosette; corolla 3 mm. wide, white and tinged with lavender, with rounded lobes. **L. subulàta** Ives. Fig. 1. Tidal shores and pools near the sea; Nfd. to Md. (*L. aquatica* var. *tenuifolia*, G; *L. aquatica*, in part, B.)

Leucóspora

L. multífida (Michx.) Nutt. Fig. 2. Much-branched annuals, covered with fine sticky hairs; flowers in the axils of most of the leaves, small, with greenish-white corolla scarcely longer than the calyx. Streams, ponds, damp fields, etc.; Ohio to Iowa & Kans., s. to Ala. & Tex. (*Conobea multifida*, G, B.)

SPEEDWELL   Verónica

Corolla mostly blue with darker lines, 2-lipped, the upper lip with 3 lobes, the lower with 2 lobes which are united so that the corolla appears to be nearly equally 4-lobed; stamens 2; pod strongly flattened. The aquatic members of this genus are all more or less reclining, often half in and half out of the water, with opposite leaves, the flowers in long racemes in the axils of the upper leaves. *V. connata* is often found completely submersed and sterile making light green patches in streams.

Limosella subulata: 1. Plant, × 1. Leucospora multifida: 2. Branch, × 2. **V. americana:**
3. Upper part of plant, × 1. **V. Beccabunga:** 4. Leaf, × 1.

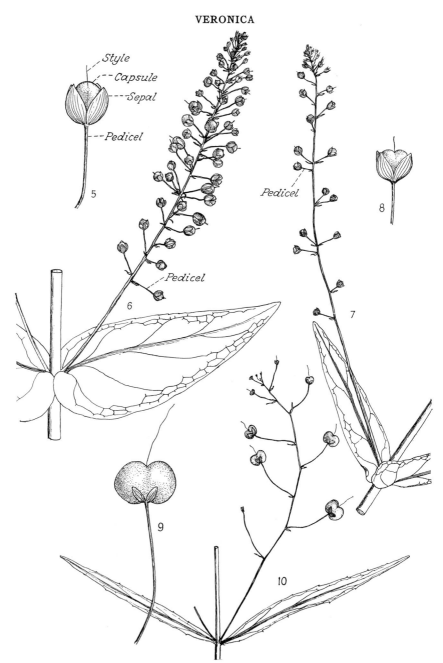

V. **Anagallis-aquatica**: 5. Fruit, × 5. 6. Leaves and inflorescence, × 1. V. **connata**: 7. Leaves and inflorescence, × 1. 8. Fruit, × 5. **V. scutellata**: 9. Fruit, × 5. 10. Leaves and inflorescence, × 1.

302

1. Leaves petioled...2
1. Leaves not petioled...3
  2. Leaves widest near the base, pointed at tip. **V. americàna** (Raf.) Schwein. American Brooklime. Fig. 3. Swamps, springs, brooks, etc.; Nfd. to Alaska, s. to N.C., Mo., Mex. & Calif.
  2. Leaves widest near the middle, rounded at tip (Fig. 4). **V. Beccabúnga** L. European Brooklime. Ditches and beaches; Que. to Mich. & W.Va.; adventive from Europe.
3. Leaves rounded at base and somewhat clasping (Fig. 6), usually more than 1 cm. wide; capsule about as wide as long, not strongly 2-lobed (Figs. 5, 8)...........4
3. Leaves with nearly parallel sides, 1 cm. or less wide, not clasping (Fig. 10); capsule much wider than long, strongly 2-lobed (Fig. 9)...........................7
  4. Racemes 30–60-flowered, with pedicels 4–8 mm. long (Fig. 6); capsule not wider than long, scarcely notched (Fig. 5)...............................5
  4. Racemes 15–30-flowered, with pedicels 3–6 mm. long (Fig. 7); capsule slightly wider than long (Fig. 8)..6
5. Pedicels without glands. **V. Anagállis-aquática** L. Water Speedwell. Fig. 6. Streams and ditches, shores, etc.; common from Conn. to Mich., and occasional elsewhere; naturalized from Europe.
5. Pedicels covered with minute gland-tipped hairs. **V. glandífera** Pennell. Streams; Pa. to Ind. & N.C.
  6. Pedicels and upper part of stem with minute gland-tipped hairs. **V. connàta** Raf. Water Speedwell. Fig. 7. In streams; Mass. & Ont. to N.D., s. to Pa., Tenn. & Okla. This and its variety are often completely submersed and sterile.
  6. Pedicels and stem without gland-tipped hairs. **V. connata** var. **glabérrima** (Pennell) Fassett. Wis. & Mo. to the Pacific Coast.
7. Plants without hairs. **V. scutellàta** L. Marsh Speedwell. Figs. 9, 10. Swamps and swales; Nfd. to Mackenzie, s. to Va., Colo. & Calif.
7. Plants with white hairs on stem and leaves. **V. scutellata** var. **villòsa** Schumacher. Ohio & Ont. westward.

TURTLEHEAD   **Chelòne**

The name Turtlehead (or Snakehead) comes from the resemblance of the flower to a reptile's head. The species and varieties are somewhat difficult to distinguish, but the only common ones are *C. glabra* and its var. *linifolia*. They all grow on shores of streams and in wet woods, and rarely venture into the water.

1. Leaves scarcely petioled except in *C. glabra* var. *elatior;* corolla not purple throughout except in one rare form; bracts at the bases of the flowers not fringed......2
1. Leaves definitely petioled; corolla bright purple; bracts fringed with fine hairs..10
  2. Leaves 1–2 cm. wide; lips of corolla white within and greenish-yellow without..3
  2. Leaves 3–4 cm. wide; lips of corolla purplish within........................4
3. Leaves not hairy beneath. **C. glàbra** var. **linifòlia** Coleman. Figs. 11, 12. Swales, lake shores and bogs; s. Ont. to Man., s. to Ohio & Ill.
3. Leaves hairy beneath. **C. glabra** var. **linifolia** forma **velùtina** Pennell & Wherry. With the last.
  4. Corolla somewhat yellowish toward the end; spikes elongating so that the flowers tend to become isolated. **C. glabra** var. **ochróleuca** Pennell & Wherry. Sandy swamps; Md. to N.C.
  4. Corolla not yellowish; spikes dense.......................................5

C. glabra var. **linifolia**: 11. Leaf, × 1.   12. Top of plant, × 1.   **C. glabra**: 13. Leaf, × 1.

5. Uppermost leaves nearly as large as those on the middle portion of the stem, rounded at base. **C. glabra** var. **dilatàta** Fern. & Wieg. Bogs and streams; Nfd. & Que. to n. Maine & N.Y.; N.C.

5. Uppermost leaves smaller than the others, narrowed to the base.............6

    6. Petioles 0.5–2 cm. long; corolla purple toward the end. **C. glabra** var. **elàtior** Raf. Stream banks and bogs; Pa. to Ind. & Ala.

    6. Petioles obscure or absent; corolla not purple on the outside except in one form
....7

7. Leaves firm, 2–4 cm. wide; spikes 2–6 cm. long............................8

7. Leaves thin, 1.3–2.5 cm. wide; spikes 6–8 cm. long. **C. glabra** var. **elongàta** Pennell & Wherry. Stream banks and swamps; Ohio to s. Ill. & Tenn.

    8. Corolla white except at the end..9

    8. Corolla purple. **C. glabra** forma **ròsea** Fernald. N.H.; rare.

9. Leaves minutely woolly beneath. **C. glabra** forma **tomentòsa** (Raf.) Pennell. Occasional.

9. Leaves not woolly beneath. **C. glabra** L. Fig. 13. The commonest form, in rich swamps, wet woods, etc.; Nfd. & n. Ont. to Minn., s. to Ga. & Ala.

    10. Bracts sharp; corolla 2.5–3.2 cm. long; petioles 0.5–1.5 cm. long. **C. obliqua** L. Wet woods and cypress swamps; Md. & w. Tenn. to Ala.; rare.

    10. Bracts blunt; corolla 3.0–3.7 cm. long; petioles 0.5–3.0 cm. long. **C. obliqua** var. **speciòsa** Pennell & Wherry. Wet woods and swamps; Ind. & s. Miss. to n.e. Ark.

MONKEY FLOWER   **Mímulus**

Flower deeply 2-lipped, the upper lip erect or spreading, the lower spreading; flower somewhat resembling a grinning face, hence the common name.

1. Flowers blue (rarely white); plants erect...................................2

1. Flowers yellow; plants usually creeping, except *M. guttatus*...................5

    2. Leaves not petioled; stems not winged; pedicels mostly longer than the calyx (Fig. 14)...3

    2. Leaves petioled; stems winged on the 4 angles; pedicels mostly shorter than the calyx (Fig. 15)..4

3. Flowers blue. **M. ríngens** L. Fig. 14. Common in wet places; N.S. to Ont. & s. Sask., s. to N.C., Ala., La., Okla. & Colo.

3. Flowers white. **M. ringens** forma **Péckii** House. Rare.

    4. Flowers blue. **M. alàtus** Ait. Fig. 15. Conn. & N.Y. to Mich., Ill. & Neb., s. to Fla. & e. Tex.

    4. Flowers white. **M. alatus** forma **albiflòrus** House. Rare.

5. Stems with long white sticky hairs. **M. moschàtus** Dougl. Wet soil along streams, perhaps introduced from the Rocky Mts.

5. Stems not sticky, with few or no hairs....................................6

    6. Flowers 2–4 cm. long, their pedicels much longer than the leaves in the axils of which they are borne; lower leaves with petioles much longer than their blades; stems erect or sprawling. **M. guttàtus** Fisch. Fig. 16. Native in the west, and locally naturalized in brooks and meadows in Conn. & N.Y. (*M. Langsdorfii*, G, B.)

M. ringens: 14. Top of plant, × 1. **M. alatus:** 15. Leaves and fruiting calyces, showing cross section of stem above, × 1. **M. guttatus:** 16. End of branch, × 1. **M. glabratus** var. **Fremontii:** 17. End of branch, × 1. **M. glabratus** var. **michiganensis:** 18. Leaf, × 1.

6. Flowers 8–22 mm. long, with pedicels shorter than or not more than twice as long as the leaves in the axils of which they are borne (Fig. 17); all petioles shorter than the blades; stems rooting at the nodes, prostrate, except sometimes toward the tip...**7**

7. Flowers 8–12 mm. long; leaves shallowly if at all wavy on the margins; stems creeping nearly throughout. **M. glabràtus** HBK., var. **Fremóntii** (Benth.) Grant. Fig. 17. Cold springs and lake shores; Ont. to Mo., Tex. & Mex., w. to s. Sask., Wyo., Utah & Ariz. (*M. glabratus* var. *Jamesii*, G; *M. Geyeri*, B.)

7. Flowers 12–22 mm. long; leaves with strongly wavy margins (Fig. 18); stems ascending toward the tip. **M. glabratus** var. **michiganénsis** (Pennell) Fassett. Springy beaches about the north end of Lake Michigan.

HEDGE HYSSOP   Gratiola

Flowers in the axils of the leaves, yellow or yellowish.

1. Leaves not toothed or very slightly toothed; flowers golden yellow.............**2**
1. Leaves toothed; flowers whitish or yellowish..................................**3**
  2. Leaves 1 cm. or more long, rounded at tip, covered with minute dark glands. **G. lùtea** Raf. Fig. 22. Pond shores and rivers; Nfd. to n. Va., w. to Ill. & Minn. (*G. aurea*, G, B.) Pale-flowered individuals occur rarely.
  2. Leaves mostly 5 mm. or less long, sharply pointed, without glands. **G. lutea** forma **pusílla** (Fassett) Pennell. Dwarf Hyssop. Fig. 23. A submersed dwarfed sterile form.
3. Plants with sticky hairs at least toward the tip.............................**4**
3. Plants without sticky hairs. **G. virginiàna** L. Figs. 25, 26. Wet loam, in the shade, usually along streams, N.J. to n. Ind. & s. Iowa, s. to the Gulf of Mexico. (*G. sphaerocarpa*, G, B.)
  4. Leaves 0.5–2.5 cm. long, not over twice as long as wide. **G. viscídula** Pennell. Fig. 29. Along streams; Del. to n. Ga. & e. Tenn. (*G. viscosa*, G, B.)
  4. Leaves longer, several times as long as wide. **G. neglécta** Torr. Fig. 27. Que. & Maine to s. B.C., s. to Ga., Tex., Ariz. & Calif. (*G. virginiana*, G, B.)

WATER HYSSOP   Bacòpa

Creeping fleshy herbs; corolla only slightly longer than the calyx; leaves with a few low teeth or none.

1. Leaves with no veins in addition to the midrib. **B. Monnièria** (L.) Wettst. Fig 19. Wet sand, river banks, etc.; Md. to Fla. & Tex. (*Bramia Monnieria*, B.)
1. Leaves with several veins radiating from the base............................**2**
  2. Stem covered with straight hairs on the upper part, without hairs below the middle; flowers shorter than their stalks; corolla white, yellow within. **B. rotundifòlia** (Michx.) Wettst. Fig. 24. Usually growing submersed with terminal tufts of leaves floating; Ind. & Tenn. to Mont., Colo. & Tex. (*Bramia rotundifolia*, B.)
  2. Stems covered with crinkled hairs except toward the base; flowers longer than their stalks; corolla blue. **B. caroliniàna** (Walt.) Robinson. Fig. 28. In the water; Va. to Fla. & Tex.

FALSE PIMPERNEL   Lindérnia

Small herbs 1–3 dm. high; flowers in the axils of the upper leaves, purplish; those coming late in the season not opening. The two following species are not always easy to distinguish.

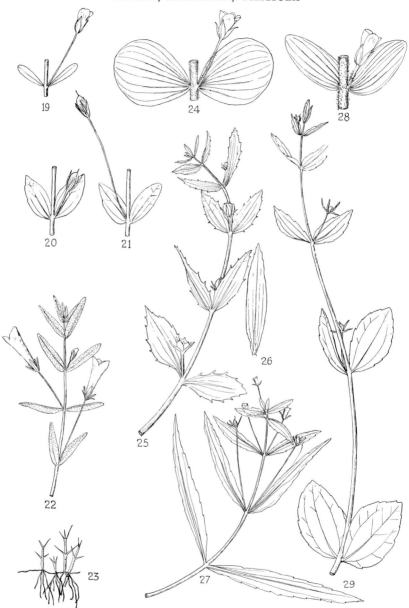

B. Monnieria: 19. Leaves and flower, × 1.   **B. rotundifolia:** 24. Leaves and flower, × 1.
B. caroliniana: 28. Leaves and flower, × 1.   **L. dubia:** 20. Leaves and fruit, × 1.   **L
anagallidea:** 21. Leaves and fruit, × 1.   **G. lutea:** 22. Top of plant, × 1.   **G. lutea** f
pusilla: 23. Plants, × 1.   **G. virginiana:** 25. Branch, × 1.   26. Leaf, × 1.   **G. neglecta**
27. Branch, × 1.   **G. viscidula:** 29. Branch, × 1.

1. Stalks of flowers stout, shorter than or slightly exceeding the leaves in the axils of which they are borne; lobes of calyx about as long as the pod; seeds pale-yellow, about 0.4 mm. long, 2–3 times as long as wide. **L. dùbia** (L.) Pennell. Fig. 20. Shores of lakes and streams; N.B. to Minn., s. to Fla., Tex. & Mex.; Wash. & Ore. (*Ilysanthes dubia*, G; *I. attenuata*, B.)

1. Stalks of flowers slender, several times as long as the leaves in the axils of which they are borne; lobes of the calyx mostly shorter than the pod; seeds brownish-yellow, about 0.3 mm. long, 1.5–2 times as long as wide. **L. anagallídea** (Michx.) Pennell. Fig. 21. N.H. to N.D. & Colo., s. to Fla., Tex. & Mex.; Wash. to Calif. (*Ilysanthes anagallidea*, G; *I. dubia*, B.)

References: Pennell, The Scrophulariaceae of eastern temperate North America, Acad. Nat. Sci. Phila. Monographs 1, 1–650 (1935)—all species except those mentioned below. Fernald, Rhodora, **37**, 440–442 (1935)—*Bacopa;* Fassett, Rhodora, **41**, 524–525 (1939)—*Veronica connata* var. *glaberrima* and *Mimulus glabratus* var. *michiganensis.*

BLADDERWORT FAMILY   LENTIBULARIACEAE

BLADDERWORT   Utricularia

Small herbs growing in the water or in damp places, with bladders (Fig. 49) to catch small animal life, except in some terrestrial species. Flowers yellow or purple, 2-lipped, the upper lip erect, the lower with a spur (Fig. 57) and usually with a prominent palate (Fig. 49); scapes (Fig. 49) 1- to several-flowered; fruit a pod with several seeds. Those species with long stems and usually profuse foliage which grow in regions of severe winters generally survive that season by more or less compact winter buds (Figs. 30–37).

1. Stems erect from a base definitely anchored in sand, mud or bog (Fig. 57), without bladders or with poorly developed bladders; bracts (Fig. 49) in pairs and united to form a tube (Fig. 38), or 3-parted (Fig. 44), or broad-based with a pair of inner bracts (Fig. 45), or attached above the base (Figs. 39, 40)..................2

1. Stems sometimes anchored at one end, but mostly drifting free in the water (Figs. 49, 53, 54), usually with several scapes; bracts 2-lobed at base (Figs. 41, 42), or clasping (Fig. 43), or attached above the base (Figs. 39, 40)...........6

  2. Bracts in pairs and united to form a tube (Fig. 38); erect stems several from a creeping one; leaves with a few spineless lobes; flower purple, solitary, facing upward (Fig. 47). **U. resupinàta** B.D. Greene. Sandy pond margins; N.B. to Fla., w. to w. Ont. & Ill. (*Lecticula resupinata*, B.)

  2. Bracts not united in pairs; stems solitary; leaves strap-like (Fig. 57) and rarely seen; flowers 1–5, not pointing upward...................................3

3. Stem very slender, hair-like, with a zigzag inflorescence (Fig. 48); pedicels much longer than the bracts (Fig. 48); bracts attached above the base (Figs. 39, 40); flowers yellow or purplish..4

3. Stem stout (Fig. 57); pedicels shorter than the bracts so that the flowers appear scarcely stalked (Fig. 57); bracts 3-parted (Fig. 44) or with a pair of inner bracts (Fig. 45); flowers yellowish..5

  4. Flowers yellow, 6–12 mm. long. **U. subulàta** L. Fig. 48. Sandy swamps and pine barrens; Mass. to Fla. & Tex. (*Setiscapella subulata*, B.)

  4. Flowers purplish, 1 mm. long, not opening. **U. subulata** forma **cleistógama** (Gray) Fernald. (*U. cleistogama*, G; *S. cleistogama*, B.)

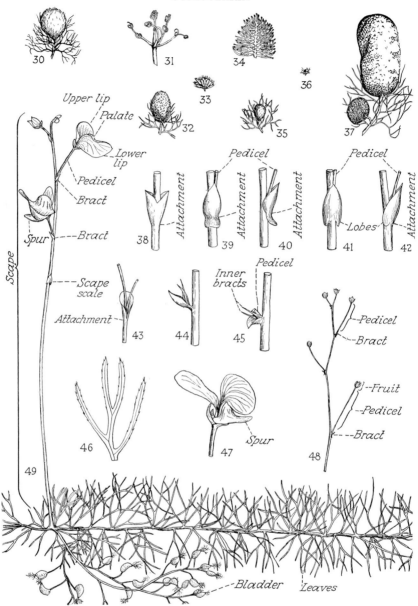

U. geminiscapa: 30. Winter bud, × 1.   U. purpurea: 31. Winter bud, × 1.   39. Bract, back view, × 4.   40. Bract, side view, × 4.   U. intermedia: 32. Winter bud, × 1.   33. Single leaf of winter bud, × 5.   41. Bract, back view, × 4.   42. Bract, side view, × 4.   46. Portion of leaf, × 8.   49. Plant, × 1.   U. vulgaris var. americana: 34. Single leaf of winter bud, × 5.   37. Winter bud, × 1.   U. minor: 35. Winter bud, × 1.   U. inflata var. minor: 36. Winter bud, × 1.   U. resupinata: 38. Bract, × 4.   47. Flower, × 2.   U. gibba: 43. Bract, × 4.   U. virgatula: 44. Bract, × 4.   U. cornuta: 45. Bracts, × 4.   U. subulata: 48. Inflorescence, × 1.

5. Corolla much longer than the calyx; bract broad-based and with a pair of inner bracts (Fig. 45). **U. cornùta** Michx. Fig. 57. Peat bogs and sandy shores; Nfd. to Ont. & Minn., s. to Fla. & Tex. (*Stomoisia cornuta*, B; including *U. juncea*, G, & *S. juncea*, B.)

5. Corolla barely longer than the calyx; bract 3-parted (Fig. 44). **U. virgátula** Barnh. Sandy pond margins, local; N.Y., N.J., Va., Fla., Miss. & Cuba. (*Stomoisia virgatula*, B.)

    6. Bladders on the scattered leaves (Figs. 51, 52) or on separate stems (Fig. 49); scape scales (Fig. 49) present or absent; bracts attached at their lower margin (Fig. 43), sometimes with basal lobes (Figs. 41, 42); flowers yellow........**7**

    6. Bladders borne at the tips of much-branched stems which are whorled and look like leaves (Fig. 50); scape scales absent; bracts attached above the base (Figs. 39, 40); flowers purple. **U. purpurea** Walt. Ponds and lakes; Que. to Minn., s. to Cuba, La. & British Honduras (*Vesiculina purpurea*, B.)

7. Divisions of leaves flattened, with a midrib, the sides parallel and abruptly tapered to a long tip from a definite point (Fig. 46); bracts with basal lobes (Figs. 41, 42)...**8**

7. Divisions of leaves hair-like (sometimes appearing flattened when pressed), without a midrib, tapering throughout (Figs. 51–53); bracts without basal lobes except in *U. vulgaris* var. *americana*...**9**

    8. Bladders on stems separate from stems bearing leaves (Fig. 49); divisions of leaves copiously spine-toothed (Fig. 46); spur of corolla almost as long as the lower lip and close to it (Fig. 49); pedicels erect in fruit. **U. intermèdia** Hayne. Shallow pools and slow streams; Greenland & Nfd. to B.C., s. to Calif., Iowa, Ill., Pa. & N.J.

    8. Bladders on the leaves; divisions of leaves rarely toothed except at tip; spur very short; pedicels curved downward in fruit. **U. mìnor** L. Shallow water and bogs; Greenland & Nfd. to Mackenzie, s. to Conn., w. N.Y., Ill., Utah & Calif.

9. Scape stout, more than 10 cm. high, with 2–20 flowers; bracts not clasping, sometimes with basal lobes; pedicels curved downward in fruit; branches 3 dm. to a meter or more in length, free-floating except sometimes for attachment at one end...**10**

9. Scape slender, less than 10 cm. high, with 1–2 flowers; bracts tapering to a clasping base (Fig. 43); pedicels ascending in fruit; branches short, creeping over mud in shallow water (Fig. 53)...**13**

    10. Bladder-bearing leaves forking repeatedly (Fig. 54); whorl of upper stems inflated and floating; bracts broad-based..............................**11**

    10. Bladder-bearing leaves forking at base, each fork consisting of a straightish (Fig. 52) or zigzag (Fig. 51) rachis with segments coming from its two sides; plants without a whorl of inflated stems; bracts various................**12**

11. Inflated floating stems 5–7 cm. long; length of scape from floats to lowest pedicel 7.5 cm. or more; flowers 3–14 on each scape; calyx lobes 4–6 mm. long. **U. inflàta** Walt. Quiet waters near the coast; Va. to w. Tenn., Fla. & Tex.

11. Inflated floating stems 4 cm. or less long; scape 5 cm. or less long, 2–5-flowered; calyx lobes 3–4 mm. long. **U. inflata** var. **mìnor** Chapm. Fig. 54. S. Maine to N.J.; Fla.; s.w. Ark.; S. A. (*U. radiata*, B.)

    12. Scapes 2–5-flowered, without scales; bracts without basal lobes; minute flowers which do not open scattered along the stem at the base of the scape (Fig. 52); leaves without spines except at the tips of the divisions. **U. geminiscàpa** Benj. Ponds and slow streams; Nfd. to Del. & Pa., rarely s. to Va., w. to Mich. & Wis. (*U. clandestina*, G.)

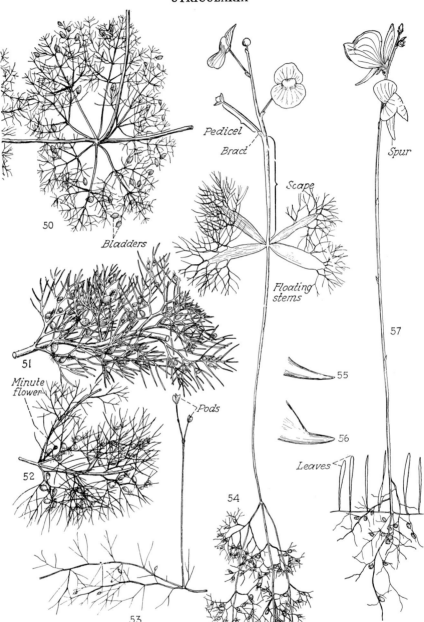

U. purpurea: 50. Branches, × 1.　U. vulgaris var. americana: 51. Branch and leaves, × 1.　U. geminiscapa: 52. Branch and leaves, × 1.　U. gibba: 53. Fruiting plant, × 1. U. inflata var. minor: 54. Plant, × 1.　U. cornuta: 57. Plant, × 1.　U. biflora: 55. Spur. × 2.　U. fibrosa: 56. Spur, × 2.

12. Scapes 6–12-flowered, with 1–5 scales; bracts with basal lobes; minute flowers absent; leaves with bristles on the margins when seen through the microscope. **U. vulgàris** L., var. **americàna** Gray. Fig. 51. Common in ponds and slow streams; Labrador & Minn. to Alaska, s. to Va., and rarely to Fla., Tex. & Mex. (*U. macrorhiza*, B.)

13. Spur shorter than the lower lip of the corolla. **U. gíbba** L. Fig. 53. Que. & N.S. to Fla., Puerto Rico & British Honduras; w. about the Great Lakes; Calif.

13. Spur equaling or longer than the lower lip............................**14**

 14. Spur tapering from base to apex (Fig. 55); all leaves bearing bladders. **U. biflòra** Lam. Mass. to Fla., Okla. & Tex. (*U. pumila*, B.)

 14. Spur conic at base, cylindric above (Fig. 56); some leaves without bladders. **U. fibròsa** Walt. Shallow pools in pine barrens; e. Mass., L.I. & N.J. to Fla. & Miss.

KEY TO WINTER BUDS

1. Bud a thickened end of the stem, 1–2 mm. in diameter, bearing crowded incurled branches (Fig. 31). **U. purpurea.**

1. Bud a rounded mass of densely crowded leaves............................**2**

 2. Divisions of leaves fringed at tip with stout gray hairs (Figs. 33, 34).......**3**

 2. Divisions of leaves not fringed, sometimes with a few hairs................**4**

3. Buds 1–2 cm. long, often one-sided or 2-lobed (Fig. 37); leaves with a central rachis (Fig. 34). **U. vulgaris** var. **americana.**

3. Buds 3–10 mm. long, oval, symmetrical (Fig. 32); leaves very small, basally parted (Fig. 33). **U. intermedia.**

 4. Buds reddish-green; divisions of leaves flattened. **U. minor.** Fig. 35.

 4. Buds green; divisions of leaves thread-like................................**5**

5. Buds 2–5 mm. in diameter. **U. geminiscapa.** Fig. 30.

5. Buds 1 mm. or less in diameter. **U. inflata** var. **minor** (Fig. 36) and **U. gibba.**

This treatment of *Utricularia* is by John W. Thomson, Jr. The key to winter buds is adapted from that of Rossbach, Rhodora, **41,** 114–115 (1939).

Reference: Fernald, Rhodora **23,** 291 (1922)—*U. subulata f. cleistogama.*

ACANTHUS FAMILY   ACANTHÀCEAE

WATER WILLOW   Dianthèra

 Plants perennial from a horizontal underground stem; leaves opposite; flowers purplish, in dense clusters which are long-stalked in the axils of the leaves; fruit a 4-seeded flattened pod.

1. Leaves long and narrow, usually shorter than the stalk supporting the flowers. **D. americàna** L. Fig. 59. Usually in shallow water; Que. & Ont. to Mich., s. to Ga. & Tex.

1. Leaves oblong, usually longer than the stalk supporting the flowers. **D. ovata** Walt. Swamps; Va. to Fla. & Tex., n. to Mo.

PLANTAIN FAMILY   PLANTAGINÀCEAE

Littorélla

L. americàna Fernald. Fig. 58. Dwarf plants with rather stiff dark thread-like tufted leaves which are dilated into papery sheaths at base; flowers very small, the staminate on a slender naked stem, the pistillate hidden at the base of the plant.

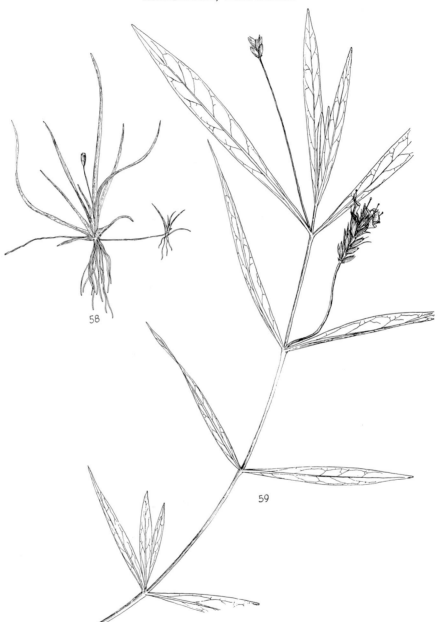

L. americana: 58. Plant, × 1.   D. americana: 59. End of branch. × 1.

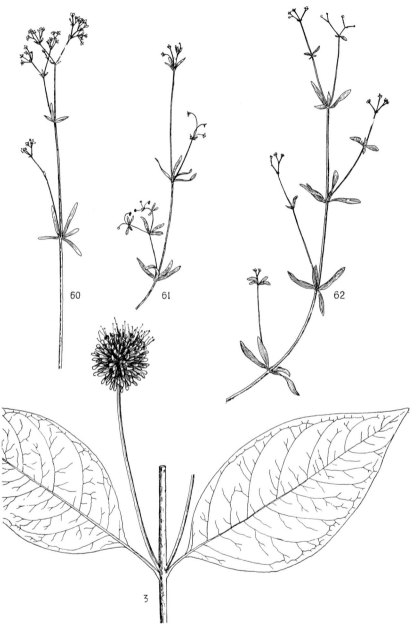

60

61

62

3

G. palustre: 60. Top of p'ant, × 1.　**G. trifidum:** 61. Top of plant, × 1.　**G. tinctorium:** 62. Top of plant, × 1.　**C. occidentalis:** 63. Leaves and head of flowers, × ½.

Submersed, Nfd. to Maine, Vt., Ont. & Minn. (*L. uniflora*, G, B.) Thought to be rare, but perhaps more abundant in the submersed sterile state than is supposed. The flowers appear only as the water recedes and are certainly very rare. The sterile plants are often difficult to distinguish from *Ranunculus reptans*.

Reference: Fernald, Rhodora, **20**, 61–62 (1918)—*L. americana.*

MADDER FAMILY RUBIÂCEAE

BUTTONBUSH Cephalánthus

C. occidentàlis L. Fig. 63. Shrub, or rarely a small tree, with leaves opposite or in 3's; flowers in spherical heads at the ends of long naked stalks in the axils of the upper leaves. Swamps and streams; N.S. & B.C. to Ont. & s. Minn., s. to Fla. & Tex.—Var. **pubéscens** Raf. differs in having the branchlets and lower leaf surfaces softly hairy. S. Ind. & s.e. Mo. to w. La. & e. Tex.

BEDSTRAW Gàlium

Low often prostrate or matted herbs with leaves in whorls of 4–6; leaf margins and often the angles of the stem minutely saw-toothed; flowers white or greenish, minute; fruit of twin spheres. Plants of swamps, wet woods, and shores, sometimes at the water's edge.

1. Flowers in much-branched clusters (Fig. 60). **G. palústre** L. Nfd.to Mich., s. to Conn. & N.Y.; perhaps partly introduced from Europe.
1. Flowers solitary or 2–3 together...2
   2. Flower stalks slender, curved, minutely roughened, rarely forking (Fig. 61). **G. trífidum** L. Nfd. & Labrador to B.C., s. to n. & w. N.E., N.Y., Ohio, Mich. & Colo.
   2. Flower stalks stout, nearly straight, smooth, often forking (Fig. 62). **G. tinctòrium** L. Nfd. & N.Y. to Mich. & Neb., s. to Fla., Mo. & Tex. (*G. Claytoni*, G, B; not *G. tinctorium*, G, B.)

   Reference: Fernald, Rhodora. **37**, 443–445 (1935)—*G. tinctorium.*

BLUEBELL FAMILY CAMPANULÂCEAE

BELLFLOWER Campánula

Corolla borne at the summit of the ovary, blue, bell-shaped, with 5 lobes.

1. Leaves several centimeters wide. **C. americàna** L. Harebell. Fig. 64. Rich woods and shores; Ont. & N.Y. to Neb., s. to Ga. & Ark.—See *Lobelia siphilitica*, with which this may be confused.
1. Leaves less than 1 cm. wide..2
   2. Calyx 1.3–3.8 mm. long, its lobes 0.7–2 mm. long; capsule 1.2–2 mm. long. **C. aparinoìdes** Pursh. Marsh Bluebell. Fig. 66. Wet grassy ground; Maine to Wis. & Neb., s. to Ga., Ky. & Ill. Submersed forms have shorter leaves and limp stems which are nearly smooth.
   2. Calyx (3–) 4–6.7 mm. long, its lobes 2–4 mm. long; capsule 3.2–5 mm. long. **C. uliginòsa** Rydb. Fig. 65. Wet meadows; N.B. to Sask., s. to N.Y., Ind. & Neb.

C. **americana**: 64. Leaves, stem and flowers, × 1. C. **uliginosa**: 65. Top of plant, × 1
C. **aparinoides**: 66. Top of plant, × 1. L. **Dortmanna**: 67. Plant, × 1.

317

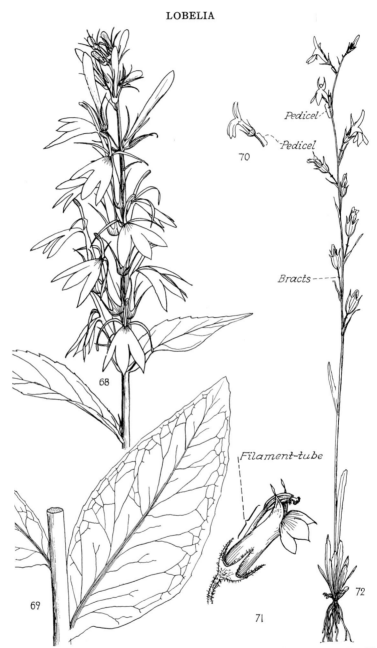

L. **cardinalis**: 68. Top of plant, × 1.   L. **siphilitica**: 69. Stem and leaves, × 1.   71. Flower, × 1.   L. **Nuttallii**: 70. Flower, × 1.   L. **Kalmii**: 72. Plant, × 1.

LOBELIA FAMILY   LOBELIÀCEAE

Lobèlia

Flowers in the axils of leaves or of bracts; corolla blue or red (rarely pink or white), with a straight tube which is 2-lipped and often split down the upper side; fruit a many-seeded pod. The only strictly aquatic species is *L. Dortmanna,* which is very different in appearance from all the others.

1. Leaves in a basal rosette, fleshy and consisting of 2 tubes united lengthwise. **L. Dortmánna** L. Water Lobelia. Fig. 67. Nfd. to n. Va., w. through N.Y. & s. Ont. to n. Wis. & Minn.; Vancouver Is. to Ore.—The rosette of leaves is nearly always submersed, and the flowers may or may not extend into the air.
1. Leaves flat, scattered on the stem..2
   2. Flowers 2–4.5 cm. long.................................. 3
   2. Flowers 7–16 mm. long...6
3. Flowers 3–4.5 cm. long, usually red. **L. cardinàlis** L. Cardinal Flower. Fig. 68. Swamps, river banks, streams, etc.; e. Que. to e. Minn., s. to the Gulf of Mexico.— Forma **ròsea** St. John, with rose-pink flowers, and forma **álba** (Eaton) St. John, with white flowers, are rare.
3. Flowers 1.8–3.3 cm. long, blue, or rarely white.............................4
   4. Filament tube (Fig. 71) 12–15 mm. long; calyx often more or less hairy......5
   4. Filament tube 8–11 mm. long; calyx without hairs. **L. elongàta** Small. Swamps and tidal marshes; s. Del. to Ga.
5. Leaves hairy beneath; calyx lobes bristly-hairy. **L. siphilítica** L. Great Lobelia. Figs. 69, 71. Moist woods, swamps and streams; Maine & n. N.Y. to n. Wis. & s.e. S.D., s. to N.C., Ala. & Mo.—This may be confused with *Campanula americana,* which has a bell-shaped 5-lobed corolla.—Forma **albiflòra** Britton, with white flowers, is rare.
5. Leaves without hairs; calyx lobes with few hairs or none. **L. siphilitica** var. **ludoviciàna** DC. Ill. & n.w. Wis. to Man., s. to e. Tex. & Colo.
   6. Stalk of flowers with a pair of minute bracts near the middle (Fig. 72); plants mostly north and west of N.J. **L. Kálmii** L. Wet meadows, bogs, lake shores, etc.; Nfd. to N.J. & s. Pa., n. to James Bay & Great Slave Lake, w. to s. Ohio, central Ill., Colo. & s. B.C.
   6. Stalk of flowers without bracts or with a pair of minute bracts nearly hidden at base; plants mostly of the Coastal Plain from L.I. southwestward..........7
7. Leaves 5 mm. or more wide; base of lower lip of corolla not hairy; plants 2–6 dm. high. **L. Nuttállii** R. & S. Fig. 70. Woods and swamps; L.I. & e. Pa. to Tenn., Ky., Ala. & Fla.
7. Leaves 4 mm. or less wide; base of lower lip of corolla hairy within; plants 7.5–13 dm. high..8
   8. Stalks of flowers not roughened, and without bracts at base. **L. Boykínii** T. & G. Swamps and cypress bogs; s. Del.; S.C. to Fla.
   8. Stalks of flowers roughened, and with a pair of little bracts at base. **L. Cánbyi** Gray. Swamps; s. N.J. to Tenn. & Ga.

References: McVaugh, Rhodora, **38**, 241–263, 276–298, 305–329, 346–362 (1936)— all species and varieties of e. N.A.; St. John, Rhodora, **21**, 217–218 (1919)—forms of *L cardinalis;* Britton, Bull. Torr. Bot. Club, **17**, 125 (1890)—*L. siphilitica* f. *albiflora.*

COMPOSITE FAMILY   COMPÓSITAE

A very large family characterized by the aggregation of the flowers into heads, each head with an involucre of bracts (Fig. 1); flowers of 2 types (of which only one may be present in some genera), those of the disk, or inner part of the head, tubular, and the marginal ones strap-shaped.

1. Leaves finely divided (Fig. 1). **Megalodonta, p. 320.**
1. Leaves not finely divided...2
    2. Leaves opposite or 3–5 at each level on the stem.....................3
    2. Leaves scattered..5
3. Leaves 3–5 at each level. **Eupatorium, p. 320.**
3. Leaves opposite...4
    4. Leaves broadest at the base, often joined (Fig. 2); flowers white or purple. **Eupatorium, p. 320.**
    4. Leaves narrowed to the base, sometimes compound; flowers yellow. **Bidens, p. 325.**
5. Leaves dotted with black glands visible with a lens. **Helenium, p. 339.**
5. Leaves not gland-dotted..—.............6
    6. Rays yellow; leaves with 3–5 parallel veins (Fig. 54). **Solidago, p. 339.**
    6. Rays white, purple or blue; veins leaving the midrib at intervals...........7
7. Leaf tips abruptly narrowed to a little point (Fig. 50). **Boltonia, p. 339.**
7. Leaves tapered to the tip. **Aster, p. 341.**

WATER MARIGOLD   Megalodónta

M. Béckii (Torr.) Greene. Fig. 1. Strictly aquatic plants, with submersed leaves cut into many thread-like divisions, the emersed leaves, if present, simple; head of flowers like that of *Bidens;* fruit a nutlet like that of *Bidens.* Ponds and streams; Que. to Man., s. to N.J. & Mo. (*Bidens Beckii,* G.)

Eupatòrium

Coarse herbs, often 1 m. or more tall, with opposite or whorled leaves; heads with disk flowers only, small, numerous, in a flat or dome-shaped cluster.

1. Leaves 3 or more in a whorl (rarely opposite), definitely petioled; flowers purple..**2**
1. Leaves opposite (rarely the upper scattered), without petioles, ordinarily joined at base so as to clasp the stem. **E. perfoliàtum** L. Boneset; Thoroughwort. Fig. 2. Wet meadows, lake shores, etc., common; N.S. & N.B. to Man., s. to Fla., Tex. & Neb.—Forma **purpùreum** Britton, with flowers more or less purplish, occurs locally, sometimes making up a large proportion of the individuals.—Forma **truncàtum** (Muhl.) Fassett, with leaves separate at base, forma **trifòlium** Fassett, with leaves in whorls of 3, and forma **laciniàtum** Stebbins, with leaves separate and cut along the margin into jagged teeth 4–9 mm. long, all occur sporadically, often in the same clump with normal plants.
    2. Leaves abruptly narrowed to the petiole, the 2 lowest side veins much the longest (Fig. 3). **E. dùbium** Willd. Joe-Pye Weed. Swamps and streams; s. N.H. to S.C.
    2. Leaves tapered to the petiole, the side veins nearly uniform (Fig. 4)........**3**

MEGALODONTA

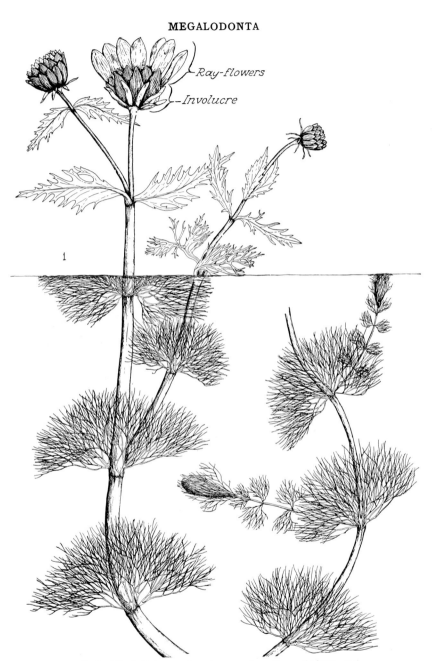

Ray-flowers

Involucre

1

M. Beckii: 1. Submersed and emersed portions of plant, × 1.

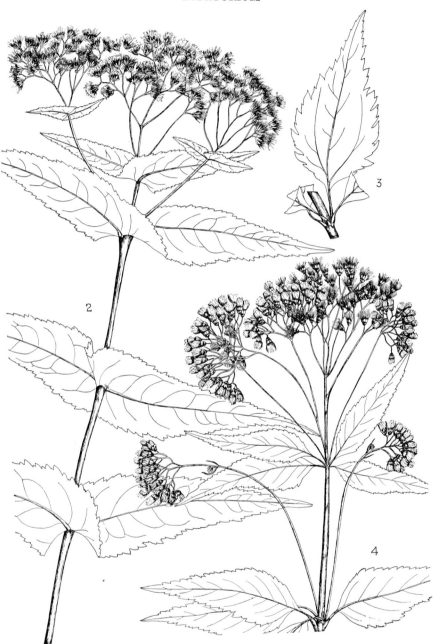

E. **perfoliatum**: 2. Top of plant, × ½.   E. **dubium**: 3. Stem and leaves, × ½.   E. **maculatum**: 4. Top of plant, × ½.

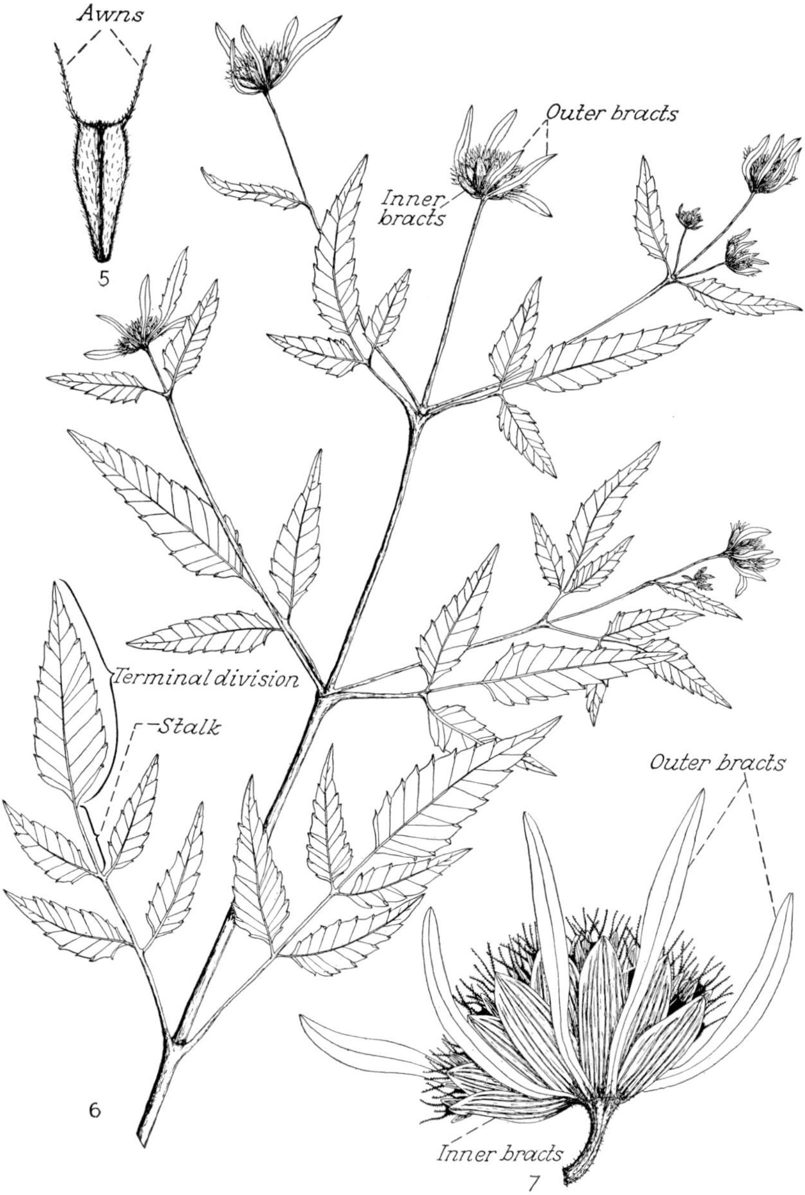

Awns

5

Outer bracts

Inner bracts

Terminal division

Stalk

6

Outer bracts

Inner bracts

7

B. discoidea : 5. Nutlet, $\times$ 3.   6. Upper part of plant, $\times \frac{1}{2}$.   7. Head, $\times$ 2.

B. frondosa: 9. Upper part of plant, $\times \frac{1}{2}$.   10. Nutlet, $\times$ 3.   11. Head, $\times$ 2.   **B. frondosa** var. **anomala**: 8. Nutlet, $\times$ 3.   **B. frondosa** var. **pallida**: 12. Leaf, $\times$ 1.

3. Uppermost leaves shorter than the inflorescence. **E. maculàtum** L. Joe-Pye Weed. Fig. 4. Common in open wet ground and on lake shores; Nfd. & Que. to Me., w. Conn.; e. Pa., w. through Minn. & Ill. to B.C. & N.M. (*E. purpureum* var. *maculatum*, G.)

3. Uppermost leaves longer than the inflorescence. **E. maculatum** var. **foliòsum** (Fernald) Wiegand. Nfd. to Me., w. to n. Mich.

References: Wiegand, Rhodora, **22,** 57–70 (1920)—taxonomic treatment of the Joe-Pye Weed group; Wiegand and Weatherby, Rhodora, **39,** 297–306 (1937)—the nomenclature of the Joe-Pye Weed group; Britton, Bull. Torr. Bot. Club, **17,** 124 (1890), Fassett, Rhodora, **27,** 55 (1925), and Stebbins, Rhodora, **32,** 133 (1930)—forms of *E. perfoliatum.*

Bur Marigold; Beggar-ticks; Pitchforks **Bìdens**

Flowers bright yellow, sometimes very conspicuous in late summer; ray flowers present or absent; head with spreading leaf-like outer bracts at base (Figs. 6, 7) in addition to the yellowish inner ones; nutlet crowned with 2–4 (in some species more) spine-like awns (Fig. 5). In the fall many members of this genus force themselves on one's attention by fastening their nutlets to clothes in great numbers.

Although most species of *Bidens* can be named in the flowering condition by one familiar with them, in some cases mature nutlets are necessary for certain identification. Like many annuals, *Bidens* is most variable in habit, and almost any species, growing on a lake shore, may range in size from dwarf unbranched individuals only a few centimeters tall to copiously branched plants a meter or more in height.

1. Leaves compound, the terminal division on a slender stalk (Fig. 6)............**2**
1. Leaves simple (Fig. 26), or deeply cleft, or even several times divided but with the terminal division on a winged stalk (Figs. 16, 40, 41)....................**7**
  2. Outer bracts 3–5, not fringed (Fig. 7). **B. discoídea** (T. & G.) Britton. Figs. 5, 6. Boggy shores, often on floating or water-soaked logs; N.S. & s. Que. to s.e. Minn., s. to Va., Ala., La. & Tex.—The barbs on the awns regularly point upward (Fig. 5); rarely individuals are found with downwardly barbed awns.
  2. Outer bracts 3–16, fringed with white hairs (Fig. 11)....................**3**
3. Outer bracts 5–8; inner bracts as long as the nutlets (Fig. 11)...............**4**
3. Outer bracts 10–16; inner bracts shorter than the nutlets (Fig. 15)...........**6**
  4. Awns barbed downward (Fig. 10)..**5**
  4. Awns barbed upward (Fig. 8). **B. frondòsa** var. **anómala** Porter. Cobblestone beaches and brackish shores; N.S. to D.C., and inland very locally through Vt., n. N.Y., s. Ont., w. Wis., Iowa, Kans. & Neb.
5. Plants bright green; branches not overtopping the main stem. **B. frondòsa** L. Figs. 9–11. Very common, from dry fields to wet shores; Nfd. & Que. to Ont., Man., Sask. & Wash., s. to Fla., La., Colo. & Calif.; adventive in Europe.
5. Plants pale-green; side branches longer than the main stem; stalk of terminal leaflet tending to be winged. **B. frondosa** var. **pállida** Wiegand. Fig.12. Pond shores, local; N.S., N.Y., Ill. & Wis. A poorly understood plant, sometimes appearing as a hybrid of *B. frondosa* with *B. comosa.*
  6. Plants not hairy. **B. vulgàta** Greene. Figs. 13–15. Fields, waste land in cities, and damp places; weedy and less apt to venture into the water than the 2 preceding species; N.S. & Que. to Alberta, s. to N.C., Mo., Nev. & Calif.

B. vulgata: 13. Nutlet, × 3.   14. Top of plant, × ½.   15. Head, × 2.

Awns

16   17

B. coronata : 16. Top of plant, $\times$ ½.   17. Nutlet, $\times$ 3.

Awns

18

19

20

21

Inner bract

Outer bracts

B. aristosa : 19. Nutlet, × 3.   20. Top of plant, × ½.   21. Head, × 2.   **B. aristosa** var. **mutica :** 18. Nutlet, × 3.

6. Plants finely white-hairy. **B. vulgata** forma **pubérula** (Wiegand) Fernald. Occurring sometimes with *B. vulgata*.

7. Specimens with mature nutlets...**8**

7. Specimens with flowers only................................**32**

  8. Awns triangular in cross section and of the same texture as the rest of the nutlet (Fig. 17). **B. coronàta** (L.) Britton. Tickseed Sunflower. Fig. 16. Swales; Mass., s. Ont. & Wis., s. to Va., Ky. & Neb. (*B. trichosperma*, G, B.) The size of the nutlets and the width of leaf segments are variable.

  8. Awns round in cross section and of firmer texture than the rest of the nutlet (Fig. 19), or rarely absent...**9**

9. Nutlets with crinkled somewhat corky or winged margins (Figs. 19, 25).......**10**

9. Nutlets with bristly but not crinkled margins (Fig. 27)....................**15**

  10. Leaves divided into several leaflets (Fig. 20); nutlet flat (Figs. 18, 19)....**11**

  10. Leaves not divided (Fig. 31); nutlets 4-angled (Fig. 28)...............**17**

11. Outer bracts 8–10, shorter than the inner (Fig. 21).......................**12**

11. Outer bracts 12–20, longer than the inner (Fig. 22).......................**14**

  12. Awns lacking (Fig. 18). **B. aristòsa** var. **mùtica** Gray. Mass. & Va., w. to Ill. & Mo.; probably adventive in the eastern part of its range.

  12. Awns present...**13**

13. Awns barbed downward (Fig. 19). **B. aristòsa** (Michx.) Britton. Fig. 20. Swamps; Maine to Minn., s. to Va., n. Ala., Miss., & s.e. Tex; probably adventive in the eastern part of its range.

13. Awns barbed upward. **B. aristosa** var. **Fritchèyi** Fernald. Ind. & Ky. to Ill. & Mo.; adventive in D.C., Mass., Maine and probably elsewhere.

  14. Awns absent or reduced to little triangular teeth (Fig. 25). **B. polylèpis** Blake. Fig. 23. Swamps; w. Ill. to Iowa, Kans., Colo. & Tex.; adventive in e. U.S. (*B. involucrata*, B, G.)

  14. Awns long, barbed downward (Fig. 24). **B. polylepis** var. **retròrsa** Sherff. Ind. & Mo.

15. Heads nodding at fruiting time (Figs. 26, 31) except in dwarfed individuals; nutlets with a firm rounded summit (Figs. 27, 28)................................**16**

15. Heads erect at flowering time; nutlets without a firm rounded summit (Fig. 32).**19**

  16. Nutlets straight and flat, not strongly keeled, without pale corky margins (Fig. 27); rays 1.5–3 cm. long. **B. laèvis** (L.) BSP. Brook Sunflower. Fig. 26. Swamps; N.H. to Fla. near the coast, and w. from Fla. to Calif.— The lower part of the stem is often prostrate and trailing in the water and soft mud.

  16. Nutlets curved, strongly keeled, with pale corky margins (Fig. 28); rays wanting or at most 1.7 cm. long...................................**17**

17. Stems branching, 1–9 dm. high; heads many, nodding in fruit; leaves not petioled **18**

17. Stems simple or nearly so, 2–20 cm. high; heads solitary or few, nearly or quite erect in fruit; leaves petioled. **B. cérnua** forma **mínima** (Huds.) Larss. Fig. 30. A dwarfed form on boggy shores; Gulf of St. Lawrence to s.N.H., w. to N.Y., Wis., etc.

  18. Leaves long-pointed, with 2–24 teeth on each side (Fig. 31). **B. cérnua** L. Bur Marigold. Abundant in wet places, N.S. to B.C., s. to N.C., Tenn., N.M. & Calif.; also in Europe.

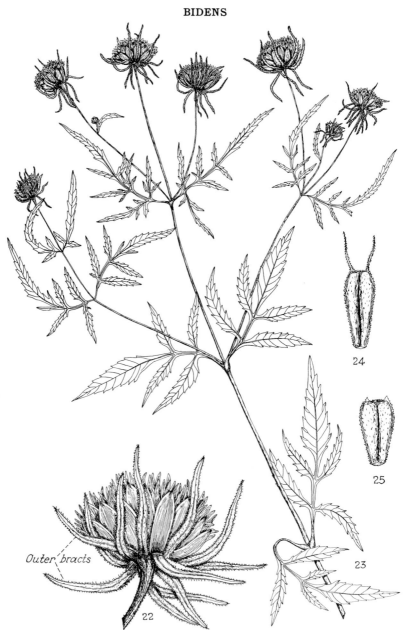

Outer bracts

B. polylepis: 22. Head, × 2. 23. Top of plant, × ½. 25. Nutlet, × 3. **B. polylepis** var. **retrorsa:** 24. Nutlet, × 3.

Ray-flowers

Disk

Rounded summit

26

27

B. laevis: 26. Top of plant, $\times$ $\frac{1}{2}$.   27. Nutlet, $\times$ 3.

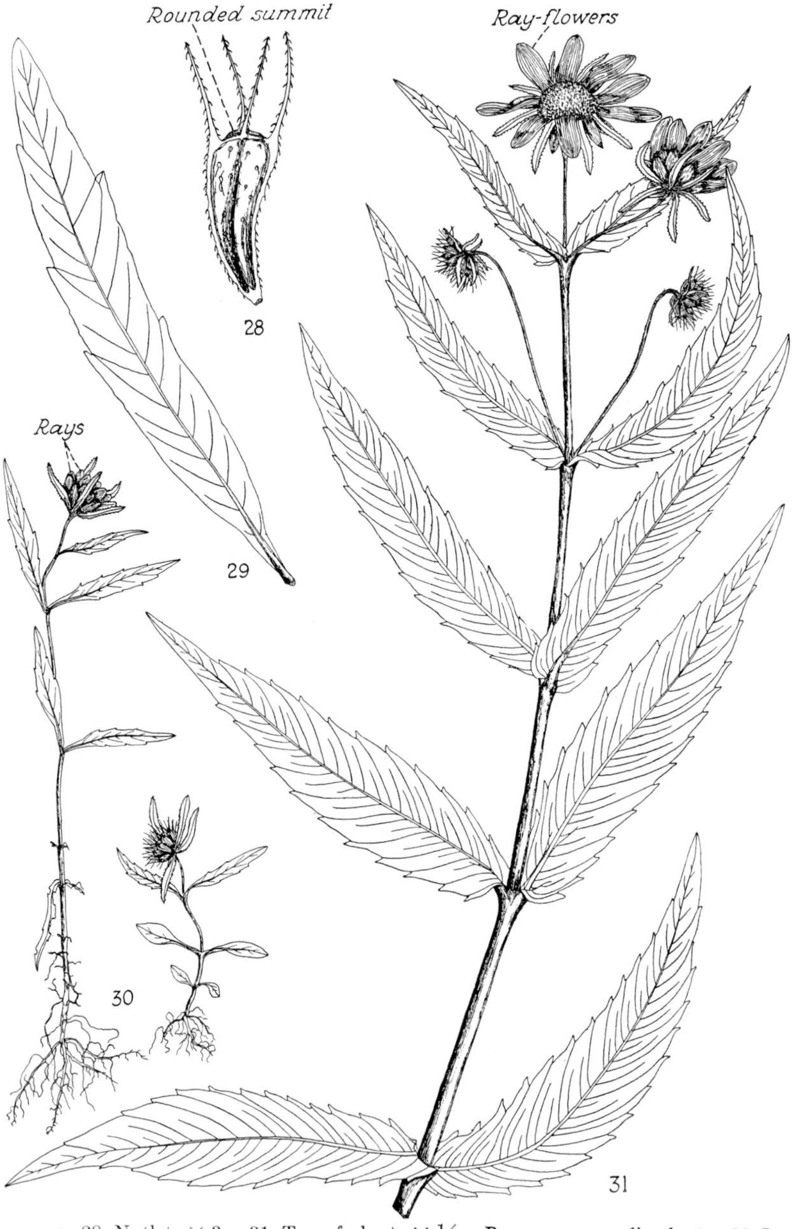

Rounded summit

Ray-flowers

Rays

28

29

30

31

B. cernua: 28. Nutlet, × 3.   31. Top of plant, × ½.   B. cernua var. oligodonta : 29. Leaf, × ½.   B. cernua f. minima : 30. Two plants, × ½.

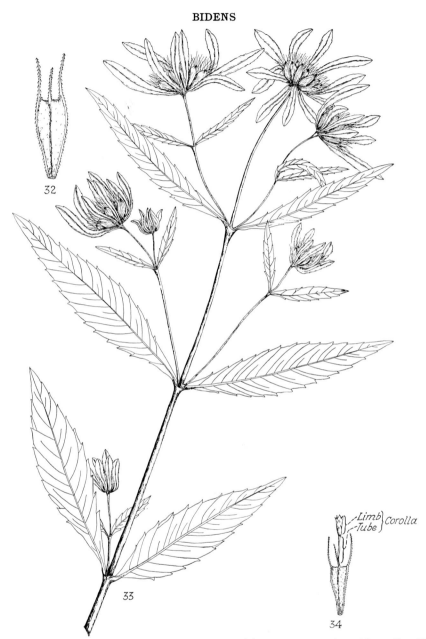

B. comosa: 32. Nutlet, $\times$ 3.   33. Top of plant, $\times$ ½.   34. Young nutlet, with corolla still present, $\times$ 3.

333

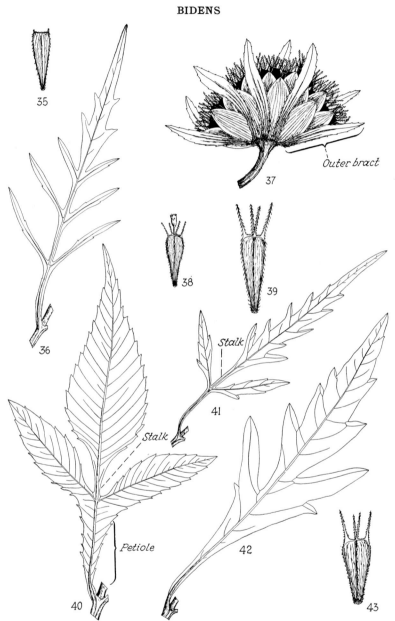

B. connata: 37. Head, × 2. 38. Young nutlet, with corolla still present, × 3. 39. Nutlet, × 3. 40. Leaf, × 1. **B. connata** var. **submutica:** 35. Nutlet, × 3. **B. connata** var. **pinnata:** 36. Leaf, × 1. **B. connata** var. **ambiversa:** 42. Leaf, × 1. 43. Nutlet, × 3. **B. connata** var. **gracilipes:** 41. Leaf, × 1.

18. Leaves blunt, with 1–6 teeth on each side (Fig. 29). **B. cernua** var. **oligodónta** Fernald & St. John.   Gulf of St. Lawrence to Mass. & w. N.Y.

19. Margins of nutlets barbed downward for their entire length, the faces barbed downward or smooth (Fig. 32); disk flowers light-yellow. **B. comòsa** (Gray) Wiegand. Fig. 33. Swales and wet shores; Maine & Que. to N.D , s. to N.C., Tenn., N.M. & Utah.

19. Margins of most nutlets barbed upward at least near the base, the faces barbed upward (Fig. 39); disk flowers dark-yellow or orange......................**20**

   20. Mature nutlets 4-angled (Fig. 44) except those near the margin of the disk **21**

   20. Mature nutlets flat (Fig. 46)..**28**

21. Leaves simple or with 2–3 broad divisions...............................**22**

21. Leaves with 3–7 very narrow divisons (Fig. 36). **B. connàta** var. **pinnàta** Wats. Sandy lake shores; n.w. Wis. & adjacent Minn.—Grades into most of the other varieties.

   22. Margin of nutlet barbed downward except at base (Fig. 39).............**23**

   22. Margin of nutlet barbed upward along its entire length (Fig. 44).......**27**

23. Awns well-developed, 4 in number on the central nutlets in each head.......**24**

23. Awns none or but 2 in number and 0.25–1 mm. long (Fig. 35). **B. connata** var. **submùtica** Fassett. Shores of Lake Nipissing, Ont.

   24. Outer bracts 1.5 cm. or less long (Fig. 37)...........................**25**

   24. Outer bracts 3–6 cm. long (Fig. 49). **B. connata** var. **fállax** (Warnst.) Sherff. Local, Que. to R.I. & Va., w. to Ind. & Minn.; introduced in Germany.

25. Petiole broadly winged (Fig. 40); blade usually 3-parted. **B. connàta** Muhl. Swamp Beggarticks. Figs. 38–40. Very common in wet places; N.S. & Que. to N.J., w. to Minn., Mo. & Ala.

25. Petiole not winged or very narrowly winged (Figs. 41, 45).................**26**

   26. Leaves not divided, except in extremely large individuals. **B. connata** var. **petiolàta** (Nutt.) Farwell. Fig. 45. Common in wet places; N.S. & Que. to Ont. & Minn., s. to Va., Tenn., Mo. & Kans.; adventive in Europe.

   26. Leaves 3-, rarely 5-, lobed. **B. connata** var. **gracílipes** Fernald. Fig. 41. Sandy beaches; Maine to Conn.; Wis.; and perhaps elsewhere.

27. Awns with all barbs pointing upward (Fig. 44); leaves not usually deeply divided. **B. connata** var. **anómala** Farwell. Local, Md., Mich., Wis. and perhaps elsewhere.

27. Awns with barbs pointing variously (Fig. 43); leaves with lobes below grading into coarse teeth above (Fig. 42). **B. connata** var. **ambivérsa** Fassett. Ditches and bogs; n. Wis.

   28. Face of nutlet with longitudinal lines (Fig. 46)......................**29**

   28. Face of nutlet without longitudinal lines (Fig. 10). See **B. frondosa** var. **pallida, 5.**

29. Leaves with winged petioles, the simple leaves or terminal segments of divided leaves narrow and sharply toothed....................................**30**

29. Leaves with slender petioles, the simple leaves or terminal segments of divided leaves a third to a half as broad as long and bluntly toothed (Fig. 48)......**31**

   30. Awns barbed upward. **B. heterodóxa** Fernald & St. John. Fig. 46. P.E.I. —a poorly understood plant, somewhat resembling *B. connata.*

   30. Awns barbed downward. **B. heterodoxa** var. **orthodóxa** Fernald & St. John. Magdalen Isls., Que.

31. Awns barbed downward (Fig. 47). **B. heterodoxa** var. **monardaefòlia** Fernald. Lake Pocotopaug, Conn.

31. Awns smooth. **B. heterodoxa** var. **agnóstica** Fernald. Lake Pocotopaug, Conn.

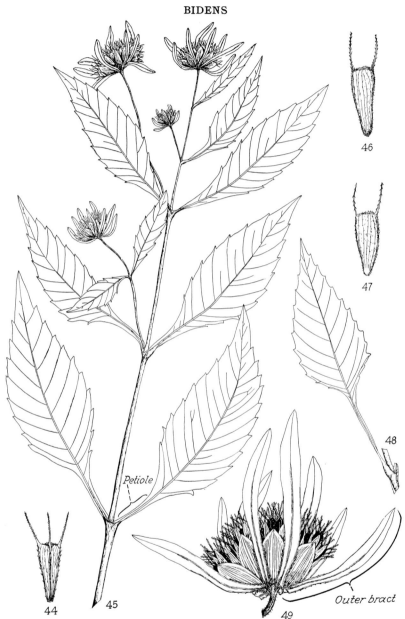

B. connata var. petiolata: 45. Top of plant, × ½. B. connata var. anomala: 44. Nutlet, × 3. B. connata var. fallax: 49. Head, × 2. B. heterodoxa: 46. Nutlet, × 3. B. heterodoxa var. monardaefolia: 47. Nutlet, × 3. 48. Leaf, × ½.

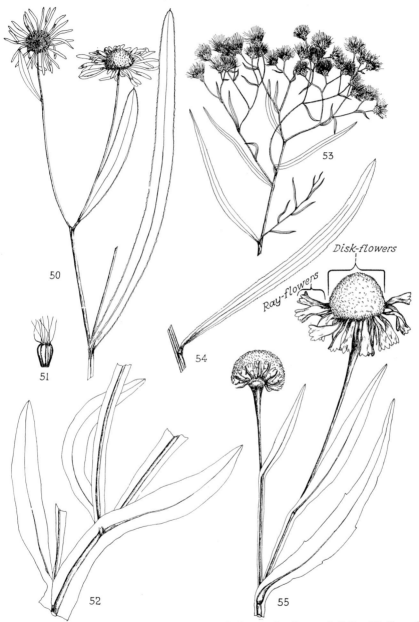

B. asteroides: 50. Top of plant, × 1. 51. Nutlet, × 5. **S. graminifolia:** 53. Top of plant, × 1. 54. Leaf, × 1. **H. nudiflorum:** 52. Stem and leaves, × 1. **H. autumnale:** 55. Top of plant, × 1.

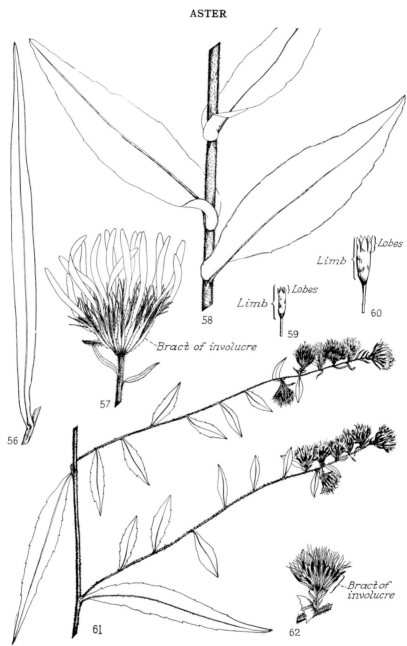

Labels visible in figure: *Limb*, *Lobes*, *Bract of involucre*

A. junceus: 56. Leaf, × 1. 57. Head, × 2. A. puniceus: 58. Stem and leaves, × 1. A. praealtus: 59. Corolla of disk flower, × 3. A. paniculatus: 60. Corolla of disk flower, × 3. A. lateriflorus var. pendulus: 61. Part of plant, × 1. 62. Head, × 2.

32. Rays conspicuous, longer than the diameter of the disk (Figs. 16, 26).....**33**
32. Rays absent (Figs. 20, 22) or inconspicuous, and shorter than the diameter of the disk (Fig. 30)..**35**
33. Leaves cut into many divisions (Figs. 16, 20, 23)........................**34**
33. Leaves simple...**17**
    34. Stalk that supports the head provided with close white hairs. **B. aristosa, 12–13,** and **B. polylepis, 14**
    34. Stalk that supports the head lacking hairs, or with a very few spreading hairs. **B. coronata, 8.**
35. Leaves not petioled or tapered at base. **B. cernua, 18.**
35. Leaves petioled, or at least tapered, at base..............................**36**
    36. Corolla of disk flowers light-yellow, the gradually flaring limb about half as long as the slender tube (Fig. 34). **B. comosa, 19.**
    36. Corolla of disk flowers (at least in *B. connata*) dark orange-yellow, the abruptly flared limb about as long as the slender tube (Fig. 38). The common plant keying to this point is **B. connata, 21–27.** An occasional individual of the local **B. frondosa** var. **pallida, 5,** may have a sufficiently well-developed wing on the stalk of the terminal leaflet to key here. The rare **B. heterodoxa, 30–31,** will also key to this point.

References: Sherff, Field Mus. of Nat. Hist., Bot. Ser., **16,** pt. 1, 1–346, pt. 2, 347–709 (1937)—a monograph of *Bidens;* Fernald, Rhodora **40,** 352 (1938)—*B. vulgata* f. *puberula;* Fassett, Rhodora, **35,** 391 (1933)—*B. connata* var. *submutica.*

SNEEZEWEED **Helènium**

Tall plants, 0.2–2 m. high; leaf bases running down as wings on the stem; ray flowers yellow (or rarely purple-brown), conspicuous.

1. Leaves mostly toothed; disk flowers yellow. **H. autumnàle** L. Fig. 55. River banks, etc.; Que. & w. Mass. to Man. & Ore., s. to Fla. & Ariz.—Hybridizes with the next.
1. Leaves not toothed (Fig. 52), or the basal ones somewhat toothed; disk flowers brownish. **H. nudiflòrum** Nutt. Ill. & Mo. to N.C. & Tex.; established in e. Pa.

GOLDENROD **Solidàgo**

Of this large and well-known genus, one species occurs commonly on wet shores of lakes and streams and is occasionally found in shallow water.

S. graminifòlia (**L.**) Salisb. Figs. 53, 54. Leaves ribbon-like, with a central vein and 1–3 fainter ones on each side of it; upper part of stem branched so that the inflorescence is flat-topped; flowers yellow, in small numerous heads which are themselves crowded in little clusters. E. Que. to Sask., s. to N.J., N.C., Ill. & Mo.—**S. graminifolia** var. **Nuttállii** (Greene) Fernald. Branches finely hairy. Nfd. to n. Wis., s. to Ala.

Boltònia

B. asteroìdes (**L.**) L'Her. Fig. 50. Tall plants, 0.2–2.5 m. high, often much-branched above and looking much like an *Aster;* nutlets (Fig. 51) flat, about 1 mm. long, crowded with several slender bristles. Along streams, river bottoms, etc.; Conn. to Minn. & Neb., s. to Fla. & La.—Var. **decúrrens** (T. & G.) Engelm. has the bases of the leaves running as 2 wings on the stem. Ill. & Mo.

A. praealtus: 63. Leaf and branch, × 1.   64. Part of leaf, × 5.   **A. paniculatus:** 66. Leaf and branch, × 1.   **A. Tradescanti:** 65. Leaf and part of stem, × 1.   **A. coerulescens:** 67. Part of leaf, × 5.   **A. interior:** 68. Leaf and branch, × 1.

Áster

Heads with yellow disk flowers and blue, purple or white ray flowers. A large genus the members of which are found in woods, dry open ground, etc. The few species listed below occur in swales and on wet shores, sometimes at the water's edge. Asters are notorious for the difficulty with which many of the species are identified; frequent hybrids add to the confusion.

1. Stem leaves clasping (Fig. 58). **A. puníceus** L. Thickets and swamps; Nfd. to Man., s. to Ga., Tenn., Ohio & Mich.
1. Stem leaves not clasping..2
  2. Bracts of the involucre nearly equal in length, mostly tapered from below the middle to a long tip, without a definitely marked elongate diamond-shaped green portion (Fig. 57); leaves very long and narrow. **A. júnceus** Ait. Fig. 56. Wet meadows, etc.; Que. to B.C., s. to N.E., n. Pa., Wis. & Neb.
  2. Bracts unequal in length, mostly tapered from above the middle and with a definitely marked elongate diamond-shaped green portion (Fig. 62); leaves seldom more than 12 times as long as wide.............................3
3. Heads borne along one side of branches (Fig. 61)...........................4
3. Heads scattered on the branches...5
  4. Branches with leaves that vary in size and are not abruptly smaller than those on the main stem. **A. lateriflòrus** (L.) Britton. Thickets and shores; Gulf of St. Lawrence to Conn. & n. Pa., and on the mountains to N.C.; Ont. & n. Mich.
  4. Branches with leaves that are uniform and are abruptly smaller than those on the main stem. **A. lateriflorus** var. **péndulus** (Ait.) Burgess. Figs. 61, 62. Maine to Fla., w. to Wis., Ill., Mo., Miss. & perhaps Tex.
5. Disk flowers with lobes a sixth to a third as long as the limb (Fig. 59)..........6
5. Disk flowers with lobes a third to a half as long as the limb (Fig. 60)...........7
  6. Veinlets of leaves conspicuous, forming a network with meshes that are not elongate (Fig. 64). **A. praeáltus** Poir. Fig. 63. Low ground; Ohio & Ky. to Mich. & Wis., s. to Ark., Ariz. & n. Mex. (*A. salicifolius*, G, B.)
  6. Veinlets obscure, forming elongate meshes (Fig. 67). **A. coeruléscens** DC. Shores of rivers and ditches; Wis. to Alberta, s. to Kans., Tex., n. Mex. & s. Calif.
7. Bracts of the involucre 4.5–8 mm. long (Fig. 66); heads, including spread of rays, 12–25 mm. wide...8
7. Bracts of the involucre 3.3–4.5 mm. long (Fig. 68); heads 10–16 mm. wide......9
  8. Leaves 12 or more times as long as wide. **A. paniculàtus** Lam. Fig. 66. Meadows, thickets and shores; N.B. to N.J. & e. Pa., w. to Wis., n. Ill. & Mo.
  8. Leaves less than 12 times as long as wide. **A. paniculatus** var. **símplex** (Willd.) Burgess. N.B. to Va., w. to S.D., Neb. & perhaps Tex.
9. Heads few on each branch; leaves soft, not very sharp-pointed, 3–10 cm. long, rarely toothed. **A. Tradescánti** L. Fig. 65. River shores; N.S. to Maine & Vt. (*A. vimineus* var. *saxatilis*, G; not *A. Tradescanti*, G, B.)
9. Heads numerous on each branch; leaves firm, sharp-pointed, up to 15 cm. long, often toothed. **A. intèrior** Wiegand. Fig. 68. W. N.Y. & Ohio to Wis., Ill., Mo. & La. (*A. Tradescanti*, in part, G.)

References: Wiegand, Rhodora, **30**, 161–179 (1928)—*A. lateriflorus*, and Rhodora, **35**, 16–38 (1933)—*A. praealtus* and following species; Fernald, Rhodora, **35**, 312–314 (1933)—*A. Tradescanti*.

APPENDIX

USE OF AQUATIC PLANTS BY BIRDS AND MAMMALS

Available information concerning the use of plants by birds and mammals consists of data of heterogeneous nature, some general and some local in application. The following generalizations are from statements compiled by Frederick and Frances Hamerstrom, from the publications listed in the bibliography. The kinds of animals have been lumped for convenience into the following arbitrary groups: marsh birds, wildfowl, shore birds, upland game birds, song birds and mammals. References to specific animals appear in the animal index (p. 351), but much special information concerning season, locality and relation of the food to other foods is omitted and may be found only by referring to the original papers, indicated by numbers following each statement. Many of the original papers are local in their import, whereas others refer to a large territory; to determine whether text statements are correspondingly local or general each should be checked from the annotated bibliography. For example, "72" refers only to Nebraska, whereas a statement followed by "65" is usually of wide application. A plant here described as "unimportant" may have been listed by some of the authors as of local importance.

All the plant names have been reduced to the nomenclature of the main text of this book. When an animal name is listed without further comment it should be understood that the plant is eaten by that animal.

1. ACANTHACEAE, *Dianthera:* pinnated grouse eat seeds.[49]
2. ALGAE, general: marine algae and *Characeae* most often mentioned, but some others may be important, often as food for wildfowl[58,59,65,76,91]; deer[89]; support aquatic animals[77]; may smother other plants.[77]
3. *Chara* and *Nitella:* often important for wildfowl[7,11,52,54,56,58,59.65,67,72,87,91], perhaps chiefly because of minute aquatic animals[77]; may grow too luxuriantly[77]; moose.[81]
4. *Spirogyra:* wildfowl[54]; sometimes preferred food of deer.[4,89]
5. *Vaucheria:* deer.[4]
6. ALISMACEAE, general: wildfowl eat mostly tubers and nutlets.[52,54,58,95]
7. *Alisma:* nutlets eaten by wildfowl[52,54,58,65], usually not important.[72]
7a. *A. gramineum:* unimportant for ducks.[72]
7b. *A. Plantago-aquatica:* wildfowl[52,54,58,95], usually not important[72,91]; pheasant.[30]
8,9. *Echinodorus* spp. and *Lophotocarpus calycinus:* wildfowl eat small quantities of nutlets.[58.65]
10. *Sagittaria:* wildfowl eat many tubers, some nutlets and other parts[52,54,58,59,60,65,88,91]; stems, roots and tubers food for muskrats[77,79], porcupine[79], and beaver.[14]
10a. *S. cristata:* lesser importance for ducks.[13,72]
10b. *S. cuneata:* wildfowl eat many submersed tubers and some other parts.[13,57,67,91]

10*c. S. graminea:* good duck food.[57]

10*d. S. latifolia:* wildfowl eat many submersed tubers and some other parts[11,12,39, 52,54,57,58,72,77,87,91]; attracts marsh birds, wildfowl and song birds[39]; muskrats.[11,26, 35,46]

10*e. S. rigida:* muskrats[11]; ducks.[11,57]

10*f. S. teres:* ducks.[54,57,58,65]

11. ARACEAE, general: wood duck eats seeds.[54]

12. *Acorus Calamus:* muskrats.[35,46,79]

13. *Calla palustris:* sparingly eaten by muskrat.[35]

14. *Peltandra virginica:* seeds eaten by wildfowl, sometimes important[54,57,65,77,87,91]; rarely eaten by muskrat.[46]

15. ASCLEPIADACEAE, *Asclepias* spp. Ducks.[54,58]

16. *A. incarnata:* muskrats eat roots sparingly.[35]

17. BETULACEAE, *Alnus:* upland game birds eat buds, leaves and flowers[23,29,33,39,70, 78,83]; attractive to song birds[39] as food[60,61,62]; wildfowl[54]; deer, mostly starvation diet[2,71,84]; beaver[1,5,6,24,45,79,90]; moose.[81]

17*a. A. crispa:* moose[69]; beaver.[11]

17*b. A. incana:* upland game birds[15,83]; moose eat leaves[69,81]; inferior food for deer[22, 36,43,44 55,74]; beaver.[11,34]

17*c. A. rugosa:* deer[4,17]; beaver.[3]

18. BORAGINACEAE, *Myosotis* spp.: birds eat fruits.[60,61,62]

19. CALLITRICHACEAE, *Callitriche* spp.: ducks.[58,95]

20. CAMPANULACEAE, *Campanula americana:* deer.[4]

21. CERATOPHYLLACEAE, *Ceratophyllum demersum:* mostly seeds, sometimes foliage, important food for wildfowl[9,12,13,52,54,56,58,59,65,67,72,77,87,88,91]; shelters shrimps and other small animals[77]; muskrat[12]; sometimes crowds out other plants.[77]

22. COMPOSITAE, general: ducks[52,54,58]; only a few species are food for wildfowl.[65]

23. *Bidens:* attracts upland game birds, ducks and song birds[39]; mostly fruits eaten by upland game birds[21,37,38,39,40,42,47,48,80,86], and ducks.[52,54,58,65]

23*a. B. cernua:* ducks[58], of lesser importance.[72]

23*b. B. comosa:* upland game birds eat fruits.[30,50,94]

23*c. B. coronata:* of lesser importance for ducks.[72]

23*d. B. frondosa:* mallard[58]; ruffed grouse[49]; muskrat.[46]

24. *Eupatorium* spp.: mallard[58]; ruffed grouse eat leaves and fruits.[70]

25. *Megalodonta Beckii:* ducks[12]; of lesser importance.[72]

26. CORNACEAE, *Cornus* spp.: berries eaten by upland game birds[32,33,51,61,62,64,70], by song birds[61,62,64], and by wildfowl[52,54,58,77]; deer[2,17,25,43]; beaver.[5]

26*a. C. stolonifera:* wildfowl[58]; attractive to song birds[39] as food[60,62,75], upland game birds[39,49,93]; deer, sometimes important[2,22,36,44,55]; moose[69]; beaver[1,45], muskrat.[46]

27. *Nyssa* spp.: berries eaten by song birds[60,61,62,64]; ducks.[54,58]

27*a. N. aquatica:* ducks.[54,58]

27*b. N. sylvatica:* attractive to song birds[39] as food[60,61,62,75]; ducks eat seeds[54,58]; upland game birds eat berries[10,33,39,45,46,47,48,49], berries and buds[70]; deer browse.[4, 17,85]

28. CRUCIFERAE, general: only Water Cress has repute as wildfowl food, but related species usually not distinguished probably have equal value.[65]

29. *Nasturtium officinale:* ducks eat foliage[56,65,72,77,87,88,91]; muskrat[26]; deer[4]; shelters small aquatic life[77,91]; good on game farm[56] but may crowd other plants in the wild[56,91]; sometimes available all winter in Northern states.[77]

30. CYPERACEAE, general: nutlets favorite food for some wildfowl, less readily eaten by others[52,54,58,59,77,88], other parts also eaten[58,65]; *Cyperaceae* and *Najadaceae*

considered most important families for wildfowl[65] but extensive use may reflect abundance rather than preference[77]; muskrat[79]; beaver[79,90]; deer.[2]

31. *Carex* spp.: upland game birds[10,18,32,33,39,42,48,51]; attractive to marsh birds, wildfowl, shore birds and song birds[39]; nutlets eaten by wildfowl[11,52,54,59,65,67]; moose[69,81]; beaver[5,35]; deer[89]; muskrat.[11,35,46]

31a. *C. aquatilis:* wildfowl[65]; moose.[81]

31b. *C. comosa:* lesser importance for ducks.[72]

31c. *C. lacustris:* wildfowl.[7,65,91]

31d. *C. lasiocarpa:* fair for wildfowl.[91]

31e. *C. lenticularis;* moose.[81]

31f. *C. oligosperma:* sharp-tailed eats nutlets.[83]

31g. *C. rostrata:* roots and sprouts important muskrat food[35]; wildfowl[65]; lesser importance for ducks.[72]

31h. *C. stricta:* roots and sprouts sparingly eaten by muskrats[35]; bobwhite eats nutlets[47,48]; lesser importance as duck food.[72]

31i. *C. Tuckermani:* upland game birds.[12]

32. *Cladium* spp.: wildfowl.[58,59,65]

32a. *C. mariscoides:* ducks.[52,54,58,59]

33. *Cyperus* spp.: tuber-bearing species of local value to wildfowl[58,65], nutlets also eaten[54,58,65]; ducks[7,52,59]; upland game birds[32,47,49]; attractive to marsh birds, wildfowl, shore birds and song birds.[39]

33a. *C. dentatus:* valuable for wildfowl.[65]

33b. *C. erythrorhizos:* nutlets and spikelets eaten by ducks.[14]

33c. *C. esculentus:* of local importance for wildfowl, especially where flooded in winter.[9,54,57,58,65,72,87,91]

33d *C. ferruginescens:* ducks[52,58], of lesser importance[72]; of little value for muskrats.[82]

33e. *C. inflexus:* lesser importance as duck food.[72]

33f. *C. rivularis:* lesser importance as duck food.[72]

33g. *C. strigosus:* lesser importance as wildfowl food[65,72]; sparingly eaten by muskrats.[35]

34. *Dulichium arundinaceum:* slight importance for wildfowl[54,65] and as muskrat food.[35]

35. *Eleocharis* spp.: eaten by wildfowl[52,54,58,59,65], mostly stems and roots, since nutlets are small[65]; muskrats eat roots.[11,35,82]

35a. *E. acicularis:* turkeys eat young spikes[10]; fair wildfowl food[72,91]; muskrats.[35]

35b. *E. compressa:* lesser importance for ducks.[72]

35c. *E. obtusa:* muskrats.[35]

35d. *E. palustris:* fair food for wildfowl, whole plant eaten.[72,91]

35e. *E. parvula:* good wildfowl food, especially in brackish pools[9,91]; tubers eaten by birds.[65]

35f. *E. pauciflora:* tubers good wildfowl food[54,58], especially in northern regions.[65]

35g. *E. quadrangulata:* wildfowl food on coastal marshes, especially in the south.[9]

36. *Fimbristylis* spp.: wildfowl eat nutlets[54,58,59,65], fair food.[91]

37. *Fuirena* spp.: unimportant for wildfowl.[54,65]

38. *Psilocarya* spp.: of slight importance to wildfowl.[58,65]

39. *Rynchospora* spp.: ducks eat nutlets.[54,58,65]

39a. *R. corniculata:* ducks.[54,58,65]

40. *Scirpus* spp.: mostly nutlets, some tubers, are good duck food[7,11,52,54,58,59,63,65,76,77,87]; geese eat rootstocks and herbage[65]; geese and swans eat shoots[77]; bobwhite eats nutlets[47]; attracts marsh birds, wildfowl and song birds[39]; muskrat.[11,63,79]

40a. *S. acutus:* important wildfowl food.[7,57,65,67,72,77,87,91]

40*b*. *S. americanus:* often important wildfowl food[9,52,54,57,58,59,65,67,72,77,87]; important muskrat food.[82]

40*c*. *S. etuberculatus:* ring-necked duck eats nutlets.[52]

40*d*. *S. fluviatilis:* often important for wildfowl, nutlets mostly eaten.[7,52,54,57,58,59,65,72,91]

40*e*. *S. Olneyi:* wildfowl[65,91]; important muskrat food.[82]

40*f*. *S. validus:* often important wildfowl food.[7,54,57,65,67,72,77,91]

41. ELATINACEAE, general: ducks.[95]

42. ERICACEAE, *Chamaedaphne calyculata:* winter staple for sharp-tailed[78]; sometimes browsed by deer.[2,36,74]

43. ERIOCAULACEAE, *Eriocaulon* spp.: leaves eaten by ducks.[54,58]

44. EQUISETACEAE, *Equisetum* spp.: food for wildfowl[54,58,65], mainly geese[65]; ruffed grouse[49]; moose in winter.[79]

44*a*. *E. fluviatile:* sometimes important for moose.[69,81]

45. GENTIANACEAE, *Nymphoides* spp.: sometimes food for wildfowl.[65,95]

46. GRAMINEAE, general: principally grains, sometimes young shoots, often important food for wildfowl[52,54,58,59,63,65,76], follow *Cyperaceae* and *Najadaceae* in importance[65]; upland game birds[18,19,20,21,39,40,41,49,70,78,80,92]; mourning dove[92]; attract marsh birds, wildfowl, shore birds[39]; young grass eaten by deer[36,43,71,89]; muskrat[26,35,79]; beaver.[5,6,34,68,79,90]

47. *Agrostis* spp.: unimportant for wildfowl.[54,65]

47*a*. *A. stolonifera* var. *major:* young plants heavily grazed by deer.[36]

48. *Beckmannia Syzigachne:* unimportant for wildfowl.[54,65]

49. *Calamagrostis* spp.: young plants heavily grazed by deer[36]; muskrat.[35]

49*a*. *C. canadensis:* young sprouts eaten by moose.[69]

50. *Echinochloa* spp.: ducks[54,88]; among most important grasses[52,77]; attracts upland game birds, wildfowl, song birds.[39]

50*a*. *E. pungens:* grains eaten by some birds, stems and leaves by others, often important for wildfowl[7,9,52,54,57,58,59,65,67,72,77,87,91] and upland game birds[12,21,32,42,47,48,50,78,80,86,87,94]; attracts song birds.[60,61,62]

50*b*. *E. Walteri:* good wildfowl food[87,91]; muskrat.[82]

51. *Eragrostis* spp.: of little importance for wildfowl.[54,58,65]

51*a*. *E. hypnoides:* grains and spikelets eaten by ducks.[9]

52. *Fluminea festucacea:* wildfowl food.[67,91]

53. *Glyceria* spp.: ducks[54,65]; important muskrat food[35]; heavily grazed by deer.[36]

53*a*. *G. striata:* good duck food.[54,57,58]

54. *Leersia* spp.: ducks[52,58,65], sometimes important.[58]

54*a*. *L. oryzoides:* duck food[52,58,59,72], sometimes important[54]; occasionally eaten by muskrats.[35,46]

55. *Phalaris arundinacea:* unimportant for wildfowl[65,72]; the European race, often introduced, crowds out native vegetation valuable to wildlife.

56. *Phragmites maximus:* little or no value for wildfowl food, replacing better plants[65,77,87,91]; sometimes good for cover or preventing wave erosion[77,87]; favorite food for muskrat.[79,82]

57. *Poa* spp.: little importance for wildfowl[54,58,65]; sometimes staple food for Hungarian partridge[94]; wild turkey.[9,49]

57*a*. *P. pratensis:* young plants eaten by deer, hay not so much[4,36,71]; eaten sparingly by moose.[69]

58. *Spartina pectinata* (or possibly some salt-marsh species): may be locally important as wildfowl food, but generally of little value.[54,57,58,65,67,72,91]

59. *Zizania aquatica:* very important food for wildfowl, among most valuable grasses, grains and other parts eaten[9,11,12,16,39,52,54,57,58,65,67,72,77,87,88,91,95]; impor-

tance perhaps overestimated in view of local and seasonal availability[65]; song birds[77,95]; upland game birds[47,48]; attracts song birds[39]; muskrats[11,57,77]; deer[57,77]; moose.[57,81]

60. *Zizaniopsis milacea:* sometimes important food for wildfowl[52,58], but not generally[54,65,91]; may become weedy.[91]

61. HALORAGIDACEAE, general: seeds and sometimes leaves sparingly eaten by ducks.[52,54,65]

62. *Hippuris vulgaris:* seeds and sometimes foliage eaten by wildfowl[52,54,58,65], a fair food, harboring small animal life.[91]

63. *Myriophyllum* spp.: many wildfowl eat fruits, a few eat foliage,[52,54,58,59,65,87,95] unimportant, and may become weedy[77]; sparingly eaten by muskrats[35]; moose.[81]

63a. *M. exalbescens:* fair duck food.[12,52,67,72,87,91]

63b. *M. heterophyllum:* mallard.[58]

63c. *M. verticillatum:* American widgeon.[54]

64. *Prosperpinaca* spp.: seeds sometimes important food for ducks[52,54,58,59,65,76] and rarely swans[65]; sparingly eaten by muskrats.[35]

64a. *P. palustris:* canvasback eats seeds.[52]

64b. *P. pectinata:* mallard.[58]

65. HYDROCHARITACEAE, general: ducks.[52,54,58]

66. *Anacharis* spp.: ducks[54]; beaver.[14]

66a. *A. canadensis:* wildfowl food of variable importance[12,52,54,56,58,65,72,77,87,88,91]; shelters small aquatic life[77,91]; may suppress other plants[56,91]; sparingly eaten by muskrat.[12,35]

67. *Limnobium Spongia:* seeds eaten by wildfowl[54,58,91]; perhaps locally important [52,56,65,95]; marsh birds[95]; may become weedy.[91]

68. *Vallisneria americana:* excellent food for wildfowl, which eat all parts, especially winter buds and rootstocks[9,11,12,52,54,57,58,65,67,72,76,77,87,88,91,95]; attracts marsh birds, wildfowl, shore birds[39]; harbors minute animals[77,91]; muskrat.[10]

69. HYPERICACEAE, *Hypericum* spp.: ducks[54,58,59]; ruffed grouse.[70]

69a. *H. majus:* muskrat.[35]

69b. *H. punctatum:* deer.[4]

70. IRIDACEAE, *Iris versicolor* and *I. virginica:* sometimes muskrat food[35,46,82]; wildfowl and marsh birds[95]; seeds probably not eaten by wildfowl, but persists as cover under heavy grazing.[77]

71. ISOETACEAE, *Isoetes* spp.: sharp-tailed eats leaves.[49]

72. JUNCACEAE, *Juncus* spp.: wildfowl[54]; upland game birds[39,49]; attractive to wildfowl, marsh birds, song birds[39]; bases and roots sparingly eaten by muskrat[35]; moose.[69]

72a. *J. effusus:* sparingly eaten by muskrat.[35]

72b. *J. macer:* seeds and stems eaten by a few wildfowl[91]; deer[4]; may become weedy.[91]

73. LABIATAE, general: upland game birds[32,33]; pintail[54]; muskrat.[79]

74. *Lycopus uniflorus:* tubers important muskrat food.[35]

75. *Mentha* spp.: green-winged teal[54]; ruffed grouse[70]; muskrat.[26]

76. *Scutellaria lateriflora:* pheasant.[30]

77. LEMNACEAE, general: important food for most wildfowl[39,52,54,56,58,59,65,77,87,88,91,95]; marsh birds[95]; harbor small aquatic animals[65,77]; pheasant[42]; may smother other plants.[77]

78. *Lemna* spp.: wildfowl[7,52,59], eaten heavily by some wildfowl[54,58], little by others[52]; pheasant[42]; beaver.[14]

78a. *L. minor:* often important wildfowl food[9,52,54,58,59,67,72,77,91,95]; attracts small aquatic animals[77]; muskrat.[46]

78b. *L. perpusilla:* ducks.[12,54,95]

78c. *L. trisulca:* often important wildfowl food[58,67,72,77,91,95]; attracts small aquatic animals.[77,91]

78d. *L. valdiviana:* ducks.[95]

79. *Spirodela polyrhiza:* often good wildfowl food[9,11,12,52,54,58,72,77,91]; pheasants.[42]

80. *Wolffia* spp.: often good wildfowl food.[52,58,87,91]

80a. *W. columbiana:* ducks[58]; muskrat.[46]

80b. *W. punctata:* sometimes important for ducks[52,72]; muskrat.[46]

81. *Wolffiella floridana:* sometimes good wildfowl food.[52,91]

82. LENTIBULARIACEAE, *Utricularia* spp.: sometimes listed as food for wildfowl[95]. but has little or no value[87,91]; may harbor minute animal life[91]; sometimes becomes weedy[91]; sparingly eaten by muskrat.[35]

82a. *U. vulgaris:* little importance for ducks.[72]

83. LOBELIACEAE, *Lobelia siphilitica:* deer.[4]

84. LYTHRACEAE, *Decodon verticillatus:* ducks eat seeds[54,58,65]; sometimes important for muskrats.[46]

85. MALVACEAE, general: ducks eat seeds.[54,58,59]

86. *Hibiscus* spp.: ducks[54,58]; bobwhite.[27]

86a. *H. palustris:* black duck.[58]

88. MARSILEACEAE, *Marsilea quadrifolia:* fruiting bodies eaten by ducks.[52,54,58,65]

89. MUSCI, general: some reported as eaten[58], others refused[77], by mallards; ruffed grouse[33]; deer[2], probably little eaten.[4]

90. MYRICACEAE, *Myrica Gale:* occasional food of muskrats[46]; buds and catkins eaten by sharp-tailed[78]; deer.[44,74]

91. NAJADACEAE, general: includes the most important food for wildfowl[52,54,58,59, 63,65,95]; marsh birds.[95]

92. *Najas* spp.: stems, foliage and seeds important for ducks.[65,76]

92a. *N. flexilis:* stems, foliage and seeds important (among most important) wildfowl food[12,52,54,57,58,59,72,87,91]; heavily eaten by mallards.[35,77]

92b. *N. guadalupensis:* good wildfowl food.[9,91]

92c. *N. marina:* ducks.[52,54]

93. *Potamogeton* spp.: often favorite food for wildfowl, some eat whole plant, others prefer certain parts[7,11,52,54,57,58,59,65,76,77,88,91]; staple for ducks, all species eaten[57]; heavy use of nutlets by ducks may reflect abundance rather than preference[77]; attractive to marsh birds, wildfowl and shore birds[39]; often heavily eaten by muskrats[11,35]; beaver[14,34]; deer[36]; moose.[69,81]

93a. *P. americanus:* wildfowl[12,54,72,87], sometimes important.[9]

93b. *P. amplifolius:* desirable duck food.[12,77]

93c. *P. angustifolius:* desirable duck food.[12]

93d. *P. bupleuroides:* good duck food.[9]

93e. *P. crispus:* eaten by ducks, but may become a weed.[77]

93f. *P. diversifolius:* pintail.[54]

93g. *P. epihydrus:* ducks[13,77,95]; muskrats.[35]

93h. *P. filiformis:* good duck food.[57]

93i. *P. foliosus:* often important for wildfowl.[9,52,54,58,67,91]

93j. *P. Friesii:* ducks.[54,58]

93k. *P. gramineus:* tubers and other parts often important duck food.[54,57,72,87,91]

93l. *P. illinoensis:* duck food[95], of lesser importance.[72]

93m. *P. lucens:* mallard.[58]

93n. *P. natans:* sometimes important for ducks[12,52,54,72,87,95]; rootstocks, and nutlets held late in the season, good duck food.[77]

93o. *P. obtusifolius:* slight value for ducks.[87]

93*p*. *P. pectinatus:* nutlets and tubers make this the most important pondweed for ducks.[7,9,18,52,54,57,58,59,67,72,77,87,91]

93*q*. *P. praelongus:* ducks.[58,77]

93*r*. *P. pusillus:* good food for wildfowl.[7,9,52,54,57,58,59,72,87,91]

93*s*. *P. Richardsonii:* sometimes important duck food.[52,54,57,58,67,72,77,87,91]

93*t*. *P. Robbinsii:* ducks[12]; tough and probably not eaten by wildfowl.[77]

93*u*. *P. tenuifolius:* moose.[81]

93*v*. *P. Vaseyi:* ducks.[95]

93*w*. *P. zosteriformis:* sometimes duck food.[12,54,72,77]

94. *Ruppia maritima:* mostly in salt water, but often important as duck food within its restricted range.[9,52,54,56,58,59,63,65,66,87,91]

95. *Zannichellia palustris:* nutlets and sometimes foliage often good for wildfowl, especially in brackish pools.[9,52,54,57,58,59,65,91]

96. NYMPHAEACEAE, general: wildfowl eat seeds.[52,54,58,87]

97. *Brasenia Schreberi:* ducks[52,54,57,58,59,76,77,88,91], sometimes locally important.[52,65]

98. *Cabomba caroliniana:* slight use by waterfowl.[54,58,65]

99. *Nelumbo lutea:* wildfowl eat seeds[52,88,91,95], generally sparingly[65], but they are sometimes locally important[57]; attracts marsh birds, wildfowl, song birds[39]; roots eaten by beaver.[79]

100. *Nuphar* spp.: wildfowl eat seeds[52,54,58,65,77]; deer eat leaves, stems, flowers[79]; porcupine[79]; beaver eat roots[24]; heavily eaten by muskrats[11,12,77]; algae and insects grow under leaves[77]; may crowd out other plants.[77]

100*a*. *N. advena* and *N. variegatum:* seeds eaten by wildfowl[12,13,52,54,57,58,95], of fair food value[72,87,91]; attracts shore, marsh, and song birds[39]; leaves important for deer[89]; moose[69,81]; muskrat.[12,35,46,87]

100*b*. *N. microphyllum:* seeds eaten by wildfowl.[52,54,91,95]

101. *Nymphaea odorata* and *N. tuberosa:* seeds fair food for wildfowl, sometimes heavily eaten, also rootstocks[12,52,54,57,58,59,65,91,95]; attract wildfowl and marsh birds[39]; rootstocks and base of petioles eaten by muskrats[11,12,35,82]; "roots" eaten by beaver[5,14,24,68,79]; deer[71,89]; moose[69,79,81]; porcupine.[79]

102. OLEACEAE, *Forestiera acuminata:* seeds eaten by ducks[56,58,91], sometimes locally important.[54,65]

103. *Fraxinus* spp.: seeds eaten by bobwhite,[48] song birds[60,61,62], and occasionally wildfowl[54,58,65]; beaver.[5,24]

103*a*. *F. nigra:* lightly browsed by deer[17,22,55], or sometimes heavily[2]; beaver.[68,90]

104. ONAGRACEAE, *Jussiaea* spp.: only members of the family valuable to wildfowl, seeds eaten by ducks.[54,58,65]

105. *Ludwigia palustris:* base of stems sparingly eaten by muskrat.[35]

106. *Trapa natans:* no food value for wildfowl except as source of small aquatic animals, and shades out other plants.[91]

107. PINACEAE, *Taxodium distichum:* cone scales, seeds and galls eaten by ducks.[52,54,58,65]

108. POLYGONACEAE, *Polygonum* spp.: nutlets the only part eaten[54], and often important to wildfowl[52,54,58,59,63,65,77,87,95]; upland game birds[12,18,21,27,28,29,30,31,33,39,40,41,42,49,53,66,78,80,86,92]; shore birds[66]; song birds[92]; attractive to wildfowl, song birds[39]; heavily eaten by deer in summer[36], sparingly otherwise[36,43]; sparingly eaten by muskrat.[35]

108*a*. *P. Careyi:* nutlets eaten by upland game birds.[12,32]

108*b*. *P. coccineum:* wildfowl.[7,9,52,65]

108*c*. *P. Hydropiper:* among *Polygonum* most eaten by wildfowl[9,52,54,57,58,65,76,91,95]; nutlets occasionally eaten by upland game birds.[32,47,78]

108*d*. *P. hydropiperoides:* among *Polygonum* most eaten by wildfowl.[9,52,54,57,58,65,91,95]

108e. *P. lapathifolium:* among *Polygonum* most often eaten by wildfowl[7,9,52,54,58,65,67,91]; upland game birds.[12,32,47,48,49]

108f. *P. natans:* among *Polygonum* most often eaten by wildfowl[7,12,52,54,58,59,65,67,72,77,91]; upland game birds.[77,92]

108g. *P. opelousanum:* often important for wildfowl.[54,57,58,65]

108h. *P. pensylvanicum:* nutlets often important for wildfowl[9,54,57,58,65,72,87,91]; good for upland game birds[12,27,29,32,47,48,49,50,78,86,87,94]; muskrat[46]; deer.[4]

108i. *P. Persicaria:* nutlets for wildfowl[7,9,52,54,58], among *Polygonum* most often eaten[65]; nutlets for upland game birds.[29,30,32,50,78,86,94]

108j. *P. densiflorum:* wildfowl.[52,54,65]

108k. *P. punctatum:* wildfowl food[52,54,58,72,95], among the best[9,57,65]; upland game birds[10,50,94]; muskrat[26]; deer.[4]

109. *Rumex* spp.: nutlets eaten by wildfowl[52,54,58], but less than those of *Polygonum*[65]; upland game birds[27,29,38,39,42]; attract song birds[39]; heavily browsed by deer[36]; sparingly eaten by muskrat.[35]

109a. *R. maritimus:* wildfowl[52,54,58], of slight value.[91]

109b. *R. obtusifolius:* nutlets sparingly eaten by wildfowl[91]; occasionally by muskrat[26,46]; deer.[4]

109c. *R. verticillatus:* deer.[4]

110. PONTEDERIACEAE, general: ducks.[95]

111. *Eichornia crassipes:* no food value for wildfowl except as source of minute animal life, and shades out good food plants.[91]

112. *Heteranthera* spp.: ducks.[54]

112a. *H. dubia:* locally attractive to some wildfowl.[54,65,77]

113. *Pontederia cordata:* seeds occasionally good duck food[54,57,58,65,77,87,91]; good muskrat food.[11,77]

114. PRIMULACEAE, *Samolus floribundus:* deer.[4]

115. RANUNCULACEAE, *Caltha palustris:* seeds for upland game birds[18,92]; moose.[69]

116. *Ranunculus* spp.: upland game birds[33,49,70,86]; wildfowl.[54,58]

116a. *R. flabellaris:* nutlets and foliage for wildfowl[54,58], of fair food value.[72,91]

116b. *R. septentrionalis:* moose.[69]

116c. *R. trichophyllus* and its close relatives: nutlets and foliage for wildfowl[12,58,65], of lesser importance.[72,91]

117. RUBIACEAE, general: ducks.[54]

118. *Cephalanthus occidentalis:* often important for wildfowl, supports insects, and nutlets eaten[52,54,58,59,65,91]; attracts marsh birds, wildfowl, upland game birds, song birds[39]; beaver[45]; muskrat.[46]

119. *Galium* spp.: fruits sparingly eaten by ducks[52,54,58,59,65,72]; upland game birds[18,30,33,70]; sparingly eaten by muskrat.[35]

120. SALICACEAE, *Salix* spp.: ducks[54,58]; attract marsh birds, wildfowl, song birds[39]; upland game birds[8,23,29,32,33,49,70,78,83,93]; sometimes heavily browsed by deer[2,17,36,43,55,71]; moose[69,81]; important for beaver[5,6,11,14,24,34,45,68,79,90]; muskrat eat leaves[26,35,46]; porcupine.[81]

120a. *S. Bebbiana:* deer browse.[74]

120b. *S. discolor:* moose.[81]

120c. *S. longifolia:* deer[4]; beaver.[3]

120d. *S. longipes:* deer[4]; beaver.[3]

120e. *S. nigra:* deer[4]; beaver.[3]

121. SALVINIACEAE, general: ducks.[95] *Azolla caroliniana:* probably of little importance for ducks.[59,65]

122. SCROPHULARIACEAE, *Bacopa* spp.: seeds and plant eaten by wildfowl.[65]

122a. *B. rotundifolia:* ducks.[52,54]

123. SOLANACEAE, general: seeds eaten by pheasants.[80]

124. *Solanum* spp.: seeds eaten by killdeer.[66]

124a. *S. Dulcamara:* stems eaten by muskrats, berries rejected[46]; berries winter food for pheasants[18,19,20,21,30,93]; ruffed grouse.[33]

125. Sparganiaceae, *Sparganium* spp.: nutlets eaten by wildfowl, usually in small quantities[11,52,54 58,59,76,77], freely by some[65], essentially a cover plant[77]; attracts marsh birds and wildfowl[39]; preferred deer food[36]; muskrat.[11]

125a. *S. americanum:* nutlets eaten by wildfowl[52,54,58], of fair value[91]; whole plant heavily eaten by muskrat.[35]

125b. *S. angustifolium:* good duck food.[57]

125c. *S. chlorocarpum:* ducks[12]; whole plant heavily eaten by muskrat.[35]

125d. *S. eurycarpum:* sometimes good wildfowl food[52,54,57,58,67,72,87,91]; basal parts muskrat food.[26,46]

125e. *S. multipedunculatum:* nutlets eaten by ducks.[52]

126. Typhaceae, *Typha* spp.: stalks and roots important food for muskrats[26,52]; beaver[5]; attracts marsh birds, wildfowl, song birds.[39]

126a. *T. angustifolia:* fair food for geese, rarely important for ducks[65,91]; excellent muskrat food.[26,46,82]

126b. *T. latifolia:* fair food for some geese, rarely for ducks[65,77 91]; excellent muskrat food[11,26,35,46,82,87]; beaver.[34]

127. Umbelliferae, general: only a few are food for wildfowl.[65]

128. *Cicuta* spp.: seeds occasionally eaten in quantity by wildfowl.[54,58,65]

128a. *C. bulbifera:* moose.[69]

129. *Hydrocotyle* spp.: seeds and some leaves eaten by wildfowl.[52,54,58,65]

129a. *H. umbellata:* desirable duck food.[57]

130. Urticaceae, *Planera aquatica:* seeds are preserved under water and are locally important for wildfowl.[54,56,58,65,91]

131. Verbenaceae, *Lippia* spp.: seeds eaten by ducks.[54,58]

131a. *L. nodiflora:* ducks.[54,58]

132. Xyridaceae, *Xyris* spp.: mallard.[58]

ANIMAL INDEX

In the preceding section each plant or group of plants has been numbered. In this section, each animal name is followed by a series of numbers corresponding to the numbers of plants in the preceding section, to indicate the species reported as eaten by each animal.

Wildfowl

Coot: 3, 21, 30, 59, 66a, 67, 68, 92a, 93, 99, 101, 108, 108f, 112a.
Ducks. Baldpate (*see* **American Widgeon**). **Black Duck:** 2, 3, 6, 10, 10c, 10d, 10f, 19, 21, 23, 26, 26a, 27b, 30, 31, 32, 32a, 33, 33c, 35, 35f, 36, 39, 39a, 40, 40d, 41, 43, 45, 46, 50a, 54, 57, 58, 59, 60, 62, 63, 64, 67, 68, 69, 77, 78, 80, 82, 84, 86, 86a, 91, 92, 92a, 93, 93i, 93p, 93s, 94, 95, 96, 97, 100, 100a, 101, 107, 108, 108c, 108d, 108e, 108f, 108g, 108k, 108r, 109, 110, 113, 118, 119, 120, 121, 125, 125a, 125d, 129, 130, 131a. **Bluebills** (*see also* **Scaups** and **Ringnecks**): 3, 10b, 21, 30, 68, 92a, 93, 97, 100, 108, 125. **Bufflehead:** 2, 3, 6, 19, 21, 30, 41, 45, 59, 63 67, 68, 77, 82, 91, 92a, 93, 94, 97, 110, 121. **Canvasback:** 2, 3, 6, 10, 10b, 10d, 19, 21, 23, 26, 30, 31, 33, 33c, 40, 40b, 40d, 40e, 41, 45, 46, 50a, 54, 59, 62, 63, 64a, 65, 67, 68, 70, 77, 79, 82, 91, 92a, 93, 93p, 93s, 94, 97, 99, 100b, 101, 107, 108, 108c, 108e, 109, 110, 119, 121, 125, 125a, 125d, 125e. **Dabblers** (*see* **River Ducks**). **Deep Water Ducks** (*see* **Diving Ducks**). **Diving Ducks:** 93b. **Eider:** 91, 110. **Gadwall:** 2, 3, 4, 6, 19, 21, 22, 23, 26, 30, 31, 33, 33b, 35, 35a, 36, 40, 40b, 40d, 41, 43, 45, 46, 50a, 51a, 53, 54a, 57, 58, 59, 60, 62, 63, 64, 65, 66, 67, 68, 70, 77, 78, 82, 91, 92, 92a, 93, 93j, 93p, 93r, 94, 95, 96, 97, 100a, 100b, 101, 107, 108, 108c, 108d, 108e, 108f, 108g, 108h, 108i, 108k, 109, 110, 113, 116, 117, 118, 119, 120, 121, 125, 129. **Goldeneye, American:** 2, 3, 6, 19, 21, 30, 31, 32, 35, 41, 45, 46, 61, 62, 63, 66a, 67, 68, 70, 77, 82, 91, 92, 93, 94, 97, 101, 121, 125. **Barrow's Goldeneye:** 6, 19, 41, 45, 63, 67, 70, 77, 82, 91, 110, 121. **Harlequin:** 77. **Mallard:** 2, 3, 6, 7, 7b, 8, 10 10b, 10d, 14, 15, 19, 21, 22, 23, 23a, 23d, 24, 26, 26a, 27, 27a, 27b, 30, 31, 32, 32a, 33, 33c, 33d, 35, 36, 38, 39, 39a, 40, 40b, 40d, 40e, 41, 44, 45, 46, 50, 50a, 51, 53, 53a, 54, 54a, 58, 59,

60, 62, 63, 63b, 64, 64b, 65, 66a, 67, 68, 77, 78, 78a, 79, 80, 80a, 82, 84, 85, 86, 88, 89, 91, 92, 92a, 93, 93i, 93j, 93m, 93p, 93q, 93r, 94, 95, 96, 97, 98, 99, 100, 100a, 101, 102, 103, 104, 107, 108, 108c, 108d, 108e, 108f, 108g, 108i, 108k, 109, 109a, 110, 113, 116, 116a, 116c, 118, 119, 120, 121, 125, 125a, 125d, 128, 129, 130, 131, 132. **Merganser, Hooded:** 19, 21, 41, 45, 63, 67, 68, 77, 82, 91, 94, 110, 121. **Old Squaw:** 6, 21, 50a, 67, 68, 70, 77, 82, 91, 93, 94, 110, 121. **Pintail:** 2, 3, 6, 7, 7b, 10, 10b, 10f, 15, 17, 19, 21, 23, 27, 30, 31, 32, 32a, 33, 33b, 35, 35f, 36, 37, 39, 39a, 40, 40b, 40d, 40e, 41, 45, 46, 50, 50a, 51a, 54a, 57, 58, 59, 60, 61, 62, 63, 64, 67, 68, 72, 73, 77, 78, 82, 85, 86, 88, 91, 92, 92a, 93, 93f, 93i, 93p, 93r, 94, 95, 96, 97, 100, 100a, 101, 108, 108c, 108d, 108e, 108f, 108g, 108h, 108i, 108k, 109, 110, 112a, 113, 116, 118, 119, 121, 122a, 125, 128, 129, 131, 131a. **Redhead:** 2, 3, 6, 10, 10b, 19, 21, 26, 30, 31, 35, 40, 40b, 40d, 41, 45, 46, 50a, 54, 59, 62, 63, 64, 65, 66a, 67, 68, 70, 77, 78, 82, 91, 92, 92a, 92c, 93, 93p, 93s, 93w, 94, 95, 96, 97, 99, 100b, 101, 108, 108d, 108e, 108i, 110, 121, 125, 130. **Ringbill** (*see* **Ringneck**). **Ringneck:** 2, 3, 6, 10, 10d, 19, 21, 30, 33, 33b, 33d, 35, 40b, 40c, 41, 45, 46, 50, 50a, 51a, 57, 60, 63, 63c, 65, 67, 68, 70, 77, 78a, 81, 82, 91, 92, 92a, 93, 93i, 93p, 93r, 93s, 94, 95, 96, 97, 99, 100, 100a, 100b, 101, 107, 108, 108b, 108d, 108f, 108j, 108k, 110, 118, 119, 121, 125, 125a, 125d, 125e. **River Ducks:** 93b, 116a. **Ruddy:** 2, 3, 6, 10, 10b, 19, 21, 30, 33, 40, 41, 45, 46, 50a, 59, 63, 67, 68, 70, 77, 82, 91, 92, 93, 94, 97, 98, 100a, 101, 110, 121, 125. **Scaup, Greater:** 2, 3, 6, 7, 10, 19, 21, 26, 30, 32a, 33, 35, 40, 40b, 40d, 41, 45, 46, 50a, 59, 61, 62, 63, 63a, 65, 67, 68, 70, 77, 82, 88, 91, 92, 92a, 93, 93p, 94, 95, 96, 97, 99, 100, 100a, 100b, 101, 108, 108c, 108d, 108f, 108i, 108j, 109, 109a, 110, 118, 121, 125, 129. **Lesser Scaup:** 2, 3, 6, 7, 7b, 10, 10d, 19, 21, 22, 30, 31, 32, 32a, 33, 35, 40, 40b, 40d, 41, 45, 46, 50, 50a, 54a, 59, 61, 62, 63, 63a, 65, 66a, 67, 68, 70, 77, 80, 80b, 82, 91, 92, 92a, 93, 93n, 93o, 94, 95, 96, 97, 99, 100a, 100b, 101, 108, 108c, 108d, 108e, 108f, 110, 118, 119, 121, 125, 125a, 125d, 125e. **Scoter, American:** 19, 41, 45, 63, 70, 77, 82, 91, 110, 121. **Surf Scoter:** 2, 19, 21, 41, 45, 63, 68, 70, 77, 82, 91, 93, 94, 97, 110, 121. **White-winged Scoter:** 2, 19, 21, 41, 45, 63, 68, 70, 77, 82, 91, 93, 94, 97, 110, 121. **Shoveller:** 2, 3, 6, 10, 10b, 19, 21, 30, 31, 32, 32a, 33, 35, 36, 40, 40b, 40d, 41, 45, 46, 50a, 54a, 63, 64, 67, 69, 70, 77, 78, 78a, 82, 85, 87, 91, 92a, 93, 93p, 93r, 94, 95, 97, 100a, 100b, 101, 108, 108f, 110, 119, 121, 125. **Teal:** 33b, 51a, 59. **Blue-winged Teal:** 2, 3, 6, 7, 7b, 10, 10b, 10d, 19, 21, 23, 30, 31, 31c, 32a, 33, 34, 35, 36, 39, 40, 40a, 40b, 40d, 40f, 41, 45, 46, 50, 50a, 54a, 58, 59, 60, 61, 62, 63, 64, 66, 66a, 67, 68, 70, 77, 78, 82, 84, 86, 91, 92, 92a, 92c, 93, 93n, 93p, 93r, 94, 95, 96, 97, 100a, 100b, 101, 104, 108, 108b, 108c, 108d, 108e, 108f, 108g, 108h, 108i, 108j, 108k, 109, 110, 112, 112a, 116, 116a, 117, 118, 119, 120, 121, 125, 125a, 129, 131. **Cinnamon Teal:** 2, 3, 6, 19, 30, 31, 35, 40, 40b, 41, 45, 46, 59, 61, 62, 63, 67, 70, 77, 82, 85, 91, 93, 93p, 94, 95, 100a, 100b, 101, 108, 108b, 108c, 108d, 108e, 108f, 108i, 108k, 109, 110, 116a, 119, 121, 125, 125a. **Green-winged Teal:** 2, 3, 6, 7, 7b, 10, 10b, 17, 19, 21, 23, 26, 30, 31, 32, 32a, 33, 35, 36, 39, 39a, 40, 40b, 40d, 40f, 41, 44, 45, 46, 47, 48, 50, 50a, 54a, 58, 59, 60, 61, 62, 63, 64, 67, 68, 70, 72, 75, 77, 78, 82, 84, 91, 92, 92a, 93, 93p, 93r, 93s, 93w, 94, 95, 97, 100, 100a, 100b, 101, 107, 108, 108c, 108d, 108e, 108f, 108g, 108h, 108i, 108j, 108k, 109, 110, 113, 116, 116a, 118, 119, 121, 122, 122a, 125, 129, 131. **Whistler** and **Whistlewing** (*see* **American Goldeneye**). **Widgeon:** 3, 10b, 21, 59, 68, 77, 93, 94, 97, 101, 108c, 125. **American Widgeon:** 2, 3, 4, 6, 10, 15, 19, 22, 23, 26, 30, 31, 32, 32a, 33, 35, 36, 40, 40b, 40d, 41, 43, 44, 45, 46, 59, 61, 62, 63, 63c, 65, 66a, 67, 68, 70, 77, 78, 82, 91, 92, 92a, 93, 93p, 93r, 94, 95, 96, 97, 100, 100a, 100b, 101, 107, 108, 108c, 108d, 108e, 108f, 108i, 110, 116, 118, 119, 121, 125, 128, 131. **European Widgeon:** 68, 93, 94, 125. **Wood Duck:** 2, 3, 6, 10, 10d, 11, 14, 19, 21, 22, 23, 26, 27, 27a, 27b, 30, 31, 33, 33c, 35, 36, 39, 39a, 40, 40d, 41, 45, 46, 50a, 51, 53, 53a, 54a, 59, 60, 62, 63, 64, 65, 66a, 67, 68, 69, 70, 77, 78, 78a, 78b, 79, 82, 84, 86, 91, 92a, 93, 93a, 93k, 93m, 93p, 96, 97, 98, 99, 100, 100a, 100b, 101, 102, 103, 104, 107, 108, 108c, 108d, 108e, 108f, 108g, 108h, 108i, 108k, 109, 110, 112, 112a, 113, 116, 117, 118, 119, 120, 121, 125, 125a, 125d, 129, 130, 131. **Ducks,** unspecified: 3, 6, 7a, 7b, 10, 10a, 10b, 10c, 10d, 10e, 10f, 14, 21, 23, 23a, 23c, 25, 29, 30, 31, 31b, 31g, 31h, 32, 33, 33c, 33d, 33e, 33f, 33g, 35, 35a, 35b, 35d, 36, 39, 39a, 40, 40a, 40b, 40d, 40e, 40f, 44, 46, 50, 50a, 50b, 52, 53, 53a, 54, 54a, 55, 56, 58, 59, 62, 63, 63a, 64, 66a, 67, 68, 72, 77, 78a, 78b, 78c, 78d, 79, 80, 80b, 82a, 84, 87, 88, 91, 92, 92a, 93, 93a, 93b, 93c, 93e, 93g, 93h, 93i, 93k, 93l, 93n, 93o, 93p, 93r, 93s, 93t, 93v, 93w, 94, 95, 97, 99, 100, 100a, 101, 104, 107, 108, 108c, 108d, 108e, 108f, 108g, 108h, 108k, 110, 112a, 113, 116a, 116c, 117, 118, 119, 120, 125, 125b, 125c, 125d, 126, 126a, 126b, 129, 129a, 130.

Geese. Blue Goose: 35. **Brant:** 2, 93, 94. **Canada Goose:** 2, 10b, 46, 50a, 59, 68, 77, 93, 94. **Snow Goose:** 50a, 59, 68, 93. **White-fronted Goose:** 2, 50a, 93. **Geese,** unspecified: 10, 10d, 30, 31, 32, 33, 35, 40, 40a, 40b, 40d, 40e, 44, 46, 50, 50a, 58, 59, 62, 66a, 68, 77. 93, 94, 108, 125, 126a, 126b, 129.

Grebes: 10d, 31, 33, 40, 46, 59, 68, 93, 99, 100a, 101, 118, 125, 126.

Swans. Whistling Swan: 10, 21, 68, 93, 94. **Whooper Swan:** 106. **Swans,** unspecified: 31, 32, 40, 40d, 44, 64, 66a, 68, 94, 125.

Wildfowl, unspecified: 7, 7b, 8, 9, 10, 10b, 10d, 10f, 14, 21, 22, 28, 29, 30, 31, 31a, 31c, 31d, 31g, 33, 33a, 33c, 33g, 34, 35, 35a, 35d, 35e, 35f, 35g, 36, 37, 38, 40, 40a, 40b, 40d, 40e, 40f, 45, 46, 47, 48, 50a, 50b, 51, 52, 55, 56, 57, 59, 60, 61, 62, 63a, 64, 66a, 67, 68, 72b, 78a, 79, 80, 81, 92a, 92b, 93, 93d, 93i, 93p, 93r, 93s, 94, 95, 96, 99, 100, 100a, 100b, 101, 102, 103, 104, 108, 108b, 108c, 108d, 108e, 108f, 108g, 108h, 108i, 108j, 108k, 109, 109a, 109b, 111, 113, 116a, 116c, 118, 122, 125a, 125d, 126, 127, 128, 130.

Marsh Birds

Bittern: 33, 40, 99, 100a, 101, 118, 125, 126.
Gallinule, Florida: 59, 67, 70, 77, 91. **Gallinule, Purple:** 59, 67, 70, 77, 91.

Heron: 10*d*, 33, 40, 93, 99, 100*a*, 101, 125, 126.

Rail, Black: 59, 67, 77, 91. **King Rail:** 10*d*, 31, 33, 40, 46, 68, 72, 93, 99, 100*a*, 101, 118, 120, 125, 126. **Sora Rail:** 10*d*, 31, 33, 40, 46, 59, 67, 68, 72, 77, 91, 93, 99, 100*a*, 101, 118, 120, 125, 126. **Virginia Rail:** 10*d*, 31, 33, 40, 46, 68, 72, 93, 99, 100*a*, 101, 118, 120, 125, 126. **Yellow Rail:** 59, 67, 77, 91.

Shore Birds

Killdeer: 124.

Plover: 31, 33, 46, 68, 93.

Sandpiper: 31, 33, 46, 68, 93.

Snipe, Wilson's: 68, 93.

Shore birds, unspecified: 31, 33, 46, 68, 93.

Upland Game Birds

Bobwhite: 23, 26, 26*a*, 27*b*, 30, 31, 31*h*, 33, 40, 46, 50*a*, 59, 72, 86, 103, 108, 108*c*, 108*e*, 108*h*, 109, 115, 118.

Grouse, Pinnated: 1, 17, 31, 46, 108, 108*c*, 108*e*, 108*h*, 116*c*, 120. **Ruffed Grouse:** 17, 17*b*, 23, 23*d*, 24, 26, 26*a*, 27*b*, 30, 31, 33, 44, 46, 50, 69, 73, 75, 89, 108, 108*h*, 109, 116, 116*c*, 119, 120, 124*a*. **Sharp-tailed Grouse:** 17, 17*b*, 26*a*, 31*f*, 42, 46, 71, 72, 90, 108, 120.

Partridge, Hungarian: 23, 23*b*, 31, 46, 50, 50*a*, 57, 108, 108*h*, 108*i*, 108*k*, 109.

Pheasant: 7*b*, 17, 23, 23*b*, 26*a*, 30, 31, 46, 50, 50*a*, 76, 77, 78, 79, 108, 108*h*, 108*i*, 109, 115, 116, 119, 120, 123, 124*a*.

Prairie Chicken (*see also* **Grouse, Pinnated** and **Sharp-tailed**): 26, 30, 31, 33, 50*a*, 73, 108*a*, 108*c*, 108*e*, 108*h*, 108*i*, 120.

Turkey, Eastern Wild: 27*b*, 31, 35*a*, 57, 108*k*. **Wild Turkey:** 30, 57.

Upland Game Birds, unspecified: 31*i*, 50*a*, 108*a*, 108*f*.

Mammals

Beaver: 10, 17, 17*a*, 17*b*, 17*c*, 26, 26*a*, 30, 31, 40, 46, 66, 78, 93, 99, 100, 100*a*, 101, 103, 103*a*, 118, 120, 120*c*, 120*d*, 120*e*, 126, 126*b*.

Deer: 4, 5, 17, 17*b*, 17*c*, 20, 26, 26*a*, 27*b*, 29, 30, 31, 42, 46, 47*a*, 49, 53, 57*a*, 59, 69*b*, 72*b*, 83, 89, 90, 93, 100, 100*a*, 101, 103*a*, 108, 108*h*, 108*k*, 109, 109*b*, 109*c*, 114, 120, 120*a*, 120*c*, 120*d*, 120*e*, 125.

Moose: 3, 17, 17*a*, 17*b*, 26*a*, 31, 31*a*, 31*e*, 44, 44*a*, 49*a*, 57*a*, 59, 63, 72, 93, 93*u*, 100, 100*a*, 101, 115, 116*b*, 120, 120*b*, 128*a*.

Muskrat: 10, 10*d*, 10*e*, 12, 13, 14, 16, 21, 23*d*, 26*a*, 29, 30, 31, 31*g*, 31*h*, 33*d*, 33*g*, 34, 35, 35*a*, 35*c*, 40, 40*b*, 40*e*, 46, 49, 50*b*, 53, 54*a*, 56, 59, 63, 64, 66*a*, 68, 69, 69*a*, 70, 72, 72*a*, 73, 74, 75, 78*a*, 80*a*, 80*b*, 82, 84, 90, 92*a*, 93, 93*g*, 100, 100*a*, 101, 105, 108, 108*h*, 108*k*, 109, 109*b*, 113, 118, 119, 120, 124*a*, 125, 125*a*, 125*c*, 125*d*, 126, 126*a*, 126*b*.

Porcupine: 10, 100, 101, 120.

BIBLIOGRAPHY

1. ALDOUS, SHALER E.: Beaver Food Utilization Studies. Journ. Wildlife Mangt., **2**, No. 4; 215–222 (1938). Pertains especially to Minnesota.
2. ALDOUS, SHALER E., and CLARENCE F. SMITH: Food Habits of Minnesota Deer as Determined by Stomach Analysis. Trans. Third N. Amer. Wildlife Conf., 756–767 (1938).
3. ATWOOD, EARL L., JR.: Some Observations on Adaptability of Michigan Beavers Released in Missouri. Journ. Wildlife Mangt., **2**, No. 3; 165–166 (1938).
4. ATWOOD, EARL L., JR., and JULIAN A. STEYERMARK: The White-tailed Deer in Missouri. Mimeographed. Clark National Forest, U.S. Dept. Agric. (1937).
5. BAILEY, VERNON: Beaver Habits and Experiments in Beaver Culture. U.S. Dept. Agric., Tech. Bull. 21 (1927).
6. BAILEY, VERNON: Trapping and Transporting Live Beavers. U.S. Dept. Agric., Farmers' Bull. 1768 (1937).
7. BENNETT, LOGAN J.: The Blue-winged Teal, Its Ecology and Management. George Banta Publishing Company, Menasha, Wis. (1938). Pertains principally to Iowa.
8. [BIOLOGICAL SURVEY] Winter Food of the Ruffed Grouse in the Northeast. U.S. Dept. Agric., Bur. Biol. Surv. Bi-1297 (1933).
9. [BIOLOGICAL SURVEY] Natural Plantings for Attracting Waterfowl to Marsh and Other Water Areas. U.S. Dept. Agric., Bur. Biol. Surv., Wildlife Research and Management Leaflet BS-125 (1939).
10. BLAKEY, HAROLD L.: The Wild Turkey on the Missouri Ozark Range. Preliminary report. U.S. Dept. Agric., Bur. Biol. Surv., Wildlife Research and Mangt. Leaflet BS-77 (1937).

11. BORDNER, JOHN S., WILLIAM W. MORRIS, LAMAR M. WOOD, and JOHN H. STEENIS: Land Economic Inventory of Northern Wisconsin: Sawyer County. Wis. Dept. Agric. and Markets, 138 (1932). Contains material on animal ecology.

12. BORDNER, JOHN S., H. R. ALDRICH, WILLIAM W. MORRIS, and JOHN H. STEENIS: Land Economic Inventory of the State of Wisconsin: Juneau County. State of Wisconsin Executive Council, 1 (1934). Contains material on animal ecology.

13. BORDNER, JOHN S., WILLIAM W. MORRIS, JOHN H. STEENIS, and EARL D. HILBURN: Land Economic Inventory of the State of Wisconsin: Rusk County. State of Wisconsin Executive Council, 2 (1935). Contains material on animal ecology.

14. BRADT, GLENN W.: A Study of Beaver Colonies in Michigan. Journ. Mammalogy, **19**, No. 2; 139–162 (1938).

15. BUMP, G., F. C. EDMINSTER, JR., R. W. DARROW, W. C. RITTER, and A. A. ALLEN: Progress Report on the Ruffed Grouse Investigation. N.Y. State Conserv. Dept. (1932).

16. CHAMBLISS, CHARLES E.: Wild Rice. U.S. Dept. Agric., Dept. Circ. 229 (1922).

17. CLEPPER, HENRY E.: The Deer Problem in the Forests of Pennsylvania. Commonwealth of Penna., Dept. Forests and Waters, Bull. 50 (1931).

18. DALKE, PAUL D.: Dropping Analysis as an Indication of Pheasant Food Habits. Trans. Twenty-first American Game Conf., 387–391 (1935). Pertains to Michigan.

19. DALKE, PAUL D.: Carrying Capacity of Pheasant Range. Amer. Game, March–April, **23**, 31–32 (1935). Pertains to Michigan.

20. DALKE, PAUL D.: Food of Young Pheasants in Michigan. Amer. Game, May–June, **36**, 43–46 (1935).

21. DALKE, PAUL D.: Food Habits of Adult Pheasants in Michigan Based on Crop Analysis Method. Ecology, **18**, 199–213 (1937).

22. DAVENPORT, L. A.: Find Deer Have Marked Food Preferences. Michigan Conservation, **7**, No. 4; 4–7 (1937). Pertains to Michigan.

23. DERY, D. A.: Preliminary Report on the Migration in Quebec of the Northern Sharptailed Grouse. Que. Zool. Soc. Bull. 1 (1933).

24. DUGMORE, A. RADCLYFFE: The Romance of the Beaver. J. B. Lippincott Company, Philadephia (1914).

25. EAST, BEN.: Must They Starve? Amer. Forests, **42**, No. 10; 463–465, 483 (1936).

26. ENDERS, R. K.: Muskrat Propagation in Ohio. Ohio Dept. Agric. Bull. 19 (1931).

27. ERRINGTON, PAUL L.: The Bob-white's Winter Food. Amer. Game, **20**, 75–78 (1931). Pertains to Wisconsin.

28. ERRINGTON, PAUL L.: Management of the Bob-white Quail in Iowa. Iowa State College, Extension Bull. 186 (1933).

29. ERRINGTON, PAUL L.: Emergency Values of some Winter Pheasant Foods. Trans. Wis. Acad., **30**, 57–68 (1937). Pertains to Iowa.

30. GIGSTEAD, GILBERT: Habits of Wisconsin Pheasants. Wilson Bull. **49**, 28–34 (1937).

31. GREEN, WILLIAM E., and WATSON E. BEED: Iowa Quail and Pheasants in Winter. Amer. Wildlife, November–December, **83–84**, 90–92 (1936).

32. GROSS, ALFRED O.: Progress Report of the Wisconsin Prairie Chicken Investigation. Wis. Conserv. Comm., Madison (1930).

33. GROSS, ALFRED O.: Food of the Ruffed Grouse. Game Breeder and Sportsman, **41**, 142–144 (1937). Pertains to New England.

34. HAMERSTROM, F. N., JR., and JAMES BLAKE: A Fur Study Technique. Journ. Wildlife Mangt., **3**, No. 1; 54–59 (1939). Pertains to Wisconsin.

35. HAMERSTROM, F. N., JR., and JAMES BLAKE: Central Wisconsin Muskrat Study. Amer. Midl. Nat., **21**, 514–520 (1939).

36. HAMERSTROM, F. N., JR., and JAMES BLAKE: Winter Movements and Winter Foods of White-tailed Deer in Central Wisconsin. Journ. Mammalogy, **20**, 206–215 (1939).

37. HAWKINS, ARTHUR S.: Winter Feeding at Faville Grove, 1935–36. Am. Midl. Nat., **18**, 417–425 (1937). Pertains to southern Wisconsin.

38. HAWKINS, ARTHUR S.: Winter Feeding at Faville Grove, 1935–37. Journ. Wildlife Mangt., **1**, Nos. 3–4; 62–69 (1937). Pertains to southern Wisconsin.

39. HICKS, LAWRENCE E.: Ohio Game and Song Birds in Winter. Ohio Dept. Agric., Div. Conserv., Bur. Sci. Research, Bull. 1, No. 2 (1932).

40. HICKS, LAWRENCE E.: Food Habits of the Hungarian Partridge in Ohio. Ohio Dept. Agric., Div. Conserv., Bur. Sci. Research, Bull. 104 (without date).

41. Hicks, Lawrence E.: Winter Food of the Ohio Bob-white. Ohio Dept. Agric., Div. Conserv., Bur. Sci. Research, Bull. 105 (without date).
42. Hicks, Lawrence E.: The Food Habits of the Ring-necked Pheasant. Ohio Dept. Agric., Div. Conserv., Bur. Sci. Research, Bull. 107 (1936). Pertains to Ohio.
43. Hosley, N. W., and R. K. Ziebarth: Some Winter Relations of the White-tailed Deer to the Forests in North Central Massachusetts. Ecology, 16, 535–553 (1935).
44. Howard, William Johnston: Notes on the Winter Foods of Michigan Deer. Journ. Mammalogy, 18, No. 1; 77–80 (1937).
45. Johnson, Charles Eugene: The Beaver in the Adirondacks; Its Economics and Natural History. Roosevelt Wild Life Bull. 4, 501–641 (1927).
46. Johnson, Charles Eugene: The Muskrat in New York: Its Natural History and Economics. Roosevelt Wild Life Bull. 3, 205–320 (1925).
47. Judd, Sylvester D.: The Economic Value of the Bobwhite. U.S. Dept. Agric. Yearbook, 193–304 (1903).
48. Judd, Sylvester D.: The Bobwhite and Other Quails of the United States in Their Economic Relations. U.S. Dept. Agric., Bur. Biol. Surv. Bull. 21 (1905).
49. Judd, Sylvester D.: The Grouse and Wild Turkeys of the United States, and Their Economic Value. U.S. Dept. Agric., Bur. Biol. Surv. Bull. 24 (1905).
50. Kelso, Leon: A Note on the Food of the Hungarian Partridge. The Auk, 49, No. 2; 204–207 (1932).
51. Kelso, Leon: Winter Food of Ruffed Grouse in New York. U.S. Dept. Agric., Bur. Biol. Surv., BS-1 (1935).
52. Kubichek, W. F.: Report on the Food of Five of Our Most Important Game Ducks. Iowa State Coll. Journ. Sci., 8, No. 1; 107–126 (1933). Discusses redhead, ring-necked, canvas-back, greater scaup, lesser scaup.
53. Leffingwell, Dana J.: The Ring-necked Pheasant—Its History and Habits. State College of Washington; Occasional Papers of the Charles R. Connor Museum, No. 1 (1928).
54. Mabbott, Douglas C.: Food Habits of Seven Species of American Shoal Water Ducks. U.S. Dept. Agric. Bull. 862 (1920). Discusses gadwall, baldpate, European widgeon, blue-winged teal, pintail, cinnamon teal, wood duck.
55. Maynard, L. A., Gardiner Bump, Robert Darrow, and J. C. Woodward: Food Preferences and Requirements of the White-tailed Deer in New York State. N.Y. State Conserv. Dept. and State Col. Agric. Bull. 1 (1935).
56. McAtee, W. L.: Eleven Important Wild-duck Foods. U.S. Dept. Agric. Bull. 205 (1915).
57. McAtee, W. L.: Propagation of Wild-duck Foods. U.S. Dept. Agric. Bull. 465 (1917).
58. McAtee, W. L.: Food Habits of Mallard Ducks in the United States. U.S. Dept. Agric. Bull. 720 (1918).
59. McAtee, W. L.: Notes on the Food Habits of the Shoveller or Spoonbill Duck (Spatula clypeata). The Auk, 39, No. 3; 380–386 (1922).
60. McAtee, W. L.: How to Attract Birds in the Northeastern United States. U.S. Dept. Agric. Farmers' Bull. 621, reprinted (1922).
61. McAtee, W. L.: How to Attract Birds in the Middle Atlantic States. U.S. Dept. Agric. Farmers' Bull. 844, rev. (1924).
62. McAtee, W. L.: How to Attract Birds in the East Central States. U.S. Dept. Agric. Farmers' Bull. 912, revised (1926).
63. McAtee, W. L.: Wildlife of the Atlantic Coast Salt Marshes. U.S. Dept. Agric., Bur. Biol. Surv., Wildlife Research and Mangt. Leaflet BS-17 (1935).
64. McAtee, W. L.: Fruits Attractive to Birds—Northeastern States Region No. 4. U.S. Dept. Agric., Bur. Biol. Surv., Wildlife Research and Mangt. Leaflet BS-44 (1936).
65. McAtee W. L.: Wildfowl Food Plants. Collegiate Press, Inc., Ames, Iowa (1939).
66. McAtee, W. L., and F. E. L. Beal: Some Common Game, Aquatic, and Rapacious Birds in Relation to Man. U.S. Dept. Agric. Farmers' Bull. 497, revised (1924).
67. Metcalf, Franklin P.: Wild Duck Food of North Dakota Lakes. U.S. Dept. Agric. Tech. Bull. 221 (1931).
68. Morgan, Lewis H.: The American Beaver and His Works. J. B. Lippincott Company, Philadelphia (1868).

69. Murie, Adolph: The Moose of Isle Royale. Univ. Mich., Mus. Zool., Misc. Pub. 25 (1934).
70. Nelson, A. L., Talbott E. Clarke, and W. W. Bailey: Early Winter Food of Ruffed Grouse on the George Washington National Forest. U.S. Dept. Agric. Circ. 504 (1938). Pertains to Virginia and West Virginia.
71. Newsome, William Monypeny: Whitetailed Deer. Charles Scribner's Sons, New York (1926).
72. Oberholzer, Harry C., and W. L. McAtee: Waterfowl and Their Food Plants in the Sandhill Region of Nebraska. U.S. Dept. Agric. Bull. 794 (1920).
73. Olson, Herman F.: An Analysis and Interpretation of Deer Study Data on Superior National Forest, 1934–1937 inclusive. Typed report (Oct. 15, 1937). Original not seen; quoted in No. 2.
74. Pearce, John: The Effect of Deer Browsing on Certain Western Adirondack Forest Types. Roosevelt Wildlife Bull. **7**, No. 1; 1–61 (1937).
75. Peterson, Roger T.: Song-bird Sanctuaries, with Tables of Trees, Shrubs and Vines Attractive to Birds. Natl. Assoc. Audubon Soc., Circ. 19 (1937).
76. Phillips, J. C.: Ten Years of Observation on the Migration of Anatidae at Wenham Lake, Massachusetts. The Auk, **28**, No. 2; 188–200 (1911).
77. Pirnie, Miles David. Michigan Waterfowl Management. Dept. of Conserv., Game Div., Lansing (1935).
78. Schmidt, F. J. W.: Winter Food of the Sharp-tailed Grouse and Pinnated Grouse in Wisconsin. Wilson Bull. **48**, 186–203 (1936).
79. Seton, Ernest Thomson: Lives of Game Animals. Doubleday, Doran & Company, Inc., Garden City, New York (1929).
80. Severin, H. C.: A Summary of an Economic Study of the Food of the Ring-neck Pheasant in South Dakota. Proc. S. D. Acad. Sci., **16**, 44–58 (1936).
81. Shelford, V. E., and Sigurd Olson: Sere, Climax and Influent Animals with Special Reference to the Transcontinental Coniferous Forest of North America. Ecology, **16**, No. 3; 375–402 (1935).
82. Smith, Frank R.: Muskrat Investigations in Dorchester County, Md., 1930–34. U.S. Dept. Agric. Circ. 474 (1938).
83. Snyder, L. L.: A Study of the Sharp-tailed Grouse. Univ. of Toronto Studies, Biol. Ser. 40 (1935). Pertains to Ontario and Quebec.
84. Spiker, Charles J.: Some Late Winter and Early Spring Observations on the White-tailed Deer of the Adirondacks. Roosevelt Wild Life Bull. **6**, No. 2; 327–385 (1933).
85. Stegeman, LeRoy C.: A Food Study of the White-tailed Deer. Trans. Second North Amer. Wildlife Conf., 438–445 (1937). Pertains to North Carolina.
86. Swenk, M. H.: The Food Habits of the Ring-necked Pheasant in Central Nebraska. Col. Agric., Univ. Neb., Agric. Exper. Sta. Research Bull. 50 (1930).
87. Terrell, Clyde B.: Wild Fowl and Fish Attractions for South Dakota. Game and Fish Commission, Pierre, S.D. (1930).
88. Tiffany, L. H.: Importance of Aquatic Plants to Animal Life. Ohio Dept. of Agric., Div. Conserv., Bur. of Sci. Research, Bull. 18 (without date).
89. Townsend, M. T., and M. W. Smith: The White-tailed Deer of the Adirondacks. Roosevelt Wild Life Bull. **6**, No. 2; 161–325 (1933).
90. Warren, Edward Royal: The Beaver. Williams & Wilkins Company, Baltimore (1927).
91. Waterfowl Food Plants. More Game Birds in America, Inc., New York. Revised (1936).
92. Wilson, Kenneth A., and Ernest A. Vaughn: The Resettlement Administration and Its Wildlife Program on the Eastern Shore of Maryland. 13-page pamphlet, publisher not stated (1937).
93. Wight, Howard M.: Suggestions for Pheasant Management in Southern Michigan. Dept. of Conserv., Lansing (1933).
94. Yeatter, Ralph E.: The Hungarian Partridge in the Great Lakes Region. Univ. of Mich., School of Forestry and Conserv., Bull. 5 (1934).
95. Yorke, F. Henry: Our Ducks. The American Field Publishing Co., Chicago (1899).

The two following publications have appeared too recently to be included.

Martin, A. C., and F. M. Uhler. Food of Game Ducks in the United States and Canada. U.S. Dept. Agric., Tech. Bull. 634 (1939). Contains descriptions, photographs and drawings of plants, maps of their ranges, and discussions of their uses as food.
Cottam, Clarence: Food Habits of North American Diving Ducks. U.S. Dept. Agric., Tech. Bull. 643 (1939).

THE RELATION OF PLANTS TO FISH

Aquatic plants may serve as food, shade, or protection for fish, or they may support algae or small animals which are directly or indirectly food for game fish, or they may form habitats for the deposition of eggs, or they may aid animal life by oxygenating the water. The relations are complex, and most statements in the literature are very general. The following compilation is by Carl Leopold and Roderick Huff. The superior numerals refer to the bibliography below.

Alisma Plantago-aquatica: shade and shelter for young fish.[12]

Anacharis: shelter and support of insects.[5,12]

Brasenia Schreberi: shade and shelter.[2]

Cabomba caroliniana: cover and valuable food producer.[1,12]

Ceratophyllum demersum: good shelter for young fish, and supports insects valuable as fish food.[1,3,9,12]

Chara: fair shelter, and excellent producer of fish food, especially for young trout, largemouthed and smallmouthed black bass; has a softening effect on water, abstracting lime and carbon dioxide and depositing marl.[2,3,6,10,12]

Cladophora: best food producer in fast water, especially for rainbow trout.[10]

Eleocharis acicularis: forms spawning ground for largemouthed black bass.[12]

E. palustris var. *major:* supports insects.[4]

Heteranthera dubia: food and shelter.[12]

Juncus: forms spawning grounds for rock bass, bluegills, and other sunfish.[5]

Lemnaceae: food and cover.[11]

Lemna: poor food producer and excessively shady, so not beneficial.[5,12]

Wolffia: good food and cover.[11]

Marsilea quadrifolia: excellent shade and shelter.[12]

Myriophyllum: shelter, and valuable food producer supporting many insects.[1,5,9,11,12]

M. exalbescens: roots preferred by black bass for nesting.[12]

Nasturtium officinale: excellent food producer for trout.[10]

Najas: good food producer and shelter.[5,9,12]

Nelumbo lutea: good shade and fair shelter.[12]

Nuphar: shade and shelter; leaves harbor insects.[5]

N. advena: shade and shelter, but poor food producer.[12]

N. variegatum: shade and cover[2]; food and cover.[11]

Nymphaea odorata: negative value when too dense.[12]

N. tuberosa: shade and shelter.[2,12]

Nymphoides cordatum: excellent shelter and fair food producer.[12]

Phragmites maximus: supports insects.[4]

Polygonum: good food and cover.[11]

Pontederia cordata: shade and shelter[12]; slight value for food and cover.[11]

Potamogeton: food and shelter; leaves eaten by bluegills; softens water, removing lime and carbon dioxide and depositing marl.[1,3]

P. americanus: supports insects, but only fair food and shelter.[5,11,12]

P. amplifolius: supports insects; good food supply.[5,8,9]

P. angustifolius: good for fish environments.[5]

P. crispus: good food, shelter, and shade; valuable for early spawning fish like carp and goldfish.[8,12]

P. filiformis: starch in tubers and rootstocks.[8]

P. foliosus: good food producer and shelter.[12]

P. gramineus: starch in rootstocks[8]; food and cover.[11]

P. natans: good for fish environments.[5,11]

P. obtusifolius: supports insects[8]; fair food and cover.[11]

P. pectinatus: starch in tubers; food and shelter for young trout and other fish.[2,6,8,10, 11,12]

P. praelongus: feeding grounds for muskellunge; also good food producers for trout.[2, 8,10]

P. pusillus: good food and cover.[11]

P. Richardsonii: supports insects[8]; good food and cover.[11]

P. Robbinsii: food and shelter, particularly for northern pike.[2,8]

P. zosteriformis: does not generally support insects.[8]

Ranunculus: fair food producer for trout.[10]

Ruppia: excellent food and cover.[11]

Sagittaria australis: shade and shelter for young fish.[12]

Scirpus: used for nesting by bluegills and largemouthed black bass[5]; good food and cover.[11]

S. acutus: support for insects and shelter for young fish[4,5]; slight value for food and cover.[11]

S. americanus: supports insects[4]; slight value for food and cover.[11]

S. validus: support for insects and shelter for young fish.[4,5]

Sparganium eurycarpum: no value for food or cover.[11]

Spirodela polyrhiza: not desirable due to excess shading[5]; poor food producer.[12]

Trapa natans: poor food producer and shelter.[12]

Typha: supports insects.[4]

T. angustifolia: spawning ground for sunfish, and shelter for young fish.[5]

Utricularia: good food and cover.[11]

U. purpurea: fair food producer and cover.[2,5]

Vallisneria americana: good shade and shelter, supports insects, and is valuable fish food.[2,5,9,11,12]

Zannichellia palustris: fair food producer for trout.[10]

Zizania: of doubtful value[5]; good food and cover[10]; eaten by carp.[7]

BIBLIOGRAPHY

1. Aldrich, A. D.: Construction and Management of Small Lakes and Fish Rearing Ponds. Parks and Recreation, **20**, No. 7; 335–341 (1937).
2. Bordner, J. S., Earl D. Hilburn, Wm. W. Morris, Lewis Posekany, and Russell Sanford: Inventory of Northern Wisconsin Lakes. State of Wisconsin, State Planning Board, Bull. 5 (1939).
3. Evermann, B. W., and H. W. Clark: Lake Maxinkuckee. Indiana Dept. Conserv., Pub. 7, 2 vols. (1920).
4. Frohne, W. C.: Limnological Rôle of Higher Aquatic Plants. Trans. Amer. Micros. Soc., **57**, No. 3; 256–268 (1938).
5. Hubbs, C. L., and R. W. Eschmeyer: Improvement of Lakes for Fishing. Mich. Dept. Conserv., Institute for Fisheries Research, Bull. 2 (1938).
6. Hubbs, C. L., J. R. Greeley, and C. M. Tarzwell: Methods for the Improvement of Michigan Trout Streams. Mich. Dept. Conserv., Institute for Fisheries Research, Bull. 1 (1932).
7. McAtee, W. L.: Propagation of Wild-duck Foods. U.S. Dept. Agric. Bull. 465 (1917).
8. Moore, Emmeline: The Potamogetons in Relation to Pond Culture. Bull. U.S. Bur. Fisheries, **33**, 255–291 (1913).
9. Muttkowski, R. A.: The Fauna of Lake Mendota. Trans. Wis. Acad., **19**, 374–482 (1918).
10. Needham, P. R.: Trout Streams. Comstock Publishing Company, Inc., Ithaca, N.Y. (1938).
11. Terrell, Clyde B.: Wild Fowl and Fish Attractions for South Dakota. Game and Fish Commission, Pierre, S.D. (1930).
12. Titcomb, J. W.: Aquatic Plants in Pond Culture. U.S. Bur. Fisheries, Doc. 643 (1909).

GLOSSARY

Aborted fruits. Fruits which have not developed or "filled out."

Acute. Ending in an angle of less than 90 deg. (Fig. 4, p. 23).

Adventive. Of foreign origin; deliberately or accidentally introduced, and of sporadic occurrence.

Alternate leaves. Leaves borne with but one at each level on the stem.

Annual. Living one season.

Anther. The upper, usually enlarged, portion of a stamen, containing the pollen.

Apex. The end farther from the central stem.

Appressed. Lying closely against. The scales are appressed to the axis of the spikelet in Fig. 1, p. 19.

Awn. A bristle-like appendage (Fig. 24, p. 108, and Fig. 5, p. 323).

Axil. The angle between a leaf or a bract and the stem on which it is borne (Fig. 14 and explanation, p. 52).

Berry. A pulpy or juicy fruit, like a grape.

Blade. The expanded, usually flattened, portion of a leaf.

Bract. A leaf reduced to a scale, often with a flower in its axil (Fig. 8, p. 80).

Bractlet. A very small bract.

Bulblet. A small bulb-like structure usually borne in the axil of a leaf (Fig. 70, p. 273).

Calyx. The outer, usually green, whorl of floral parts, made up of sepals which may be separate (Fig. 35, p. 236) or united (Fig. 23, p. 291); if of united sepals, it often consists of a *tube* (the united portion) and *lobes* (the free tips).

Calyx lobes. See *calyx.*

Calyx tube. See *calyx.*

Capsule. A pod; a dry fruit with many seeds.

Carpel. A segment of a pistil; Fig. 4, p. 253, shows a pistil composed of three carpels; Fig. 18, p. 242, shows pistils composed of two carpels each; Fig. 46, p. 68, shows a pistil composed of one carpel.

Catkin. A dense spike of flowers with conspicuous bracts, often flexible and drooping at maturity.

Cell. One of the units of which most plants are built, usually too small to be seen by the naked eye. In most of the figures on pp. 37–39 plants are shown as composed of cells.

Centimeter (abbreviated *cm.*). $\frac{1}{100}$ m.; about $\frac{2}{5}$ in. *See* scale on inside back cover.

Chloroplast. The green body or bodies within cells capable of manufacturing food.

Cm. See *centimeter.*

Compound leaves. Leaves whose blades are divided into several distinct *leaflets* (Figs. 1, 4, 6, 9, p. 33); the leaflets may themselves be subdivided (Fig. 7, p. 33).

Cone. A group of seed- or spore-bearing scales arranged on a usually elongate axis, as a pine cone.

Corm. An enlarged base of a stem.

Corolla. The whorl of floral parts next within the calyx, usually of white or colored *petals*, which may be separate (Fig. 35, p. 236), or united (Fig. 75, p. 277). If the petals are united, the corolla may be divided into *tube* and *lobes*, like the calyx, or into *tube* and *lips* (Fig. 19, p. 290).

359

Crisped.   With wavy surface toward the margin.

Decimeter (abbreviated *dm.*). $\frac{1}{10}$ m.; about 4 in.   *See* scale on inside back cover.

Disk-flower.   In heads of Composites, the flowers on the central portion (Fig. 55, p. 337).

Dissected.   Cut into many small divisions (Fig. 11, p. 7).

Dm.   See *decimeter.*

Elliptical.   Oval or oblong, with the ends equally narrowed or rounded (Fig. 9, p 29).

Embryo.   The undeveloped plant contained within a seed.

Emersed.   Extending out of the water.

Filament.   The usually thread-like stalk of a stamen (Fig. 1, p. 78; Fig. 6, p. 125). The filaments may be united into a *filament tube* (Fig. 71, p. 318).

Floret.   A small flower; in grasses, the pistil and stamens plus lemma and palea (Fig. 2, p. 124).   A grass spikelet may contain one (Fig. 58, p. 116) or several florets (Fig. 4, p. 124).

Frond.   In the *Lemnaceae*, the green leafless stem.

Gland.   A small swelling, often stalked (Fig. 25, p. 206), and frequently secreting resin, wax or other substances.

Globular.   Globe-shaped.

Glumes.   The lowest pair of scales in a grass spikelet (Fig. 20, p. 104; Figs. 3 and 4, p. 124).

Grain.   The mealy fruit of a grass.

Head.   A short dense cluster of flowers without stalks (figures on p. 295).

Immersed.   Under water.

Inflorescence.   The flowering portion of a plant.

Internode.   The part of a stem between successive points where leaves are borne.

Introduced.   Brought intentionally from another region.

Involucel.   A small bract at the base of a flower or flowering branch, generally in the same inflorescence with an *involucre*, composed of large bracts of similar nature (Fig. 128, p. 148).

Involucre.   A cluster of bracts in an inflorescence (Fig. 128, p. 148; Fig. 1, p. 321), usually at the base of a head of flowers.

Lateral.   Along the sides.

Leaflet.   A division of a compound leaf; on p. 33, Fig. 6 shows a leaf with 10 leaflets, Fig. 8 shows a leaf with 3 leaflets, and Fig. 9 shows a leaf with 5 leaflets.

Lemma.   The outer scale of a grass floret (Figs. 2–4, p. 124).

Ligule.   The collar-like appendage, membranous or composed of hairs, at the junction of the blade and the sheath in the grass leaf (Fig. 1, p. 124).

Limb.   In a tubular corolla, the upper expanded portion (Fig. 34, p. 333).

Linear.   Long and narrow, with parallel sides.

Littoral.   Growing on shores.

M.   See *meter.*

Megaspore.   The female or larger type of spore in *Isoetes.*

Meter (abbreviated *m.*).   A unit of measurement in the metric system, equal to about 1 yd. in the English system, or 39.37 in.

Micron. $\frac{1}{1000}$ mm.

Microspore.   The male or smaller type of spore in *Isoetes.*

Midrib.   The main or central vein of a leaf or leaflet.

Millimeter (abbreviated mm.). $\frac{1}{1000}$ m.; $\frac{1}{100}$ dm.; $\frac{1}{10}$ cm.; about $\frac{1}{25}$ in.   *See* scale on inside back cover.

Mm.   See *millimeter.*

Naturalized.   Of foreign origin but now well established in a wild state.

Node. The level of a stem at which one or more leaves arise.

Nutlet. A small hard-coated one-seeded fruit.

Oblong. Longer than broad and with nearly parallel sides.

Obtuse. Blunt or rounded at the end.

Oögonium. The egg-bearing organ of some Algae.

Opposite leaves. Leaves borne two at a node (Fig. 1, p. 23).

Ovary. The basal swollen portion of a pistil, in which the seeds develop.

Ovate. Shaped like an egg with the broader end downward.

Palate. A rounded projection in the mouthlike opening of some flowers (Fig. 49, p. 310).

Palea. The inner scale of a grass floret (Figs. 2–4, p. 124).

Panicle. A loose open cluster of flowers, more or less branched (Fig. 60, p. 94).

Pedicel. The stalk of a single flower or fruit (Fig. 4, p. 229).

Peduncle. A stalk bearing a flower (Fig. 58, p. 314) or a group of flowers (Fig. 11, p. 202).

Perennial. Living for several years.

Perianth. General term for all the flattened structures making up the outer whorls of a flower; a perianth generally consists of calyx and corolla.

Perigynium. In *Carex*, the special sac which encloses the nutlet.

Petal. See *corolla*.

Petiole. The stalk of a leaf (Fig. 3, p. 21; Figs. 7 and 8, p. 23; Fig. 9, p. 33).

Pinnate. Compound, with leaflets arranged on two sides of an axis (Fig. 6, p. 33).

Pistil. The organ in which seeds develop, usually in the center of a flower (Fig. 2, p. 124; Fig. 6, p. 254); made up of stigma, style, and ovary.

Pistillate flower. Bearing pistils and no stamens; female.

Pith. The light spongy central portion of a woody stem.

Pod. A thin-walled dry fruit containing several or many seeds.

Pollen. The usually spherical, often yellow or brown, powder-like bodies contained in the anthers.

Raceme. An inflorescence of stalked flowers along a common axis (Fig. 6, p. 302).

Rachilla. A small central axis (Fig. 27, p. 127).

Rachis. The central axis of a raceme or a compound leaf, etc.

Ray-flower. The strap-like flower on the margin of a Composite head (Fig. 55, p. 337).

Recurved. Curving downward or backward.

Reflexed. Sharply bent downward or backward.

Reticulate. Marked with lines forming a network.

Rootstock. An underground stem, usually bearing scale leaves, and with erect stems or leaves at intervals (Fig. 8, p. 176).

Rosette. A cluster of radiating leaves (Fig. 8, p. 17; Fig. 2, p. 21).

Scape. A leafless or nearly leafless stem bearing flowers (Fig. 49, p. 310).

Segment. A subdivision or lobe of a deeply cut or several times compound leaf.

Sepal. See *calyx*.

Sheath. A collar-like outgrowth at a node (Fig. 9, p. 45), or the basal part of a leaf wrapped about the stem (Fig. 1, p. 124; Fig. 50, p. 132).

Simple leaf. A leaf which is not divided into distinct leaflets. See *compound leaves*.

Spadix. A fleshy spike of flowers (figures on p. 166).

Spathe. An involucral leaf more or less enwrapping flowers or groups of flowers or their bases (Fig. 71, p. 97; Fig. 1, p. 165).

Spike. An elongate inflorescence with flowers growing close together on a common axis (Fig. 11, p. 202).

Spikelet.   The small spike of grasses (Fig. 20, p. 104; Fig. 3, p. 124) and some sedges (Fig. 8, p. 125; Fig. 30, p. 128).

Spore.   A usually microscopic one-celled structure serving to reproduce lower plants as seeds do higher plants.

Spur.   A more or less elongate saclike organ, usually part of a calyx or corolla (Fig. 47, p. 310).

Stamen.   The male organs of a flower, usually in a whorl just inside the petals (Fig. 35, p. 236); made up of an anther and a filament.

Staminate.   Bearing stamens and no pistils; male.

Stigma.   The uppermost part of a pistil, usually feathery or sticky to receive the pollen (Fig. 2, p. 124; Fig. 6, p. 125).

Stipe.   A little stalk (Fig. 30, p. 190).

Stipules.   The appendages at the base of a petiole (Fig. 3, p. 21).

Stolon.   A stem which trails on the ground and often takes root; a runner (Fig. 1, p. 301).

Style.   The tapered upper portion of a pistil (Fig. 6, p. 125; Fig. 17, p. 232).

Supra-axillary.   Borne at a distance above the leaf next below (Fig. 18 and explanation, p. 52).

Taproot.   A central root growing straight down as a continuation of the stem (Fig. 33, p. 208).

Terminal.   At the end.

Thallus.   The flat green body of some lower plants (Fig. 36, p. 41).

Tuber.   A thickened underground branch (Fig. 6, p. 56).

Tubercle.   A swollen appendage, especially on the nutlets of some sedges (Figs. 63 and 64, p. 134).

Umbel.   A group of flowers on stalks radiating from the same point (Fig. 69, p. 96); an umbel may be compound, each larger stalk terminating in a little umbel (Fig. 64, p. 272).

Venation.   The system of veins in a leaf; venation is ordinarily either parallel (Fig. 3, p. 29) or netted (Figs. 2, 6, 7, 9, and 10, p. 29).

Whorled.   Arranged in a circle at one level on the stem (Figs. 4 and 9, p. 7; Fig. 2, p. 23).

Winter bud.   A small hardened branch with close leaves that survives the rest of the plant in winter and renews growth in spring (Fig. 66, p. 72; Figs. 30–37, p. 310).

REVISION APPENDIX

P. 6   Line 15: (*Armoracia*) for (*Neobeckia*).

P. 7   Fig. 11 is **Armoracia.**

P. 10   Line 9: *Peplis diandra* for *Didiplis diandra* forma *aquatica.*

P. 11   Fig. 8 is **Peplis diandra.**

P. 22   *Lippia*, p. 286, will key to the first **6.**

P. 24   Line 8: (*Elodea*) for (*Anacharis*).

          Line 9: (*Justicia*) for (*Dianthera*).

          Line 29: (*Elodea*) for (*Anacharis*).

          Line 2 from bottom of page: (*Peplis diandra*) for (*Didiplis diandra* forma *terrestris*).

P. 25   Figs. 1 and 8 are **Elodea.** Fig. 9 is **Peplis diandra.**

P. 26   *Potamogeton crispus*, p. 57, will key to the first **9.**

P. 27   Figs. 4 and 8 are **Armoracia.**

P. 28   First 3: Buckwheat Family (*Polygonaceae*) for Smartweed (*Polygonum*).

          Some members of the *Onagraceae*, p. 255, will key out to the second **12.**

P. 30   Some species of *Salix*, p. 182, will key to the second **4** as well as the first **4.**

P. 33   Fig. 2 is **Armoracia.** Fig. 11 is **Rorippa islandica** var. **Fernaldiana.**

P. 34   Some species of *Echinodorus*, p. 93, will key to the first **10.**

P. 36   To **Reference** add: Second Edition (1950). G. W. Prescott, How to Know the Fresh-water Algae, published by Wm. C. Brown Co. (1954).

P. 40   To **References** on mosses (bottom of page) add: T. C. Frye and Lois Clark, Hepaticae of North America, Univ. of Washington Publ. Biol., **6**, 163–334 (1943); Elizabeth M. Dunham, How to Know the Mosses, The Mosher Press (1951); H. S. Conard, How to Know the Mosses, Wm. C. Brown Co. (1944); E. T. Bodenburg, Mosses, Burgess Publishing Co. (1954).

P.42   Three species of *Azolla* are found in our area. The characters that separate them are highly technical. For these characters see the treatment by Svenson, Am. Fern Journ., **34**, 69–84 (1944). The species with ranges in our area are: **A. caroliniàna** Willd.; Mass. & N.Y. to La. **A. mexicàna** Presl.; Wis., Ill. & westward. **A. filiculoìdes** Lam.; not native to our area but found as an introduction in N.Y.

P. 44   **E. palustre** L. var. **americanum** Vict. forma **verticillàtum** Milde for **E. palustre** var. **americanum** forma **luxurians**.

P. 47   **I. melanopoda** Gay & Dur. for **I. melanopoda** J. Gay.

       I. echinóspora Dur. var. **Braúnii** (Dur.) Engelm. for **I. Braunii**.

       I. echinospora var. **robústa** Engelm. for **I. Braunii** forma **robusta**.

Pp. 48   **T. angustifolia** var. **elongata** may be a hybrid between **T.**
& 49   **angustifolia** and **T. latifolia**. It is treated as a species by some: **T. glaùca** Godr.

       Add to **Reference**: Hotchkiss and Dozier, Am. Midl. Nat., **41**, 237–254 (1949)—*Typha* in N. Am.; Fassett and Calhoun, Evolution, **6**, 367–379 (1952)—Introgression between *T. latifolia* and *T. angustifolia*.

P. 53   **S. minimum** (Hartm.) Fries for **S. minimum** Fries.

P. 55   Supplementary key to *Potamogeton*:

     1. Stipules united with the base of the leaf for a distance of 10 mm. or more (Figs. 2, 27); floating leaves absent......................2
     1. Stipules free from the leaf (Fig. 39) or united for a distance of less than 10 mm. (Fig. 71)..8
       2. Leaves 3–8 mm. wide, auricled at base (Fig. 33)..............3
       2. Leaves 0.2–2 mm. wide, not auricled at base.................4
     3. Leaves minutely toothed on the margins (Fig. 33). **P. Robbínsii** Oakes.
     3. Leaves not toothed. **P. Robbinsii** forma **cultellàtus** Fassett.
       4. Leaves acute and sharp-pointed at apex (Fig. 3); fruits 2.5–4 mm. long, with a short beak (Fig. 5). **P. pectinàtus** L.
       4. Leaves obtuse and blunt at apex (Fig. 29), or sometimes minutely apiculate; fruits 2–3 mm. long, beakless (Fig. 28)..............5
     5. Primary stipules with sheath loose, inflated (Fig. 27); stems thick; flowers in 5–12 whorls; fruits about 3 mm. long. **P. vaginàtus** Turcz.
     5. Stipules with sheath tight about the stem; stems slender; flowers in 2–5 whorls; fruits about 2 mm. long...........................6
       6. Spikes 1.5–5 cm. long; whorls mostly separated, the lower more than 7 mm. apart. **P. filifòrmis** Pers.
       6. Spikes 0.5–2.5 cm. long; whorls mostly contiguous, the lower less than 7 mm. apart...7
     7. Leaves 0.25–0.5 mm. wide, obtuse. **P. filiformis** var. **boreàlis** (Raf.) St. John.
     7. Leaves 0.75–2 mm. wide, obtuse or minutely apiculate. **P. filiformis** var. **Macoùnii** Morong. Que to Alberta, s. to Wis., Colo., & Calif.
       8. Submersed leaves ribbon-like with parallel sides, flaccid, 2–10 mm. wide, with a prominent cellular median band (Fig. 34); fruits 2.5–4 mm. long, embryo coil more than one revolution.........9
       8. Submersed leaves not as above; embryo coil less than one revolution, or if more, then fruits less than 2.5 mm. long (*P. tennesseensis* may have some fruits up to 3 mm. long)...................10
     9. Submersed leaves 5–10 mm. wide, 7–13-nerved; fruits 3–4.5 mm. long (Fig. 35). **P. epihỳdrus** Raf.

9. Submersed leaves 2–8 mm. wide, 3–7-nerved; fruits 2.5–3.5 mm. long (Fig. 36). **P. epihydrus** var. **ramòsus** (Peck) House (*P. epihydrus* var. *Nuttallii*).

   10. Submersed leaves linear, less than 6 mm. wide, more than 20 times as long as wide (usually absent at maturity in *P. natans*)..**11**

   10. Submersed leaves lanceolate to ovate or oblong..............**39**

11. Floating leaves present (Fig. 1)...........................**12**

11. Floating leaves absent (Fig. 51); embryo coil less than one complete revolution...**21**

   12. Floating leaf blades mostly more than 2 cm. wide; fruits mostly more than 2.5 mm. long; embryo coil less than one revolution...**13**

   12. Floating leaf-blades mostly less than 2 cm. wide; fruits mostly less than 2.5 mm. long..................................**14**

13. Floating leaf blades 2.5–6 cm. wide, usually heart-shaped (Figs. 1, 24); fruits mostly 3.5–5 mm. long, keels obscure, fruit coat wrinkled; apex of embryo pointing toward basal end. **P. nàtans** L.

13. Floating leaf blades 1–3 cm. wide, rounded or wedge-shaped at base; fruits mostly 2.5–3.5 mm. long, keels prominent; apex of embryo pointing a little above basal end. **P. Oakesiànus** Robbins (Fig. 24 is not this species, as labelled, but *P. natans*).

   14. Embryo coil less than one revolution, scarcely visible through the fruit wall; stipules free from leaf base (Fig. 39)..........**15**

   14. Embryo coil more than one revolution, often visible through the thin fruit wall; dorsal keel thin, wing-like, and often toothed; stipules of some or all of submersed leaves adnate to base of leaf (Fig. 71)...**16**

15. Fruiting plants with floating leaves (Fig. 65); sterile plants with submersed leaves only; submersed leaves 0.1–0.5 mm. wide and tapering to a bristle tip (Fig. 67); fruit keeled and distinctly beaked (Fig. 69). **P. Vàseyi** Robbins.

15. Fruiting plants with submersed leaves only; plants with floating leaves sometimes with flowers but not with fruits; submersed leaves 0.4–1 mm. wide, merely pointed (Fig. 68); fruits not keeled and only slightly beaked (Fig. 70). **P. lateràlis** Morong.

   16. Submersed leaves obtuse to acute (Figs. 74, 78) but not tapering to bristle tips; floating leaves rounded at apex (Fig. 72)........**17**

   16. Submersed leaves acute to long tapering (Fig. 79); floating leaves somewhat pointed at apex (Fig. 77)......................**18**

17. Submersed leaves blunt or rounded at tip (Fig. 74); fused portion of stipule much longer than the free portion (Fig. 71); fruits 1.3–2.2 mm. wide, wings on the sides scarcely developed, beak lacking (Fig. 73). **P. Spiríllus** Tuckerman.

17. Submersed leaves acute (Fig. 78); fused portion of stipule about half as long as the free portion (Fig. 75); fruits 1–1.5 mm. wide, with low fine wings on the sides and a minute beak. **P. diversifòlius** Raf.

   18. Submersed leaves 0.1–0.6 mm. wide, with 1–3 nerves; floating leaves 1–10 mm. wide, with 3–7 nerves; peduncles 0.1–1.5 cm. long; wall of fruit thin, the coiled form of the embryo clearly visible (Fig. 73)..**19**

   18. Submersed leaves 0.2–2 (–3) mm. wide, with 1–5 nerves; floating leaves 5–35 mm. wide, with 9–23 nerves; peduncles 3–8 cm. long; wall of fruit thick, the form of the embryo usually not visible. **P. tennesseénsis** Fernald. W.Va. & Tenn.

19. Fruits 1–1.5 mm. long, dorsal keel narrow, entire or minutely toothed, lateral keels low and nearly toothless (Fig. 80)..................**20**

19. Fruits 1.6–2.2 mm. long, dorsal keel broad, coarsely toothed, lateral keels thin, sinuate or coarsely toothed (Fig. 81). **P. bicupulàtus** Fernald.

    20. Rhizome pale and soft; submersed leaves long and flaccid, **P. capillàceus** Poir.

    20. Rhizome black and woody; submersed leaves short and firm. **P. capillaceus** var. **átripes** Fernald (*P. capillaceus*). Acid clay pool; s.e. Va.

21. Leaves 15–35-nerved; fruits 3–3.5 mm. wide. **P. zosterifòrmis** Fernald.

21. Leaves 1–9-nerved; fruits 1–3 mm. wide (unknown in *P. longiligulatus*)..**22**

    22. Stipules coarsely fibrous, often whitish....................**23**

    22. Stipules delicate, greenish (sometimes delicately fibrous in *P. obtusifolius*)...**28**

23. Leaves tapering to bristle tips...............................**24**

23. Leaves obtuse or acute at apex but not ending in a bristle.......**25**

    24. Leaves 5–9-nerved; midrib not bordered by loosely cellular tissue; peduncles 1.5–3 cm. long; spike more than 1 cm. long, of several separated whorls of flowers. **P. longiligulàtus** Fernald. This species is unknown with fruit and may be a sterile hybrid.

    24. Leaves 3-nerved; midrib bordered by loosely cellular tissue (Fig. 55); peduncles 0.5–1.5 cm. long; spike very short, of 1 whorl of flowers; fruits usually present. **P. Híllii** Morong.

25. Stipules with margins united into a tube which sheaths the stem (Fig. 43; see remarks under key entry No. 33); leaves 3–7-nerved; fruits 1.2–2 mm. wide, 2–3 mm. long, obscurely keeled................**26**

25. Stipules not united to form a tube; margins may overlap but are not fused (Fig. 53); leaves 3-nerved; fruits 2.2–2.7 mm. wide, 3.4–4 mm. long, prominently keeled. **P. Pòrteri** Fernald.

    26. Leaves 5–7-nerved, 1.5–3.5 mm. wide, apex obtuse; peduncles flattened. **P. Frièsii** Rupr.

    26. Leaves 3 (–5)-nerved, 0.5–2.5 mm. wide, apex obtuse to acute; peduncles thread-like, enlarged toward the tip..............**27**

27. Leaves obtuse, rounded or abruptly contracted to a small point (Fig. 49). **P. strictifòlius** Benn.

27. Leaves acute, tapering to the apex (Fig. 50). **P. strictifolius** var. **rutiloìdes** Fernald.

    28. Leaves 0.1–0.5 mm. wide, 1-nerved, apex bristle-tipped......**29**

    28. Leaves 0.3–4 mm. wide, 1–5 nerved, apex obtuse to acute but not bristle-tipped..**30**

29. Rootstock long and creeping; peduncles terminal, 1.5–24 cm. long (Fig. 37); fruits 1.7–2.8 mm. wide, keels obscure. **P. confervoìdes** Reichenb.

29. Stems from a winter bud; peduncles mostly lateral, 1–3.5 cm. long; fruits about 1.5 mm. wide, keel low but prominent. **P. gemmíparus** Robbins.

    30. Fruits with a thin toothed keel (Fig. 42); stipules united into a tube which sheaths the stem (Fig. 43; see remarks under key entry No. 33); stems usually without glands at the nodes **31**

30. Fruits with keels low or obscure; stems often with a pair of translucent glands at the nodes (Fig. 55) **32**

31. Stems simple to branched; leaves 1.4–2.7 mm. wide, 3–5-nerved, the midrib with 1–3 rows of loosely cellular tissue on each side near the base (Fig. 43); winter buds on very short branches (Fig. 41). **P. foliòsus** Raf.

31. Stems bushy-branched; leaves 0.3–1.5 mm. wide, 1–3 nerved, the midrib without loosely cellular tissue or with a single row on each side (Fig. 44); winter buds on long branches. **P. foliosus** var. **macéllus** Fernald.

    32. Body of fruits 2–4 mm. long, keels low and acute; leaves 2–4 mm. wide, apex rounded; stipules not united to form a tube. **P. obtusifòlius** M. & K.

    32. Body of fruits less than 3 mm. long, keels rounded and obscur e; leaves 0.3–3 mm. wide **33**

33. Young stipules with margins united into a tube which sheaths the stem (Fig. 43; this character, which must be resorted to with sterile specimens, is determined by selecting a young branch tip, boiling it in water if necessary to soften, and with a sharp blade cutting several wafers where the internodes are short and the stipules overlap. The stipule sections are then sorted out and examined, at 10 to 100 diameters, to determine if any of the sections make complete unbroken links. Care must be taken to distinguish from margins that merely overlap and are stuck together); peduncles 1.5–8 cm. long, spikes 0.6–1.2 cm. long, of 3–5 separated whorls; leaves 1–3 mm. wide **34**

33. Stipules not united to form a tube, margins may overlap but are not fused (Fig. 53); peduncles rarely more than 3 cm. long; spikes 0.2–0.8 cm. long, of 1–3 contiguous whorls; leaves 0.3–2.4 mm. wide **35**

    34. Main leaves 1–3 mm. wide. **P. pusíllus** L. (*P. panormitanus*).

    34. Main leaves 0.3–1 mm. wide. **P. pusíllus** var. **mìnor** (Biv.) Fernald (*P. panormitanus* var. *minor*).

35. Apex of leaves mostly obtuse (Fig. 64) **36**

35. Apex of leaves mostly acute (Figs. 59, 60) **37**

    36. Mature leaves 3–7 cm. long. **P. Berchtòldii** Fieber (*P. pusillus* var. *mucronatus*).

    36. Mature leaves 0.8–2.5 cm. long. **P. Berchtoldii** var. **polyphýllus** (Morong) Fernald (*P. pusillus* var. *polyphyllus*).

37. Mature leaves with 1 or 2 rows of loose cellular material each side of midrib (Fig. 60) **38**

37. Mature leaves with 3–5 rows of loose cellular material each side of midrib (Fig. 59). **P. Berchtoldii** var. **lacunàtus** (Hagstr.) Fernald (*P. pusillus* var. *lacunatus*).

    38. Mature leaves 0.5–1.5 mm. wide, with 2 rows of lacunae each side of midrib, at least in the lower half of the leaf. **P. Berchtoldii** var. **acuminàtus** Fieber (*P. pusillus* var. *typicus*).

    38. Mature leaves 0.3–1 mm. wide, with 1 row of lacunae each side of midrib. **P. Berchtoldii** var. **tenuíssimus** (M. & K.) Fernald (*P. pusillus* var. *tenuissimus*).

39. Stem flattened; leaves all submersed, margins toothed; stipules slightly fused to base of leaf; fruits with beak 2–3 mm. long; winter buds hard. **P. críspus** L.

39. Stem round in cross-section; submersed leaves not toothed; fruits with beak not more than 1 mm. long; winter buds rare, not hard......**40**

   40. Submersed leaves petioled or tapering to a nonpetioled base, scarcely clasping; floating leaves present or absent...........**41**

   40. Leaves all submersed, without petiole, heart-shaped or rounded at base and clasping the stem (Figs. 10, 11)................**47**

41. Submersed leaves without petiole; floating leaves (usually absent) delicate, blade tapering without sharp distinction into the petiole; fruit wall hard and smooth...................................**42**

41. Submersed leaves with or without petiole; floating leaves leathery, blade distinct from petiole; fruit wall soft and spongy...........**43**

   42. Submersed leaves usually more than 8 times as long as broad, tapering to the acutish apex. **P. alpìnus** Balbis var. **tenuifòlius** (Raf.) Ogden (*P. tenuifolius*).

   42. Submersed leaves usually less than 8 times as long as broad, apex rounded. **P. alpinus** var. **subellípticus** (Fernald) Ogden (*P. tenuifolius* var. *subellipticus*).

43. Submersed leaves with 11–37 nerves; floating leaves heart-shaped to rounded at base, nerves mostly more than 21...................**44**

43. Submersed leaves with 3–29 nerves; floating leaves wedge-shaped to rounded at base, nerves mostly less than 21...................**45**

   44. Stem not conspicuously black-spotted; submersed leaves usually curved, 2.5–7.5 cm. wide; floating leaves rounded at base, mostly with more than 30 nerves; fruits wedge-shaped at base, 3.5–4.5 (–5) mm. long. **P. amplifòlius** Tuckerm.

   44. Stem usually conspicuously black-spotted; submersed leaves not curved, 1–2.5 cm. wide; floating leaves mostly heart-shaped at base, mostly with less than 30 nerves; fruits rounded or lobed at base, 3–3.5 (–4) mm. long. **P. púlcher** Tuckerm.

45. Submersed leaves with petioles 2–13 cm. long (Fig. 13), apex acutish but not sharp-pointed; fruits 3.5–4 mm. long, usually reddish, keels mostly rough with short hard points. **P. nodòsus** Poir. (*P. americanus* and var. *ncvaebcracensis*).

45. Submersed leaves nonpetioled or with petioles up to 4 cm. long, apex acutish or sharp-pointed, sometimes rounded but with a short sharp point at the tip; fruits 1.7–3.5 mm. long, usually greenish, keels evident but scarcely rough..................................**46**

   46. Stem simple or once-branched (Fig. 12), (1–) 1.5–5 mm. in diameter; submersed leaves 1.5–4 cm. wide, mostly 3–5 times as long as wide, with or without petiole, nerves mostly 9–17; floating leaves 4–13 cm. long, 2–6.5 cm. wide, petioles mostly shorter than the blades; fruiting spikes (2.5–) 3–6 cm. long; fruits (2.5–) 2.7–3.5 mm. long, (2.1–) 2.2–3 mm. wide. **P. illinoénsis** Morong. (including also *P. lucens* and *P. angustifolius*).

   46. Stems usually much branched (Fig. 21), 0.5–1 mm. in diameter; submersed leaves 0.2–1 (–1.5) cm. wide, 7–12 (–30) times as long as wide, without petiole, nerves mostly 3–9; floating leaves 1.5–5 (–7) cm. long, 1–2 (–3) cm. wide, petioles mostly longer than the blades, fruiting spikes 1–2.5 cm. long; fruits 1.7–2.5 (–2.8) mm. long, 1.6–2 (–2.3) mm. wide. **P. gramíneus** L. A variable species that hybridizes freely with many other species, especially *P.*

illinoensis and *P. perfoliatus*. The following poorly marked varieties may be recognized.....................................**47**

47. Principal submersed leaves narrowly elliptical, sides not parallel, 5–10 times as long as wide, or if more than 10 times then not less than 6 cm. long, with (3–) 5–9 nerves........................**48**

47. Principal submersed leaves linear, sides essentially parallel for most of their length, 10–20 (–30) times as long as wide, less than 5 cm. long, with 3 nerves. **P. gramineus** var. **myriophýllus** Robbins.

    48. Principal submersed leaves 1.5–4.5 (–6.5) cm. long, with 5–7 nerves. **P. gramineus** L. This includes all varieties and forms of *P. gramineus* as they are treated on pages 57, 59, 60, and 61.

    48. Principal submersed leaves 5–9 (–13) cm. long, with 7–9 (–11) nerves. **P. gramineus** var. **máximus** Morong ex Benn.

49. Rhizomes spotted with rusty red; leaves ovate-oblong, (5–) 10–20 (–25) cm. long, apex often boat-shaped (Fig. 8); stipules usually persistent and conspicuous (Fig. 10); peduncles (5–) 15–60 cm. long; fruits more than 4 mm. long and more than 3 mm. wide, the keel on the back strongly developed. **P. praelóngus** Wulf.

49. Rhizomes unspotted; leaves roundish, ovate or elongate-ovate, 1–10 cm. long, apex not boat-shaped; stipules usually inconspicuous or disintegrated to fibers; peduncles 1–25 cm. long; fruits less than 3.5 mm. long and less than 3 mm. wide, the keel on the back weakly developed or none..**50**

    50. Leaves elongate-ovate, 3–10 cm. long, coarsely nerved; stipules coarse, disintegrating to persistent whitish fibers; peduncles often thickened upward, 1.5–25 cm. long; fruits (2.5–) 2.7–3.2 (–3.5) mm. long, with a cavity in that part of the wall that is between the two sides of the curved embryo. (To determine this last character, the fruit must be cut through the middle parallel to the curve of the embryo.) **P. Richardsònii** (Benn.) Rydb.

    50. Leaves orbicular to ovate, 1–6 cm. long, delicately nerved; stipules delicate, disappearing with age; peduncles of same thickness throughout the length, 1–9 cm. long; fruits (2.3–) 2.5–2.7 (–3) mm. long, without a cavity in the wall................**51**

51. Stems 1–2 mm. in diameter; leaves 1.5–3 cm. wide, with 11–21 nerves, 5–7 of them prominent. **P. perfoliàtus** L. S. Labrador, Que., & N.B.

51. Stems 0.4–1.5 mm. in diameter; leaves 0.5–2 cm. wide, with 7–17 nerves, 1–5 of them prominent. **P. perfoliatus** var. **bupleuroìdes** (Fernald) Farwell. (*P. bupleuroides*). Brackish or fresh water, Nfd. to Fla., w. to Ont., Ohio, & La.

P. 57   **P. illinoensis** Morong includes also **P. lucens** and **P. angusti-folius.**

    P. nodòsus Poir. for **P. americanus** and var. **novaeboracensis.**

    P. gramineus L. includes also **P. gramineus** var. **graminifolius** forma **terrestris.**

P. 58   Fig. 12 is **P. illinoensis.**

P. 59   Fig. 13 is **P. nodosus.** Figs. 14 and 15 are **P. gramineus.** Fig. 17 may be **P. gramineus** var. **maximus.** Fig. 18 may be **P. illinoensis.**

P. 60   Figs. 19–21 are **P. gramineus.**

P. 61   Fig. 22 is **P. alpinus** var. **tenuifolius.** Fig. 23 is **P. alpinus** var. **subellipticus.**

P. gramineus L. includes **P. gramineus** var. **spathulaeformis, P. gramineus** var. **graminifolius** forma **longipedunculatus,** and **P. gramineus** var. **graminifolius.**

P. gramineus var. **máximus** Morong ex Benn. for **P. gramineus** var. **graminifolius** forma **maximus.**

P. gramineus var. **myriophy̆llus** Robbins for **P. gramineus** var. **graminifolius** forma **myriophyllus.**

P. alpìnus Balbis var. **tenuifòlius** (Fernald) Ogden for **P. tenuifolius.**

P. alpinus var. **subellípticus** (Fernald) Ogden for **P. tenuifolius** var. **subellipticus.**

P. 63   **P. epihydrus** var. **ramòsus** (Peck) House for **P. epihydrus** var. **Nuttállii.**

P. 64   Fig. 36 is **P. epihydrus** var. **ramosus.**

P. 67   Line 1: 1.5 for 15.

P. pusíllus L. for **P. panormitanus.**

P. pusillus var. **mìnor** (Biv.) Fernald for **P. panormitanus** var. **minor.**

P. 68   Figs. 46 and 47 are **P. pusillus.**

P. 70   Fig. 59 is **P. Berchtoldii** var. **lacunatus.** Figs. 60 and 63 are **P. Berchtoldii** var. **acuminatus.** Figs. 61 and 64 are **P. Berchtoldii.** Fig. 62 is **P. Berchtoldii** var. **tenuissimus.**

P. 71   **P. Berchtòldii** Fieber for **P. pusillus.**

P. Berchtoldii var. **lacunàtus** (Hagstr.) Fernald for **P. pusillus** var. **lacunatus.**

P. Berchtoldii var. **acuminàtus** Fieber for **P. pusillus** var. **typicus.**

P. Berchtoldii var. **tenuissímus** (M. & K.) Fernald for **P. pusillus** var. **tenuissimus.**

P. Berchtoldii for **P. pusillus** var. **mucronatus.**

P. Berchtoldii var. **polyphy̆llus** (Morong) Fernald for **P. pusillus** var. **polyphyllus.**

P. 75   Add to **References:** Fernald, Rhodora, **38,** 167 (1936)—*P. tennesseensis;* Fernald, Rhodora, **39,** 380 (1937)—*P. capillaceus* var. *atripes;* Ogden, Rhodora, **45,** 57–105, 119–163, 171–214 (1943)—broad-leaved species; Ogden, Rhodora, **49,** 255 (1947)—*P. tennesseensis;* Martin, Journ. Wildlife Management, **15,** 253–258 (1951)—identification of pondweed fruits eaten by ducks; Ogden, N.Y. State Museum, Circular, No. 31 (1953)—key to species in North America, especially designed for use with fragmentary specimens.

P. 77   Par. 2, line 3: *Elodea* for *Anacharis*.

     Add: **Najas Muénscheri** Clausen. Freshwater estuaries, e. N.Y. and e. Va. Similar to *N. guadalupensis* and *N. flexilis* but may be distinguished from both by the slender seeds having 50–60 rows of deep pits and with a distinct ridge along one side of the seed.

     N. guadalupénsis (Spreng.) Magnus for **N. guadalupensis** (Spreng.) Morong.

     N. gracíllima (A. Br.) Magnus for **N. gracillima** (A. Br.) Morong.

     Add to **References:** Clausen, Rhodora, **39**, 57–60 (1937)— *Najas Muenscheri;* Setchell, Proc. Calif. Acad., IV, 25, 469–478 (1946)—*Ruppia.*

P. 79   Line 6: p. 9 for p. 7.

P. 92   Figs. 53–55 are **E. radicans.** Figs. 56 and 57 are **E. tenellus** var. **parvulus.**

P. 93   Replacement key for **Echinódorus:**

1. Slender plants rarely 10 cm. high; carpels 20 or fewer in a loose head; stamens 6 or 9; anthers basifixed; nutlets beakless (Fig. 56) **2**
1. Robust plants with ovate or cordate leaves (Fig. 53); carpels many in dense heads; stamens 9–30; anthers attached at the middle and moving freely; nutlets beaked (Fig. 55) **3**
   2. Leaf blades narrowed from the middle to an acute or obtuse base and tip (Fig. 57), petioled. **E. tenéllus** (Martius) Buchenau var. **párvulus** (Engelm.) Fassett. Mass. to Cuba, w. to Mo., Tex., & Mex.
   2. Plants submersed; leaf blades linear, not petioled. **E. tenellus** forma **Rándii** Fassett. Mass. & probably elsewhere.
3. Sepals with smooth veins; beak of nutlets 0.5–1 mm. long; each side of nutlet with 5 arching ribs; glands of nutlet narrowed at the upper end and entering the base of the beak; translucent lines of leaves mostly less than 1 mm. apart and often several mm. long; scape erect. **E. Berteròi** (Sprengel) Fassett var. **lanceolàtus** (Engelm.) Fassett. Ohio to S.D., s. to Tex.; Calif.
3. Sepals with bumpy ridges; beak of nutlets 0.2–0.8 mm. long; each side of nutlet with 3–4 abruptly curved and sometimes joining ribs; glands of nutlet rounded at both ends and not closely approaching the beak; translucent lines of leaves mostly 1 mm. or more apart and rarely exceeding 1 mm. in length; scape erect when young but soon procumbent. **E. cordifòlius** (L.) Griseb. Figs. 53–55. D.C. to n. Fla., w. to Mo., Kans., & Tex.

     A. triviàle Pursh for **A. Plantago-aquatica.**

     A. subcordàtum Raf. for **A. Plantago-aquatica** var. **parviflorum.**

     Add to **References:** Fassett, Rhodora, 57, 133–156, 174–188, 202–212 (1955)—*Echinodorus* in the western hemisphere.

     Add to range of **B. umbellatus:** rapidly spreading in Que., Ont., N.Y., Pa., & Ohio.

     Add **Reference** under **Butomus:** Gaiser, Rhodora, **51**, 385-

390 (1949)—distribution of *Butomus* in the Great Lakes region.

P. 94   Figs. 58–60 and 62 are **A. triviale**; Fig. 61 is **A. subcordatum**.

P. 97   **Elodea** for **Anacharis**.

Figs. 70–73 are **E. canadensis**. Fig. 74 is **E. Nuttallii**.

P. 98   **Elodea** for **Anacharis**.

P. 99   **Elodèa** for **Anacharis**.

E. canadensis Michx. for **A. canadensis**.

E. Nuttállii (Planch.) St. John for **A. occidentalis**.

E. densa (Planch.) Caspary for **A. densa**.

L. Spongia (Bosc) Steud. for **L. Spongia** (Bosc) Richard.

At bottom of page add:

Frogbit   Hydrócharis

H. mòrsus-ránae L. Small floating plants similar to *Limnobium* but with the petals much wider and longer than the sepals. Introduced from Europe, persisting and spreading in the Ottawa River near Ottawa and Montreal. See Dore, Can. Field Nat. **68**, 180 (1954).

P. 100   Manual of the Grasses of the United States, by A. S. Hitchcock, was revised by Agnes Chase in 1950. This second edition may be had from the Superintendent of Documents, Washington, D. C., for $3.00.

P. 104   Fig. 21 is **G. Fernaldii**.

P. 105   **G. Fernáldii** (Hitch.) St. John for **G. neogaea**.

P. 106   Figs. 22 and 23 are **P. communis** var. **Berlandieri**.

P. 107   **P. commùnis** Trin. var. **Berlandièri** (Fourn.) Fernald for **P. maximus** var. **Berlandieri**.

E. pungens var. **Wiegándii** Fassett for **E. pungens** var. **occidentalis**.

Add to **References** at bottom of page: Fassett, Rhodora, **51**, 1–3 (1949).

P. 108   Fig. 28 is **E. pungens** var. **Wiegandii**.

P. 111   **C. neglécta** (Ehrh.) Gaertn., Mey. & Schreb. for **C. neglecta** (Ehrh.) Gaertn.

Add to characters separating **C. canadensis** from **C. inexpansa** and **C. neglecta**:

1. leaves flat, 4–8 mm. wide; lemma thin and translucent in the upper part.

1. leaves mostly inrolled, 1–4 mm. wide; lemma firm and opaque.

P. 114   Figs. 51–54 are **A. carolinianus**.

P. 115   **A. caroliniànus** Walt. for **A. geniculatus** var. **ramosus**.

A. álba L. for **A. stolonifera** var. **major**.

A. alba var. palústris (Huds.) Pers. for **A. stolonifera** var. **compacta.**

P. 116   Figs. 58 and 59 are **A. alba.**

P. 117   Figs 60–63 are **Z. miliacea.**

P. 118   Figs. 64–66 are **Scolochloa festucacea.**

P. 120   Figs. 74–76 are **B. syzigachne.**

P. 121   Scolochlòa festucàcea (Willd.)Link for **Fluminea festucacea.**
B. syzigáchne for **B. Syzigachne.**

P. 122   Rhynchospora for **Rynchospora.**

P. 123   **C. vìrens** Michx. for **C. pseudovegetus.**

C. polystáchyos Rottb. var. **texénsis** (Torr.) Fernald for **C. filicinus** var. **microdontus.**

P. 126   Fig. 15 is **C. virens.**

P. 127   Fig. 16 is **C. polystachyos** var. **texensis.**

P. 129   **C. odoràtus** L. for **C. ferax.**

Add **E. quadrangulata** var. **crássior** Fernald, which may be distinguished from **E. quadrangulata** by having the free tips of the leaf sheaths 1.5–8 cm. long rather than 0.5–1 cm.

P. 131   **E. macrostáchya** Britt. for **E. mamillata.**

P. 134   Fig. 61 is **E. geniculata.** Fig. 64 is **E. Engelmanni** forma **detonsa.**

P. 135   **E. aciculàris** (L.) R. & S. for **E. acicularis** R. & S.

E. acicularis var. **submérsa** (Nilss.) Svenson for **E. acicularis** forma **inundata.**

E. geniculàta (L.) R. & S. for **E. caribaea** and var. **dispar.**

P. 137   Several named varieties of **E. obtusa** and **E. ovata** have not been differentiated here. See **References,** especially article by Svenson.

E. Engelmanni forma **detónsa** (Gray) Svenson for **E. Engelmanni** var. **detonsa.**

E. Wólfii for **E. Wolffii.**

E. intermedia forma **Haberèri** Fernald for **E. intermedia** var. **Habereri.**

Add to **References:** Moore, Rhodora, **52,** 54 (1950)—*E. ovata* var. *aphanactis;* Svenson, Rhodora, **55,** 1–6 (1953)—*E. obtusa, ovata, diandra,* and *Engelmanni.*

P. 140   Figs. 94–96 are **F. Drummondii** or **F. caroliniana,** which are not distinguished by spikelet and nutlet characters.

P. 141   **F. Drummóndii** Boeckler and **F. caroliniàna** (Lam.) Fernald for **F. puberula.** They may be separated by:

1. Stem bases enlarged and bulbous and when fully developed making dense tussocks; stolons absent; inland plant of acid habitats; flowering

May to July. **F. Drummóndii.** L.I. to Fla., w. to Ont., Mich., Ill., & Mo.

1. Stem bases not bulbous, in small tufts; spreading by slender stolons; Coastal plant of brackish habitats; flowering July to Oct. **F. caroliniàna.** N.J. to Fla. & Tex.

F. caroliniana forma **eucýcla** Fernald for **F. puberula** forma **eucycla.**

F. caroliniana forma **pycnostáchya** Fernald for **F. puberula** forma **pycnostachya.**

Add: **F. autumnalis** forma **brachyáctis** (Fernald) Blake which has the branches of the inflorescence very short.

Add to **Reference:** Fernald, Rhodora, **47,** 113–116 (1945)— *F. Drummondii.*

P. 143 First **4,** line 2: (Fig. 7, p. 5) for (Fig. 7, p. 2).

S. Purshiànus Fernald for **S. debilis.**

S. Purshianus forma **Williámsii** Fernald for **S. debilis** var. **Williamsii.**

S. Smithii forma **setòsus** Fernald for **S. Smithii** var. **setosus.**

P. 144 Figs. 109 and 112 are **S. Purshianus.** Figs. 110 and 114 are **S. validus** var. **creber.**

P. 145 Fig. 117 is **S. rubrotinctus.** Figs. 118 and 119 are **S. expansus.**

P. 147 **S. validus** var. **crèber** Fernald for **S. validus.**

Add: **S. validus** var. **creber** forma **megastáchyus** Fernald which has large spikes, 9–15 mm. long.

Add: **S. Steinmétzii** Fernald which is similar to **S. validus.** They may be separated by the following key:

1. Some spikelets in groups of 2 or more because of very short peduncles; scales scarcely longer than the mature fruits; bristles 4–6. **S. válidus** var. **crèber.**
1. All spikelets with long peduncles; scales much longer than mature fruits; bristles 0–2. **S. Steinmétzii.** Local, central Maine.

S. rubrotínctus Fernald for **S. microcarpus** var. **rubrotinctus.**

S. expánsus Fernald for **S. sylvaticus.**

S. expansus forma **Bisséllii** Fernald for **S. sylvaticus** var. **Bissellii.**

S. atróvirens Willd. for **S. atrovirens** Muhl.

S. polyphyllus forma **macróstachys** (Boeckl.) Fernald for **S. polyphyllus** var. **macrostachys.**

P. 149 **S. cyperinus** forma **Andréwsii** (Fernald) Carpenter for **S. cyperinus** var. **Andrewsii.**

S. rubricòsus Fernald for **S. Eriophorum.**

Add to **References:** Beetle, Am. Journ. Bot., **27,** 63–64 (1940); **28,** 469–476, 691–700 (1941); **29,** 82–88, 653–656 (1942); **30,** 395–401 (1943); **31,** 261–265 (1944)—keys and descriptions of

Scirpus spp.; Beetle, Am. Midl. Nat., **29**, 533–538 (1943)—key to species based on fruit characters; Fernald, Rhodora, **45**, 279–296 (1943)—descriptions of several new species, varieties and forms. Many named forms have not been included in this appendix.

P. 150   Rhynchospora for Rynchospora.

P. 151   Rhynchóspora for Rynchospora.

Add: **R. macrostáchya** var. **colpóphila** Fernald & Gale.

1. Plants mostly 0.4–1.2 m. tall; leaves firm; fruits 2.6–3.1 mm. wide; tubercle 1–1.8 mm. wide at base. **R. macrostáchya.**
1. Plants mostly 0.8–1.8 m. tall; leaves long and rather flaccid; fruits 3–3.8 mm. wide; tubercle 1.8–2.4 mm. wide at base. **R. macrostachya** var. **colpóphila.** Fresh tidal marshes; Md. to Va.

Add: **R. globulàris** (Chapman) Small and var. **recógnita** Gale for **R. cymosa.**

1. Leaves 1.5–2 mm. wide; branchlets of the inflorescence terminating in small clusters of spikelets; bracts scarcely evident; spikelets 2.5–3 mm. long; fruits 1.2–1.3 mm. long, 1–1.2 mm. wide. **R. globulàris.** Calif., Tex., & Fla., n. to Del.
1. Leaves 2–4 mm. wide; branchlets of the inflorescence terminating in large clusters of spikelets; bracts evident; spikelets 3–4 mm. long; fruits 1.3–1.6 mm. long, 1.2–1.5 mm. wide. **R. globularis** var. **recógnita.** Tex. to Fla. & n. to Ill. & Pa.

P. 152   Rhynchospora for Rynchospora.

Figs. 143 and 144 are **R. globularis** var. **recognita.** Figs. 146 and 147 are **R. capitellata.**

P. 153   Rhynchospora for Rynchospora.

P. 155   **R. capitellàta** (Michx.) Vahl for **R. glomerata** var. **minor.**

R. capitellata forma **controvérsa** (Blake) Gale for **R. glomerata** var. **minor** forma **controversa.**

R. capitellata forma **discùtiens** (Clarke) Gale for **R. glomerata** var. **minor** forma **discutiens.**

Delete: **R. gracilenta** var. **diversifolia,** which apparently is not distinguishable from typical **R. gracilenta.**

Add: **R. chalarocéphala** Fernald & Gale, which is similar to **R. cephalantha.**

1. Inflorescence of 2–6 dense, globular clusters of spikelets; spikelets ascending to reflexed; fruits 1.8–2.5 mm. long, 1.1–1.6 mm. wide. **R. cephalántha.**
1. Inflorescence of 3–7 loosely hemispherical clusters of spikelets; spikelets ascending to divergent; fruits 1.4–1.7 mm. long, 0.9–1 mm. wide. **R. chalarocéphala.** Swamps, lake margins, and ditches; N.J. to Fla.

Add: **R. filifòlia** Gray which is similar to **R. fusca.**

1. Stems solitary or few; spikelets 2–3-flowered, 4.5–7 mm. long; fruits

1.3–1.7 mm. long, tubercle 1–1.3 mm. long. **R. fúsca.**

1. Stems many, close together; spikelets 3–6-flowered, 3–4 mm. long; fruits 1–1.3 mm. long, tubercle 0.4–0.6 mm. long. **R. filifòlia.** Lake margins and damp pockets in sand; N.J. & Del., s. to Fla., Cuba, & Tex.

Add to **References:** Fernald, Rhodora, **42,** 421 (1940)—*R. macrostachya* var. *colpophila;* Gale, Rhodora, **46,** 89–134, 159–197, 207–249, 255–278 (1944)—all species of *Rhynchospora* in our area except *R. macrostachya, R. inundata,* and *R. corniculata.*

P. 158   **C. hystricina** for **C. hystericina.**

P. 159   **C. hystricìna** for **C. hystericina.**

        C. aquatilis var. **áltior** (Rydb.) Fernald for **C. aquatilis** var. **subtricta.**

P. 160   Figs. 185 and 186 are **C. aquatilis** var. **altior.**

P. 168   Add: **L. mínima** Philippi which is similar to **L. minor.**

1. Joints 3-nerved, rounded on both sides. **L. mìnor.**

1. Joints 1-nerved, flat on the lower side. **L. mínima.** Ohio to Calif. & s. into S. Am.

P. 169   Line 2: "tapered from the flattened base to the slender tip" for "taper-pointed, round in cross section."

        Add: **E. decangulàre** L. and **E. compréssum** Lam.

1. Leaves 3–9-nerved; scapes 4–7-angled; heads 2.5–10 mm. wide. **E· septangulàre.**

1. Leaves with many fine nerves; scapes 10–12-angled; heads 5–15 mm. wide **2**

    2. Leaves obtuse; bractlets of the receptacle pale, acute and longer than the flowers; bractlets between the flowers acute. **E. decangulàre.** Shallow pools and wet depressions, along the coast from N.J. to Tex.

    2. Leaves sharp-pointed; bractlets of the receptacle dark, obtuse and about as long as the flowers; bractlets between the flowers obtuse. **E. compréssum.** Shallow pools and wet depressions, along the coast from N.J. to Tex.

P. 173   Add: **P. cordata** var. **lanceolàta** (Nutt.) Griseb.

1. Mature perianth essentially without hairs, very rarely glandular; mature fruits 6–10 mm. long. **P. cordàta.**

1. Mature perianth with short glandular hairs; mature fruits 5–6 mm. long. **P. cordata** var. **lanceolàta.** Del. to Fla., W.I., Tex., & S. Am.

Key to **Heteranthera,** first **2,** line 3: "Maine" for "w. N. E."

P. 174   First and second **6:** delete characters referring to leaves.

P. 177   Figs. 11 and 12 are **J. tenuis.**

P. 179   **J. ténuis** Willd. for **J. macer.**

P. 181   **I. brevicaùlis** Raf. for **I. foliosa.**

        Line 3 from bottom of page: "long" for "slong."

P. 182   S. intèrior Rowlee for S. longifolia.

S. interior var. Wheèleri Rowlee for S. longifolia forma Wheeleri.

P. 183   Fig. 1 is S. caroliniana. Figs. 2 and 5 are S. interior. Fig. 3 is S. interior var. Wheeleri.

P. 185   S. caroliniàna Michx. for S. longipes.

S. eriocéphala Michx. for S. missouriensis.

P. 187   Figs. 12–14 are S. rigida.

P. 189   S. rígida Muhl. for S. cordata.

×S. myricoìdes (Muhl.) Carey for S. cordata var. myricoides. This is a hybrid of S. rigida and S. sericea.

Add: S. pedicellaris var. hypoglaùca Fernald which has leaves distinctly whitened on the under side. This variety is found in the same habitats and over the same range as var. pedicellaris.

P. 190   Fig. 26 is S. rigida. Fig. 33 is S. interior.

P. 191   S. caroliniàna for S. longipes.

S. intèrior for S. longifolia.

S. rígida for S. cordata.

S. eriocéphla for S. missouriensis.

P. 192   Fig. 36 is S. interior. Fig. 40 is S. caroliniana. Figs. 43 and 48 are S. rigida.

P. 193   S. caroliniàna for S. longipes.

S. intèrior for S. longifolia.

S. eriocéphala for S. missouriensis.

S. rígida for S. cordata.

P. 195   Figs. 54, 56 and 57 are A. rugosa. Fig. 55 is A. glutinosa.

P. 196   Figs. 58 and 60 are A. serrulata.

P. 197   A. glutinòsa (L.) Gaertn. for A. vulgaris.

A. rugòsa (Du Roi) Spreng. for A. incana.

A. serrulàta (Ait.) Willd. for A. rugosa.

P. 198   R. orbiculàtus Gray for R. Brittanica.

P. 199   Figs. 1 and 2 are R. orbiculatus.

P. 200   Fig. 8 is P. amphibium var. stipulaceum. Fig. 10 is P. amphibium.

P. 201   Add to second 1: (P. punctatum and P. Hydropiper may sometimes key here but both may be distinguished from annual species by having sepals with dark glands).

P. amphíbium L. for P. natans forma genuinum.

P. amphibium var. stipulàceum (Coleman) Fernald for P. natans forma Hartwrightii.

Delete: P. coccineum forma terrestre. Terrestrial forms occur in many aquatic plants and usually are scarcely worthy of recognition. If this form is to be named, it should be forma terréstris.

P. 205   **P. pensylvanicum** for **P. pensylvanicum** var. **genuinum.** A white-flowered variant is forma **albìnum** Fernald.

Add: **P. pensylvanicum** var. **eglandulòsum** Myers which lacks hairs. Local, Ohio.

Delete: **P. Hydropiper** var. **projectum** as not being separable from **P. Hydropiper.**

P. 207   Fig. 30 is **P. Hydropiper.**

P. 209   For **P. punctatum** and var. **robustius** use the following:

1. Fruiting calyx 3–4 (–5) mm. long.............................**2**
1. Fruiting calyx 1.6–2.5 (–3) mm. long.........................**7**
  2. Stems prostrate at base, rooting at the nodes; nutlets all triangular in cross section..**3**
  2. Stems from a short taproot, scarcely prostrate at base and rooting at few or no nodes; nutlets usually both triangular and lens-shaped on the same plant..**4**
3. Stems, above the rooting base, much branched; branches of the inflorescence from the axil of a blade-bearing leaf; leaves 7–13 cm. long, 1–2 cm. wide; fruiting calyx 3–3.5 mm. long. **P. punctàtum** Elliott. N.Y. to Iowa & Kans., s. to Fla. & Tex.
3. Stems, above the rooting base, simple or nearly so; branches of the inflorescence from the axil of a bladeless bract; leaves 10–17 cm. long, 2–4 cm. wide; fruiting calyx 3.5–5 mm. long. **P. punctatum** var. **màjus** (Meisn.) Fassett. N. S. to Fla., locally inland to Ind. & Mo.; W.I.; Guatemala to n. S. A.
  4. Fruiting calyx 3–4.5 mm. long; usually more than half of the nutlets lens-shaped in cross section.................................**5**
  4. Fruiting calyx 2.4–3.4 mm. long; all or nearly all nutlets triangular in cross-section. **P. punctatum** var. **littoràle** Fassett. Sandy lake shores of n.w. Wis.
5. Calyx with 50 or more well-developed glands.....................**6**
5. Calyx with 30 or fewer obscure glands. **P. punctatum** var. **pàrvum** Victorin & Rousseau. Que. to Md.
  6. Stalked base of calyx 0.75 mm. or less long. **P. punctatum** var. **confertiflòrum** (Meisn.) Fassett. Gulf of St. Lawrence to Ga., w. to s. B.C. & Okla.
  6. Stipitate base of calyx 1 mm. or more long. **P. punctatum** var. **confertiflorum** forma **longicóllum** Fassett. Local; Wis.
7. Stem from a short taproot, rooting at few nodes or none; nutlets all or nearly all triangular in cross-section. **P. punctatum** var. **littoràle.**
7. Stem prostrate and rooting at the nodes; nutlets both triangular and lens-shaped in cross-section. **P. punctatum** var. **parviflòrum** Fassett. S.e. U.S.; rare in Del., N.J., & Colo.

Add to **References:** Fassett, Brittonia, **6,** 369–393 (1949)—*P. punctatum* and its varieties and forms.

P. 210   Add: **C. echinàtum** Gray which differs from **C. demersum** as follows.

  1. Nutlets with 2 spines near the base; leaf segments strongly toothed on one side. **C. demérsum.** Que. to B.C. & s. throughout N.A.

1. Nutlets with several spines on the sides; leaf segments without teeth or essentially so. **C. echinàtum.** N.B. to Fla., w. to Minn. & Tex.

Add **Reference:** Muenscher, Am. Journ. Bot., **27,** 231–233 (1940)—*Ceratophyllum*; Fassett, Communicaciones del Institute Tropical de Investigaciones Cientificas, Año **2,** No. 2, 25–45 (1953)—*Ceratophyllum*.

Add **N. odorata** var. **gigantèa** Tricker. A larger variety having leaves with upturned margins.

P. 213   Dr. E. O. Beal considers **Nuphar** in our area to be subspecific variants of the Eurasian **N. luteum** (L.) Sibth. & Sm.

P. 217   Delete: **Nuphar advena** var. **brevifolium.** Known from a single collection, this plant differs from the variable **N. advena** only in size.

Add to **References:** Fassett, Castanea, **18,** 116–128 (1953)—*Cabomba*; Beal, Doctorate Thesis, State Univ. of Iowa (1955)—*Nuphar*.

P. 220   Figs. 15–17 are **R. Gmelini** var. **Hookeri.**

P. 221   **R. Gmélini** DC. var. **Hoòkeri** (D. Don) Benson for **R. Purshii** and **R. Purshii** forma **terrestris.**

R. Gmelini var. **prolíficus** (Fernald) Hara for **R. Purshii** var. **prolificus.**

P. 227   Delete: **C. palustris** var. **flabellifolia** and **C. palustris** var. **radicans.** These varieties appear to be but inconstant phases, at least in our area.

Add to **References:** Benson, Am. Midl. Nat., **40,** 1–261 (1948).

P. 228   **Armoracia** for **Neobeckia.**

P. 229   Figs. 1 and 3 are **R. islandica** var. **hispida.** Fig. 2 is **R. islandica.** Fig. 4 is **R. islandica** var. **Fernaldiana.**

P. 230   Figs. 5–10 are **R. islandica** var. **hispida** and var. **Fernaldiana.**

P. 231   **R. islándica** (Oeder ex Murray) Borbás for **R. palustris.**

R. islándica var. **híspida** (Desv.) Butters & Abbe for **R. palustris** var. **hispida** and forma **inundata.**

R. islandica var. **Fernaldiàna** Butters & Abbe for **R. palustris** var. **glabrata** and forma **aquatica.**

P. 233   Fig. 21 is **C. Douglassii.**

P. 234   **Armoracia** for **Neobeckia.**

Figs. 24–30 are **Armoracia aquatica.**

P. 235   **C. Douglássii** (Torr.) Britt. for **C. bulbosa** var. **purpurea.**
Armoràcia for **Neobeckia.**

A. aquática (Eaton) Wiegand for **N. aquatica.**

P. 241   **C. terréstris** Raf. for **C. deflexa** var. **Austini.**

C. vérna L. for **C. palustris.**

Add: **C. ánceps** Fernald which is similar to **C. heterophylla.** The following key is taken from Dr. Fassett's monograph noted below where detailed illustrations, descriptions and range maps are to be found. Technical terms not included in the glossary have been avoided.

1. Plants of various habit and leaf form; leaf bases connected by a narrow membranous wing..2
1. Plants all aquatic with uniform linear to lance-shaped leaves; leaf bases not connected by a wing. **C. hermaphrodítica** L. Greenland, Nfd., N.B., n. Vt., n. N.Y., w. to Colo., Calif., & Alaska.
  2. Fruits 0.6–1 mm. wide; carpels wingless or with wing at summit and narrowed or absent down the sides..........................3
  2. Fruits 1.2–1.7 mm. wide; wing equally broad all around the carpel. **C. stagnàlis** Scop. Probably introduced into e. N.A.; s. Que. to Md.; n. Wis., Wash., & Ore. Several ecological forms have been proposed; a similar organization of forms might be proposed for the other species of **Callitriche.**
3. Plants terrestrial, with essentially uniform ovate leaves but with the broader part above the middle; width of fruit exceeding the height by 0.2 mm.; anthers 0.1–0.2 mm. wide; stigmas 0.2–0.4 mm. long. **C. terréstris** Raf. Mass. to Va., w. to Mo. & s. to Ala.
3. Plants aquatic or subterrestrial, of diverse habit and foliage; width of fruit less than to barely exceeding the height; anthers 0.3–1.5 mm. wide; stigmas 0.7 mm. or more long.................................4
  4. Height of fruit exceeding the width by 0.1 mm. or less, or not at all; carpels wingless or with an obscure false wing; reticulation on fruit wall not running in vertical lines..........................5
  4. Height of fruit exceeding the width by 0.2 mm.; carpels with a wing at summit that narrows and usually disappears down the sides; reticulation on mericarps tending to run in vertical lines. **C. vérna** L. Common in N.A. n. of Mex., except s.e. U.S.
5. Foliage relatively coarse; fruits widest above the middle; linear submersed leaves with tip of vein scarcely projecting beyond the blade. **C. heterophýlla** Pursh. S.w. N.S. to Fla. & W.I., w. to Wash., & s. to S.A.
5. Foliage delicate; fruit of equal width above and below the middle; linear submersed leaves with tip of vein slightly projecting beyond the blade. **C. ánceps** Fernald. Greenland to N.E. & n. N.Y.; Ga., Utah, Wash., & Alaska.

Add to **References:** Fassett, Rhodora, **53**, 137–155, 161–182, 185–194, 210–222 (1951).—*Callitriche.*

P. 242	Figs. 15 and 16 are **C. verna.** Figs. 17 and 20 are **C. terrestris.** Fig. 18 is **C. anceps.**
P. 243	Fig. 23 and possibly Fig. 24 are **H. Moscheutos.**
P. 244	Fig. 28 is **H. ellipticum** f. **submersum.**
P. 245	Add: **Hibíscus Moscheûtos** L. which is similar to and possibly not specifically distinct from **H. palustris.**

1. Branches of exserted style densely covered with hairs; leaves mostly 1.5 times as long as wide; peduncles rarely bearing a leaf; capsule

rounded at apex, blunt or with a short beak; petals pink to purple (white, if albino). **H. palústris** L. Marshes; Mass. to N.C., w. to s. Ont. & Ill.

1. Branches of exserted style without hairs or with a few short ones; leaves mostly 2.5 times as long as wide; peduncles usually bearing a well-developed leaf; capsule tapering to a beak; petals white or creamy-white. **H. Moscheùtos** L.; Marshes; Md. to Fla., w. to Ind.

Delete: **Hibiscus incanus.** Plants in our area referred to this species are **H. Moscheutos** or **H. lasiocarpos.**

Hypericum ellípticum forma **submersum** Fassett for **H. ellipticum** forma **aquaticum.**

P. 247 Fig. 35 is **H. pyramidatum.** Fig. 36 is **H. denticulatum** var. **recognitum.** Fig. 37 is **H. denticulatum.**

P. 248 Fig. 40 is **H. adpressum** f. **spongiosum.**

Last line: "leaves 1–5-nerved" for "sepals 1–5-nerved."

P. 249 **H. pyramidàtum** Ait. for **H. Ascyron.**

H. denticulatum var. **recógnitum** Fernald & Schubert for **H. denticulatum.**

H. denticulatum Walt. for **H. denticulatum** var. **ovalifolium.**

H. adpressum forma **spongiòsum** (Robinson) Fernald for **H. adpressum** var. **spongiosum.**

Add to **References:** Fernald & Schubert, Rhodora, **50,** 208 (1948)—*H. denticulatum* var. *recognitum*; Fernald, Rhodora, **51,** 112 (1949)—*H. adpressum* forma *spongiosum.*

P. 255 **A. tères** Raf. for **A. Koehnei.**

Péplis for **Didiplis.**

Peplis diándra Nutt. for **Didiplis diandra,** including forma **aquatica** and forma **terrestris.**

P. 256 Figs. 14 and 15 are **Peplis diandra.**

P. 259 Delete: **L. arcuata.** Plants in our area are **L. brévipes.**

P. 260 Fig. 26 is **J. repens** var. **glabrescens.**

P. 261 **J. rèpens** L. var. **glabréscens** Kuntze for **J. diffusa.**

Add:

WATER CHESTNUT FAMILY   **Hydrocaryàceae** which includes **Tràpa.**

Heading before last paragraph: "**Proserpinàca**" for "**Prosperinaca.**"

P. 263 ×**P. intermèdia** Mack. for **P. intermedia** Mack.

Line 17: delete "perhaps."

Add:

MARE'S-TAIL FAMILY   **Hippuridàceae** which includes **Hippùris.**

H. vulgaris forma **fluviàtilis** (Cosson & Germain) Glück for **H. vulgaris** forma **fluviatilis** (Hoffm.) Cosson & Germain.

            M. alterniflòrum for **M. alterniflorum** var. **americanum.**

P. 264   Figs. 36 and 38 are **M. spicatum** var. **exalbescens.**

            Fig. 37 is **M. alterniflorum.**

P. 265   Figs. 42–44 are **M. verticillatum** var. **pectinatum.**

P. 268   **M. spicàtum** L. var. **exalbéscens** (Fernald) Jepson for **M. exalbescens.**

            M. verticillàtum L. var. **pectinàtum** Wallr. for **M. verticillatum.**

            Add to **References:** Fassett, Inst. Trop. Investigaciones Cientificas Univ. El Salvador, Año **2,** No. 5–6, 139–162 (1953)— *Proserpinaca*; Patten, Rhodora, **56,** 213–225 (1954)—*M. spicatum.*

P. 274   **B. pusílla** (Nutt). Fernald for **B. erecta.**

            Add **Reference:** Fernald, Rhodora, **44,** 189–191 (1942)—*B. pusilla.*

P. 278   Fig. 3 is **S. parviflorus.**

P. 279   **S. parviflòrus** Raf. for **S. floribundus.**

P. 282   Line 2: "Plant with slender . . . " for "Leaves with slender . . . ."

P. 288   Figs. 17 and 18 are **S. palustris** var. **homotricha.**

P. 289   **S. palústris** var. **homótricha** Fernald for **S. homotricha.**

P. 295   Fig. 47 is **L. amplectens.**

P. 296   Fig. 49 is **L. amplectens.**

P. 297   Fig. 51 is **M. arvensis** var. **villosa.**

P. 299   **L. ampléctens** Raf. for **L. sessilifolius.**

            M. arvensis var. **villòsa** Benth. for **M. arvensis** var. **canadensis.**

            Delete: **M. arvensis** var. **glabrata** as it is scarcely distinguishable from var. **villosa.**

            M. arvensis forma **lanàta** (Piper) Stewart for **M. arvensis** var. **lanata.**

            Add **Reference:** Stewart, Rhodora, **46,** 331–335 (1944)—several forms recognized for *Mentha arvensis.*

P. 302   Figs. 7 and 8 are **V. salina.**

P. 303   **V. Anágallis-aquática** forma **anagalliförmis** (Boreau) Beck for **V. glandifera.**

            V. salina Schur for **V. connata.**

            V. salina forma **laèvipes** (Beck) Fernald for **V. connata** var. **glaberrima.**

P. 305   **M. guttàtus** DC. for **M. guttatus** Fisch.

P. 307   **G. aùrea** Pursh for **G. lutea.**

            G. aurea forma **pusílla** Fassett for **G. lutea** forma **pusilla.**

            B. Monnièri (L.) Pennell for **B. Monnieria.**

P. 308   Fig. 19 is **B. Monnieri.** Fig. 22 is **G. aurea.** Fig. 23 is **G. aurea** f. **pusilla.**

P. 309   Add: **L. Pyxidària** L. which differs from *L. dubia* and *L. anagallidea* in having 4 anthers rather than 2 and with the lower pair of stamens fertile. It is introduced from Eurasia and now reported from N.Y., N.J., Minn., & Wash.

Add to **References**: Fernald, Rhodora, **41**, 564–569 (1939)—*Veronica salina.*

P. 310   Figs. 34 and 37 are **U. vulgaris.** Fig. 44 is **U. juncea.**

P. 311   **U. júncea** Vahl for **U. virgatula.**

P. 312   Fig. 51 is **U. vulgaris.**

P. 313   **U. vulgàris** L. for **U. vulgaris** var. **americana.**

Justícia for **Dianthera.**

J. americàna (L.) Vahl for **D. americana.**

J. ovàta (Walt.) Lindau for **D. ovata.**

P. 314   **Justicia** for **Dianthera.**

Fig. 59 is **J. americana.**

P. 324   Fig. 8 is **B. frondosa** f. **anomala.**

P. 325   **B. frondòsa** forma **anómala** (Porter) Fernald for **B. frondosa** var. **anomala.**

P. 335   To range of **B. connata** var. **submutica** add: & n. Minn.

B. heterodóxa (Fernald) Fernald & St. John for **B. heterodoxa** Fernald & St. John.

P. 337   Delete: Fig. 51 which is not **Boltonia.**

P. 338   Figs. 56 and 57 are **A. junciformis.**

Fig. 60 is **A. simplex** var. **ramosissimus.**

P. 339   Add: **S. graminifolia** var. **màjor** (Michx.) Fernald. Leaves broadly lanceolate, 7–11 times as long as wide, broadly acute or obtuse at tip. Nfd. to Que., w. to B.C. & mts. of N.M.

Replace treatment of **Boltonia** with the following:

Tall plants, 0.2–2.5 m. high, often much-branched above and looking much like an Aster; nutlets flat, with slender bristles.

1. Plants slender with loosely ascending branches; spreading by elongate stolons; leaves submembranaceous; bracts of the involucre linear, 0.5–1 mm. wide; ray flowers lilac or purplish, 0.8–1.5 cm. long; clusters of disk flowers 6–8 mm. wide; nutlets about 2 mm. long, thick-rimmed, bristles 0–0.7 mm. long. **B. asteroìdes** (L.) L'Hér, Fig. 50. Shores, stream banks and thickets; N.Y. to N.C., w. to Ohio.—Var. **glastifòlia** (Hill) Fernald has the lower leaves narrowed to a somewhat petioled base. N.J. to La.

1. Plants coarse with strongly corymbose-paniculate leafy branches, conspicuously corrugated; apparently not stoloniferous; leaves firm; bracts of the involucre oblong to obovate, 0.5–2 mm. wide; ray flowers lilac or white, 1–1.8 cm. long; cluster of disk flowers 7–15 mm. wide; nutlets 2.5–3 mm. long, broad-winged, bristles 0.7–2 mm. long. **B. latisquàma** Gray. Prairies and stream banks; Mo., Kans., & Okla.; escaped from cultivation to N.E.—Var. **decúrrens** (T. & G.)

Fernald & Griscom has the bases of the leaves running as 2 wings on the stem. Ill. & Mo.—Var. **recógnita** Fernald & Griscom has the bracts of the involucre narrowly oblong, 0.5–1.3 mm. wide. Mich. to Ky., w. to N.D., Mo., & Kans.; escaping from cultivation to N.E.—Var. **microcéphala** Fernald & Griscom is similar to var. *recognita* but with smaller heads; the bracts of the involucre are narrow–awl-shaped or lance-shaped, 0.3–0.4 mm. wide; ray flowers 5–8 mm. long, group of disk flowers 5–8 mm. wide; nutlets 1.5–2 mm. long with bristles only 1 mm. long. Wis., Minn., Ill., & Iowa.

B. latisquàma Gray var. **decúrrens** (T. & G.) Fernald & Griscom for **B. asteroides** var. **decurrens.**

Add **Reference:** Fernald, Rhodora, **42,** 482–492 (1940)— *Boltonia* in northeastern N.A.

P. 340   Fig. 66 is **A. simplex** var. **ramosissimus.** Fig. 68 is **A. simplex** var. **interior.**

P. 341   **A. juncifórmis** Rydb. for **A. junceus.**

A. símplex Willd. for **A. paniculatus** var. **simplex.**

A. simplex var. **ramosíssimus** (T. & G.) Cronquist for **A. paniculatus.**

A. simplex var. **intèrior** (Wiegand) Cronquist for **A. interior.**

INDEX

Principal references are in **bold face** figures; references to illustrations are in *italic* figures.

Juncus subtilis, 4, 174, **180**
  tenuis, **179**, 376
  Torreyi, *178*, **180**
Jussiaea, 259, **261**, 349
  decurrens, *260*, **261**
  diffusa, *260*, **261**, 381
  repens var. glabrescens, **381**
Justicia, 363, **383**
  americana, **383**
  ovata, **383**

K

Kentucky Bluegrass, **121**

L

Labiatae, 22, **286**, 347
Lady's Thumb, **205**
Lake Cress, 6, **235**
Lance-leaved Violet, 20, **251**
Large-leaf Pondweed, **57**
Leafy Liverworts, 3, **40**
Leafy Pondweed, **63**, **67**
Leatherleaf, 30, **276**
Lecticula resupinata, **309**
Leersia, 102, **111**, 346
  lenticularis, *109*, **111**
  oryzoides, *109*, **111**, 346
    f. glabra, **111**
    f. inclusa, *109*, **111**
Lemna, 164, **168**, 347, 357
  cyclostasa, **168**
  minima, **376**
  minor, *167*, **168**, 347, 376
  perpusilla, *167*, **168**, 347
  trisulca, 3, *167*, **168**, 348
  valdiviana, *167*, **168**, 348
Lemnaceae, 3, **164**, 347, 357
Lentibulariaceae, **309**, 348
Lentils, Water, **164**
Lesser Duckweed, **168**
Leucospora, 30, **300**
  multifida, **300**, *301*
Lily, 174
  Cow, **213**
  Small White Water, **210**
  Water, **210**
  Yellow Water, **213**
Limnobium, 20, *21*, 34, *35*, 98, **99**, 372
  Spongia, *98*, **99**, 347, 372
Limosella, 16, 20, **300**
  aquatica, **300**, *301*
    var. tenuifolia, **300**

Limosella aquatica, **300**
  subulata, **300**
Lindernia, 300, **307**
  anagallidea, *308*, **309**, 383
  dubia, *308*, **309**, 382
  Pyxidaria, **383**
Lippia, 26, **286**, 351, 363
  lanceolata, **286**, *287*
  nodiflora, **286**, *287*, 351
Littorella, 16, **313**
  americana, **313**, *314*, 316
  uniflora, **316**
Live-forever, 239
Liverworts, *41*, 42
  Leafy, 3, **40**
  Thallose, 3, 40, **42**
Lizard's Tail, 34, **181**
Lobelia, 18, 26, 28, **319**
  Boykinii, **319**
  Canbyi, **319**
  cardinalis, *27*, *318*, **319**
    f. alba, **319**
    f. rosea, **319**
  Dortmanna, 16, *17*, *317*, **319**
  elongata, **319**
  Kalmii, *318*, **319**
  Nuttallii, *318*, **319**
  siphilitica, 316, *318*, **319**, 348
    f. albiflora, **319**
    var. ludoviciana, **319**
Lobelia, Great, **319**
  Water, 16, **319**
Lobelia Family, **319**
Lobeliaceae, **319**, 348
Loosestrife, 22, 28, 252, **279**
  False, 22, 28, 252, **259**
  Fringed, **279**
  Spiked, 24, **252**
  Swamp, **252**
  Tufted, **279**
Loosestrife Family, 22, **252**
Lophotocarpus, 79, **91**, 93
  calycinus, **91**, *96*, 343
Lotus, 34
  American, **217**
Love Grass, **113**
Ludwigia, 22, 28, 252, **259**
  alternifolia, *258*, **259**
  arcuata, 259, 381
  brevipes, **259**, 261, 381
  glandulosa, *258*, **259**